```
615.902 P571e

Philp, Richard B.

Environmental hazards &
human health
```

Library of Congress Cataloging-in-Publication Data

Philp, Richard B.
    Environmental hazards and human health / R.B. Philp.
        p. cm.
    Includes bibliographical references and index.
    ISBN 1-56670-133-3
    1. Environmental health. I. Title.
    RA565.P48   1995
    615.9'92—dc20   94-38920
                                                                                       CIP

This book contains information obtained from authentic and highly regarded sources. Reprinted material is quoted with permission, and sources are indicated. A wide variety of references are listed. Reasonable efforts have been made to publish reliable data and information, but the author and the publisher cannot assume responsibility for the validity of all materials or for the consequences of their use.

Neither this book nor any part may be reproduced or transmitted in any form or by any means, electronic or mechanical, including photocopying, microfilming, and recording, or by any information storage or retrieval system, without prior permission in writing from the publisher.

CRC Press, Inc.'s consent does not extend to copying for general distribution, for promotion, for creating new works, or for resale. Specific permission must be obtained in writing from CRC Press for such copying.

Direct all inquiries to CRC Press, Inc., 2000 Corporate Blvd., N.W., Boca Raton, Florida 33431.

© 1995 by CRC Press, Inc.
Lewis Publishers is an imprint of CRC Press

No claim to original U.S. Government works
International Standard Book Number 1-56670-133-3
Library of Congress Card Number 94-38920
Printed in the United States of America 1 2 3 4 5 6 7 8 9 0
Printed on acid-free paper

PREFACE

There is a commonly held myth in our society that anything that is "natural" is good, wholesome and healthful, whereas anything that is "synthetic" is bad, toxic and harmful. The mere mention of the word chemical is enough to strike terror into the heart of any food faddist. This attitude is at best, naive and at worst, dangerous. Toxic substances abound in nature, ranging from inorganic heavy metals such as arsenic and mercury, through organic substances such as hydrocyanic acid, to complex enzymes and other proteins of the neurotoxins and coagulant-anticoagulants present in venoms and toxins. One potentially serious environmental hazard is natural radon gas.

Increasingly, it is necessary for students of environmental sciences to know something of toxicology and for students of toxicology to know something of the environment. This text is designed to bridge these fields by acquainting the student with the major environmental hazards, both man-made and natural, and with the risks to human health that they pose. It is designed such that topics are generally introduced in the early chapters and covered in greater detail in subsequent ones. It attempts to strike a balance between the extremes of opinion and to indicate where information is inconclusive. Examples of major accidental exposures of humans to chemical toxicants are used liberally, and case studies taken from reported incidents are provided. It is hoped that this text will assist students to acquire the information and judgmental skills necessary to differentiate between real and perceived risks, as well as acquainting them with the toxicology of important chemicals in the environment. Since most people spend 8 hours daily, 5 days a week in the workplace, it constitutes an important component of our environment and it will be considered as such.

Richard B. Philp
London, Ontario, Canada

THE AUTHOR

Richard B. Philp, D.V.M., Ph.D., is a Professor and former Chairman in the Department of Pharmacology and Toxicology at The University of Western Ontario. After graduating from the Ontario Veterinary College, he practiced veterinary medicine in Illinois and in Ontario, where he also served as a public health officer. He obtained his Ph.D. in pharmacology from U.W.O. and conducted postdoctoral studies at the Royal College of Surgeons of England. He has served on advisory committees to federal and provincial departments of agriculture regarding the use of antibiotics in the agricultural industry. He was Honorary Visiting Professor in the School of Pathology, University of New South Wales in 1986–1987. He teaches undergraduate and graduate courses in toxicology and environmental science. He has authored or co-authored over 80 scientific papers on such diverse topics as the pharmacology and toxicology of antithrombotic drugs, the carcinogenicity of bracken fern and the environmental hazards of deep-sea diving. He is the author of *Methods of Testing Proposed Antithrombotic Drugs* (CRC Press, 1981) and of several chapters in various volumes.

CONTENTS

1. **Principles of Pharmacology and Toxicology** 1
   Introduction   1
   Pharmacokinetics   3
      Absorption   4
      Distribution   5
      Biotransformation   7
      Elimination   10
   Pharmacodynamics   12
      Ligand Binding and Receptors   12
      Biological Variation and Data Manipulation   14
      Cumulative Effects   18
   Factors Influencing Responses to Xenobiotics   20
      Age   21
      Body Composition   22
      Sex   22
      Genetic Factors   22
      Presence of Pathology   25
      Xenobiotic Interactions   26
   Some Toxicological Considerations   27
      Acute vs. Chronic Toxicity   27
   Mutagenesis and Carcinogenesis   29
   The Role of Cell Repair and Regeneration in Toxic
      Reactions   32
      Response of Tissues to Chemical Insult   32
   Fetal Toxicology   33
      Teratogenesis   33
      Transplacental Carcinogenesis   35
   Review Questions   36
   Answers   39
   Essay Question   39
   Further Reading   39

2. **Risk Analysis and Public Perceptions of Risk**   41
   Introduction   41
   Assessment of Toxicity vs. Risk   42
   Predicting Risk: Workplace vs. Environment   42
      Acute Exposures   42
      Chronic Exposures   42
      Very Low-Level, Long-Term Exposures   43
      Carcinogenesis   43
   Risk Assessment and Carcinogenesis   43
      Sources of Error in Predicting Cancer Risks   46
      Extrapolation of Animal Data to Humans   49
      Natural vs. Anthropogenic Carcinogens   50
      Reliablility of Tests of Carcinogenesis   50

Environmental Monitoring 51
Setting Safe Limits in the Workplace 51
Environmental Risks: Problems with Assessment and
   Public Perceptions 54
   The Psychological Impact of Potential
      Environmental Risks 54
   Costs of Risk Avoidance 56
Some Examples of Major Industrial Accidents and
   Environmental Chemical Exposures with Human
   Health Implications 57
   Radiation 57
   Formaldehyde 57
   Dioxin (TCDD) 58
Some Legal Aspects of Risk 59
   de Minimis Concept 59
   Delaney Amendment 59
Statistical Problems with Risk Assessment 60
Review Questions 61
Answers 63
Case Study #1 64
Case Study #2 64
Further Reading 65

3. **Water Pollution** 67
Introduction 67
Factors Affecting Toxicants in Water 68
   Exchange of Toxicants in an Ecosystem
   Factors (Modifiers) Affecting Uptake of Toxicants
      from the Environment 68
Some Important Definitions 72
   Acclimation 72
   Bioconcentration 72
   Biomagnification 72
   Anthropogenic 72
Toxicity Testing in Marine and Aquatic Species 72
Water Quality 73
   Sources of Pollution 73
   Some Major Water Pollutants 75
Health Hazards of Chemical Water Pollutants 76
   Chlorinated Hydrocarbons 76
   Chlorphenoxy Acid Herbicides 77
   Organophosphates 77
   Carbamates 77
   Others 78
Acidity and Toxic Metals 78
Chemical Hazards from Waste Disposal 80
The Love Canal Story 80
   Problems with Love Canal Studies 83
Toxicants in The Great Lakes and the Implications
   for Human Health and Wildlife 83

    Evidence of Adverse Effects on Human Health 85
  Review Questions   87
  Answers   88
  Further Reading   89

4. Air-Borne Hazards   91

  Introduction   91
  Types of Air Pollution   91
    Gaseous Pollutants   91
    Particulates   91
    Smog   92
  Sources of Air Pollution   92
  Atmospheric Distribution of Pollutants   93
    Movement in the Troposphere   93
    Movement in the Stratosphere   93
    Water and Soil Transport of Air Pollutants   94
  Types of Pollutants   94
    Gaseous Pollutants   94
    Particulate Pollutants   95
  Health Effects of Air Pollution   95
    Acute Effects   95
    Chronic Effects   95
  Air Pollution in the Workplace   95
    Asbestos   96
    Silicosis   97
    Pyrolysis of Plastics 97
    Dust   97
    CO_2 and NO_2 97
  Chemical Impact of Pollutants on the Environment 98
    Sulfur Dioxide and Acid Rain   98
    The Chemistry of Ozone   99
    Chlorine   100
    The Greenhouse Effect   101
    Carbon Dioxide   101
    Methane   102
    Subtle Greenhouse Effects   102
    Global Cooling   103
    Sulfur Dioxide   103
    Motor Vehicle Exhaust   103
  Natural Factors and Climate Change   104
  Remedies   106
  Review Questions   107
  Answers   109
  Case Study #3   109
  Case Study #4   109
  Case Study #5   109
  Case Study #6   110
  Case Study #7   110
  Case Study #8   111
  Further Reading   112

5. **Halogenated Hydrocarbons**   113
   Introduction   113
     Early Examples of Toxicity from Halogenated
       Hydrocarbons   113
   Physicochemical Characteristics and Classes of
     Halogenated Hydrocarbons   114
     Antibacterial Disinfectants   114
     Herbicides   115
     Insecticides   122
     Industrial and Commercial Chemicals   122
     Solvents   124
     Trihalomethanes (THMs)   125
   Review Questions   127
   Answers   128
   Case Study #9   128
   Case Study #10   129
   Further Reading   129

6. **Toxicity of Metals**   131
   Introduction   131
   Lead (Pb)   132
     Toxicokinetics of Lead   133
     Cellular Toxicity of Lead   133
     Fetal Toxicity   134
     Treatment   135
   Mercury (Hg)   135
     Elemental Mercury Toxicity   136
     Inorganic Mercurial Salts   136
     Organic Mercurials   137
     Mechanism of Mercury Toxicity   137
     Treatment of Mercury Poisoning   138
     The Grassy Narrows Story   138
   Cadium (Cd)   140
     Cadium Pharmacokinetics   140
     Cadium Toxicity   141
     Treatment   141
   Arsenic (As)   141
     Pharmacokinetics of Arsenicals   142
     Toxicity   142
     Treatment   142
     Environmental Effects of Arsenic   143
   Chromium (Cr)   143
   Other Metals   143
   Carcinogenicity of Metals   144
   Unusual Sources of Heavy Metal Exposure   144
   Review Questions   145
   Answers   147
   Case Study #11   147
   Case Study #12   147
   Case Study #13   148
   Further Reading   148

7. **Organic Solvents and Related Chemicals** 151
   Introduction 151
   Classes of Solvents 152
     Aliphatic Hydrocarbons 152
     Halogenated Aliphatic Hydrocarbons 152
     Aliphatic Alcohols 153
     Glycols and Glycol Ethers 155
     Aromatic Hydrocarbons 155
   Solvent-Related Cancer in the Workplace 156
     Benzene 156
     Bis(chloromethyl) Ether (BCME) 157
     Dimethylformamide (DMF) and Glycol Ethers 158
     Ethylene Oxide (C_2H_4O) 158
   Factors Influencing the Risk of a Toxic Reaction 159
   Nonoccupational Exposures to Solvents 159
   Review Questions 160
   Answers 161
   Case Study #14 161
   Case Studies #15 and 16 161
   Further Reading 162

8. **Food Additives, Drug Residues and Other Food Toxicants** 165
   Food Additives 165
     Food and Drug Regulations 165
     Artificial Food Colors 168
     Emulsifiers 170
     Preservatives and Anti-Oxidants 170
     Artificial Sweeteners 171
     Flavor Enhancers 173
   Drug Residues 174
     Antibiotics and Drug Resistance 174
     Allergy 179
     Diethylstilbestrol 179
   Natural Toxicants and Carcinogens in Human Foods 183
     Some Natural Toxicants 183
     Natural Carcinogens in Foods 184
     Others 185
   Review Questions 186
   Answers 188
   Case Study #17 188
   Further Reading 189

9. **Pesticides** 191
   Introduction 191
   Classes of Insecticides 193
     Organochlorines (Chlorinated Hydrocarbons) 193
     Organophosphates 195
     Carbamate Insecticides 196
     Botanical Insecticides 197

    Herbicides   197
        Chlorphenoxy Compounds   197
        Dinitrophenols   197
        Bipyridyls   198
        Carbamate Herbicides   199
        Triazines   199
    Fungicides   199
        Dicarboximides   199
    Newer Biological Control Methods   199
    Government Regulation of Pesticides   200
    Problems Associated with Pesticides   201
        Development of Resistance   201
        Multiple Resistance 201
        Nonspecificity   202
        Environmental Contamination   202
    Balancing the Risks and the Benefits   202
    Toxicity of Pesticides for Humans   203
    Review Questions   204
    Answers   205
    Case Study #18   205
    Case Study #19   206
    Further Reading   206

10. Mycotoxins and Other Toxins from Unicellular Organisms   209
    Introduction   209
    Some Human Health Problems Due to Mycotoxins 209
        Ergotism   209
        Aleukia   211
        Aflatoxins   211
        Fumonisins   211
        Other Mycotoxic Hazards to Human Health   213
    Economic Impact of Mycotoxins   214
        Fusarium Life Cycle   214
        Zearalonone 215
        Vomitoxin (Deoxynivalenol or DON)   215
        Other Trichothecenes   215
    Detoxification of Grains   215
        Species Differences in DON Kinetics   217
    Other Toxins in Unicellular Members of the Plant Kingdom   218
    Review Questions   219
    Answers   221
    Further Reading   221

11. Animal and Plant Poisons 223
    Introduction   223
    Toxic and Venomous Animals   224
        Toxic and Venomous Marine Animals   224
        Toxic and Venomous Land Animals   228

Toxic Plants and Mushrooms 232
   Vesicants 232
   Cardiac Glycosides 232
   Astringents and Gastrointestinal Irritants
     (Pyrogallol Tannins) 233
   Autonomic Agents 233
   Dissolvers of Microtubules 233
   Phorbol Esters (e.g., Phorbol Myristate
     Acetate, PMA) 233
   Solanine and Chaconine 233
   Cyanogenic Glycosides 234
Review Questions 236
Answers 238
Case Study #20 238
Case Study #21 238
Case Study #22 238
Case Study #23 239
Case Study #24 240
Further Reading 240

12. Radiation Hazards 243
Introduction 243
Sources and Types of Radiation 244
   The Cause of Radiation 244
   Sources 244
   Types of Radioactive Energy Resulting from
     Nuclear Decay 245
Measurements of Radiation 245
   Measures of Energy 245
   Measures of Damage 246
Some Major Nuclear Disasters of Historic Importance 247
   Hiroshima 247
   Chernobyl 248
   Three Mile Island 248
   The Hanford Release 249
Radon Gas: The Natural Radiation 249
Tissue Sensitivity to Radiation 251
Microwaves 252
Ultraviolet Radiation 253
   Medical Uses of UV Radiation 253
Extra-Low Frequency (ELF) Electromagnetic
   Radiation 254
Irradiation of Foodstuffs 256
Review Questions 257
Answers 258
Further Reading 258

13. Gaia and Chaos: How Things Are Connected 261
The Gaia Hypothesis 261
Chaos Theory 261

    Some Examples of Interconnected Systems 262
        A Vicious Circle   264
        Domino Effects of Global Warming   264
        A Feedback Loop   264
    Food Production and the Environment   266
        Meat vs. Grain   267
    The Environment and Cancer   270
    Further Reading   270

14. Case Study Reviews   273
    Case Study #1   273
    Case Study #2   274
    Case Study #3   274
    Case Study #4   274
    Case Study #5   275
    Case Study #6   275
    Case Study #7   275
    Case Study #8   275
    Case Study #9   276
    Case Study #10   276
    Case Study #11   277
    Case Study #12   277
    Case Study #13   277
    Case Study #14   278
    Case Study #15   278
    Case Study #16   279
    Case Study #17   280
    Case Study #18   280
    Case Study #19   281
    Case Study #20   281
    Case Study #21   282
    Case Study #22   282
    Case Study #23   283
    Case Study #24   283

Index   285

DEDICATION

For Elizabeth, Michael, Brendan, Douglas, Danielle, Nathan, Danny, Margaret, Matthew, Gemma and kids everywhere. May they look after this place better than their forbears.

1 PRINCIPLES OF PHARMACOLOGY AND TOXICOLOGY

The right dose differentiates a poison and a remedy.
Paracelsus, 1493–1541

INTRODUCTION

The past century has seen a tremendous expansion in the number of synthetic chemicals employed by humankind as materials, drugs, preservatives for foods and other products, pesticides, cleaning agents and even weapons of war. An estimated 64,000 chemicals are currently in use commercially, with 5 billion tons being produced worldwide. Four thousand of them are used as medicinals and at least 1,200 more as household products. An estimated 700 new chemicals are synthesized each year. Add to this the numerous natural substances, both inorganic and organic, that possess toxic potential, and it is little wonder that the public expresses concern, and sometimes even panic, about the harmful effects these agents may exert on their health and on the environment. Many of these agents, perhaps 50,000 of them, have never been subjected to thorough toxicity testing.

About 500 chemicals have been evaluated for carcinogenic potential. Some 44 have been designated as possible human carcinogens on the basis of evidence, either limited or conclusive, obtained from human studies. Of these, 37 tested positive for carcinogenicity in animal tests prior to the identification of this effect in humans. There are, however, numerous other agents which have been shown to be carcinogenic in rodents but which have yet to be identified as human carcinogens. This creates significant problems regarding the legislative and regulatory decisions that need to be made about their use, and some of the areas of uncertainty that surround the extrapolation of data from the animal setting to the human one will be discussed in the following chapter. The process of extrapolation requires input from many different disciplines which may include engineering, physics, biology, chemistry, pathology, pharmacology, physiology, public health, immunology, epidemiology, biostatistics and occupational health. The field of toxicology thus depends on all of these, but perhaps draws most heavily on pharmacology, biochemistry and pathology. It is the identification of the degree of risk to which individuals or groups are exposed in a given set of circumstances which directs all of this activity.

Other forms of toxicity, for example, hepatotoxicity, nephrotoxicity and neural toxicity, may be more important in acute exposures such as may occur in the industrial setting. Reproductive and fetal toxicity have frequently been demonstrated experimentally, but their significance for the general population exposed to low levels of toxicants in the environment remains unclear.

Considerable difficulty attends efforts to extrapolate the results of toxicity tests in experimental animals to humans exposed to very low levels in their environment, especially with regard to the risk of cancer. Current legislation requires testing in two species with sufficient numbers for reliable statistical analysis. Rats and mice are generally used, as hamsters are resistant to many carcinogens and primates are too expensive and, in the case of some species, too environmentally threatened. For statistical purposes, cancer includes all tumors whether benign or malignant. A 2-year carcinogen study employing two species cost, in 1991, at least $1,000,000 plus the costs of 1 year for preparation, 1 for analysis (pathology, etc.) and 1 for documentation and statistics. Since it is not practical to test every chemical, factors to be considered in selecting test chemicals include the frequency and severity of observed effects, the extent to which the chemical is used, its persistence in the environment (examples of persistent chemicals would include chlorinated hydrocarbons) and whether transformations to more toxic agents occur.

Heavy metals, the by-products of most mining and ore extraction processes, are examples of ubiquitous toxicants with almost infinite half-lives. Mercury (Hg), for example, is present in all canned tuna at about 5 ppm, mostly from natural sources. It may be transformed by aquatic bacteria to methylmercury which has a different toxicity profile. Cadmium (Cd) enters the environment at about 7,000 tons/year, and it is concentrated by livestock because they recycle it in feces, which is used for fertilizer and which passes it on to forage grasses. Radioactive isotopes of cesium and iodine entered the food chain after Chernobyl.

The estimation of the degree of risk associated with the presence of a potentially toxic substance in the environment is the basis for all decisions relating to the legislative controls over that chemical, including its industrial use and eventual disposal. Pharmacological/toxicological principles are essential for understanding the processes involved in toxicity testing.

Pharmacology may be defined as the science of drugs. It includes a study of their sources (materia medica), their actions in the living animal organism (pharmacodynamics), the manner in which they are absorbed, moved around and excreted in the body (pharmacokinetics), their use in medicine (therapeutics) and their harmful effects (toxicology). In this context, a drug is any substance used as a medicine, but pharmacology generally includes the study of substances of abuse and in the broadest sense deals with the interactions of xenobiotics (literally, substances foreign to living organisms) whether they be natural or man-made (anthropogenic), therapeutic or toxic. In this sense, toxicology can be considered to be a branch of pharmacology. Xenobiotics may also be exploited as research tools to reveal mechanisms underlying physiological processes.

Toxicology is the study of the harmful effects of xenobiotics on living organisms, the mechanisms underlying those effects, and the conditions under which they are likely to occur.

Environmental toxicology is the study of the effects of incidental or accidental exposure of organisms, including human beings (the focus of this text), to toxins in the environment, i.e., air, water and food. While the greatest concern today centers on pollutants of human origin, it should not be forgotten that toxic substances, including carcinogens, abound in nature. The subject of environmental toxicology embraces the study of the causes, conditions, environmental impact, and means of controlling pollutants in the environment. It may also be extended to include the environment of the workplace (industrial hygiene).

Economic toxicology is the study of chemicals that are developed expressly for the purpose of improving economic gain by selectively eliminating a species (insecticides and herbicides), improving health and productivity (drugs), or preserving foodstuffs (industrial solvents, cleaning agents, etc.).

Forensic toxicology refers to the medicolegal aspects of the harmful effects of drugs and poisons administered or taken deliberately or accidentally. Detection of xenobiotics in tissues and fluids and in, or on, objects is an important aspect of this field as is the preparation of evidence for submission in court.

■ PHARMACOKINETICS

There has been a trend recently to attempt to separate toxicology from pharmacology by the use of such terms as "toxicokinetics", "toxicodynamics", etc. The distinction is largely artificial as the principles are the same in both cases. Throughout this text, "pharmacological" can be taken also to represent "toxicological".

The response of organisms to drugs and chemicals is governed by natural laws. One of these is the *"Law of mass action"* which dictates that, *in the absence of a transport system*, chemicals in solution will move from an area of high concentration to one of low concentration. If a semipermeable membrane is interposed between these areas, the chemical will move across it, assuming the chemical can penetrate the membrane. In fact, molecules will wander randomly across the barrier, but the frequency of transfers will be greater from the area of high concentration to that of the low one until equilibrium is established. Cell walls and other biological membranes function as semipermeable membranes, and this law influences the uptake of most drugs and chemicals by living organisms. The concentration of a toxicant in the environment (water, air, soil) is thus an important determinant affecting its uptake. Transport mechanisms are dealt with under "Absorption and Distribution".

Partition coefficient is the ratio of a chemical's relative solubility in two different phases. The ratio of solubility in oil (often *n*-octanol) to that in water is frequently used to predict the distribution of a xenobiotic between the aqueous and lipid phases in the body.

Absorption

Whether or not a xenobiotic is toxic, and how that toxicity is manifested, depends largely on how it is handled by the body. Substances that are not absorbed from the gastrointestinal tract have no systemic toxicity. This fact allows barium to be used as an X-ray contrast medium, despite barium's toxicity by other routes of administration. The selective toxicity of most insecticides depends solely on a greater ability to penetrate the chitin of the insect's exoskeleton than to penetrate human skin. A substance that is not readily excreted by the body, usually through the kidneys or in the feces, will accumulate to toxic levels.

The primary routes of absorption for toxicants are the skin, the lungs and the gastrointestinal tract. The latter two are important for the population at large, but the skin may be a very significant site in certain industrial settings. The site of absorption, more commonly called the *portal of entry* in toxicology, can have a significant influence on the toxicity of a substance.

Larger molecules require a degree of lipid solubility to cross biological barriers since cell membranes consist of a fluid phospholipid matrix with embedded proteins that may penetrate part way or all the way through the membrane. Factors that influence the lipophilicity of a chemical will therefore affect its absorption. Many chemicals are weak acids or bases that may exist in an ionized (polar) or a non-ionized state with an equilibrium established between them, e.g.,

$$R\text{--}H \rightarrow R^- + H^+$$
non-polar

The polar form is water-soluble whereas the non-polar form is lipid-soluble. The pH will influence the equilibrium and hence the amount of lipid-soluble form available for absorption. The *dissociation constant (pK_a)* of a substance is defined as *that pH at which 50% of it will exist in each state*. Weakly acidic drugs are shifted to the non-polar state in an acid medium and to the polar state in an alkaline medium. The reverse is true for weakly alkaline drugs. Since the pH of the stomach and upper small bowel is acidic (pH 2–4), acidic chemicals will be absorbed here. Alkaline substances tend to be absorbed in the lower small bowel and the upper colon which are more alkaline, whereas the descending colon becomes acidic again.

Lipid solubility is not essential for the passage of all molecules across membranes. There is the bulk transfer of water across the cell membrane that can carry very small (less than 200 Daltons), water-soluble molecules with it. Metallic ions such as calcium, sodium and potassium, as well as chlorine, can pass through special channels, some of which are regulated by the transmembrane potential (voltage-regulated) and others by specific receptors (receptor-activated). Specialized exchangers also exist, for example, the sodium pump.

Active transport is an energy-consuming process by which a substance may be moved against a concentration gradient. Active transport is important in the kidney and the liver. In addition to energy consumption, it is also characterized by saturability, selectivity for specific chemical configurations and the ability to move substances against an electrochemical gradient.

Facilitated diffusion is similar except that no energy is consumed, and it cannot occur against an electrochemical gradient.

Pinocytosis is a process whereby a segment of the plasma membrane of a cell invaginates to form a sack in which extracellular fluid and colloidal particles can be taken into the cell by pinching off the "mouth" of the sack. This is an important mechanism by which the mucosal cells of the intestinal tract take up nutrients and some drugs and chemicals.

Distribution

Once absorbed the agent may be distributed throughout various compartments in the body. Serum albumin possesses many non-specific binding sites for xenobiotics, especially weakly acidic ones, and it therefore becomes a transport system for many substances. The balance between dissociated (polar) and undissociated (non-polar) states also affects the distribution of a chemical as well, since pH changes from the extracellular fluid (pH 7) to the plasma (pH 7.4). The partition coefficient of a substance also influences its distribution, determining, for example, the extent to which it will be sequestered in fat. Highly lipid-soluble substances will be sequestered in body fat where they may remain for long periods. Everyone has DDT and its metabolites dissolved in their fat. The amount varies with their age and location. Use of DDT in North America was drastically reduced in the 1970s, and a complete ban was legislated in Canada in 1990. Substances like DDT, that are sequestered in fat, may be released during periods of fat loss (starvation, extreme dieting), as a result of illness and even during lactation, when lipids are transferred to milk. The released toxicant may reach concentrations at target sites sufficient to cause a toxic response. Figure 1 illustrates these relationships among storage fat, blood and target organ.

The rate of distribution of a substance is a function of the rate of blood flow through the tissues (tissue perfusion). Highly vascular organs will accumulate it first; organs that are poorly perfused will accumulate it last. The substance is thus distributed initially on the basis of tissue perfusion; then, as equilibrium states are reached, it will redistribute on the basis of its solubility. Following the intravenous injection of a chemical with a high partition coefficient, equilibrium will be established instantly with the kidney and liver because of their high vascularity, almost as quickly with the brain, with muscle in about 30 min and with fat in about 3 hr. The membranes surrounding the brain and separating it from its blood vessels constitute the *blood-brain barrier* which will only pass quite lipid-soluble agents such as all anesthetics.

Thus, tissue perfusion and partition coefficient may play important roles in determining the onset and termination of either a therapeutic or a toxic response. Sodium thiopental, an ultrashort-acting barbiturate, is used for anesthetic induction. The rate of biotransformation is so slow as to have little effect on recovery. The drug readily penetrates the blood-brain barrier because of its high lipid solubility, and the brain, which is richly perfused, rapidly takes it up and anesthesia ensues. This effect is terminated because the drug is redistributed to other tissues, including depot fat which is poorly perfused. New equilibria are established among blood, brain and other tissues so that, while initial recovery is rapid, a state of sedation may persist for several hours. In Figure 2 the effects of perfusion and partition coefficient on T1/2s of thiopental in different tissues are shown.

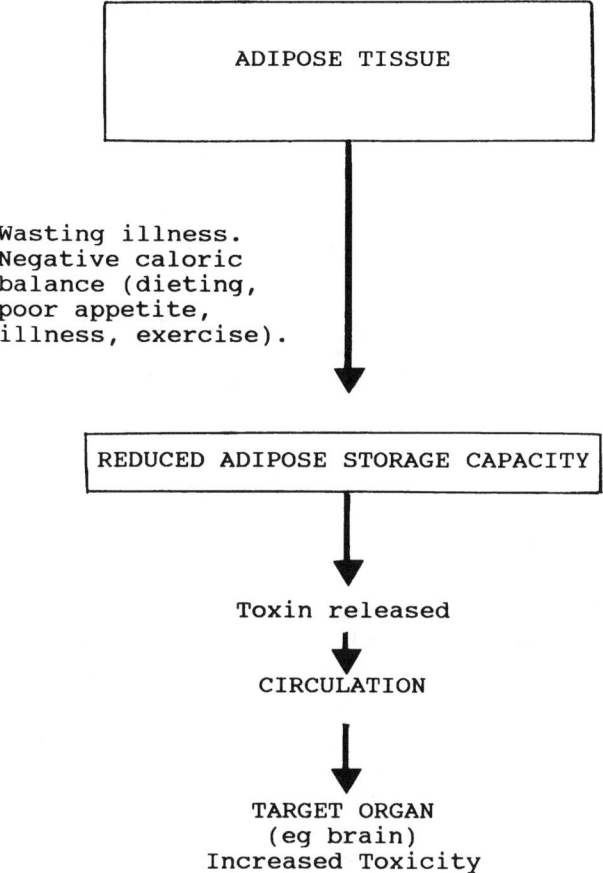

Figure 1 Disposition of lipid-soluble chemicals in adipose tissue and blood and the effect of weight loss.

Biotransformation

Biotransformations are classified as either of two types: Phase I reactions or Phase II reactions. It should be noted that these same transformations apply to therapeutic agents. Phase I reactions: in this type of reaction, also known as nonsynthetic biotransformation, a lipophilic (fat-soluble) substance is rendered more polar and hence, more water-soluble. This type of metabolite is more readily excreted by the kidneys, but it usually retains significant bioactivity. It may be more active, or less active, than the parent substance. If the parent chemical is nontoxic but the metabolite is toxic, this is toxication. A drug which requires activation is sometimes referred to as a *pro-drug*.

Phase I chemical reactions (Figure 3) include oxidation, reduction and hydrolysis and generally unmask or introduce a functional

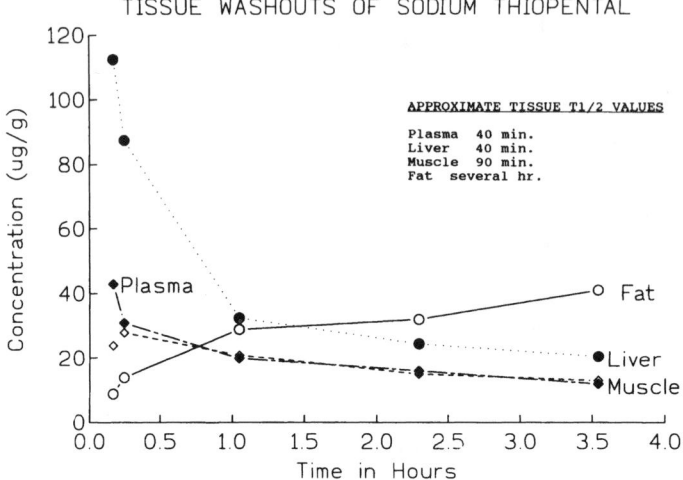

Figure 2 T1/2s of sodium thiopental, a highly lipid-soluble drug, in various tissues.

(reactive) group such as $-NH_2$, $-OH$, $-SH$, COOH. The oxidation reactions include N- and O-dealkylations, side-chain and aromatic hydroxylations, N-oxidation and hydroxylation, sulfoxide formation and desulfuration. Hydrolysis of esters and amides also occurs. Reduction reactions may involve azo (RN=NR) or nitro (RNO_2). Many oxidation reactions are under the control of a group of mixed function oxidases for which cytochrome P-450 serves as a catalyst. These are located primarily in the smooth endoplasmic reticulum (SER) of hepatic cells, but they exist in many tissues. The P-450 monooxygenases have tremendous substrate versatility, being able to oxidize lipophilic xenobiotics plus fatty acids, fat-soluble vitamins, and various hormones. This is partly because there are at least 10 variants of the enzyme (isoenzymes) and because each is capable of accepting many substrates. It should be noted that procarcinogens are converted to carcinogens by Phase I reactions. Examples of this include benzo[a]pyrene, the fungal toxin aflatoxin B_1 and the synthetic estrogen diethylstilbestrol. This process often involves the formation of an epoxide compound, as it does in the three examples given. An epoxide has the chemical configuration shown in Figure 4, making it highly nucleophilic and chemically reactive. Many epoxides are carcinogens. Figure 4 shows this chemical transformation for stilbestrol and benzo[a]pyrene, which is an example of a polyaromatic hydrocarbon (PAH), many of which are carcinogenic and are environmental pollutants. Other enzymes called epoxide hydrolases may detoxify the epoxides.

PRINCIPLES OF PHARMACOLOGY AND TOXICOLOGY 9

1. Parathion (inactive) --cytochrome P450 monooxygenase----> Paraoxon (active)

2. Pentobarbital (active)--------> Hydroxypentobarbital (inactive)

3. Codeine (poorly active)---------------> Morphine (very active)

Figure 3 Some examples of Phase I reactions. The product may be more or less active than the parent chemical, or it may be inactive.

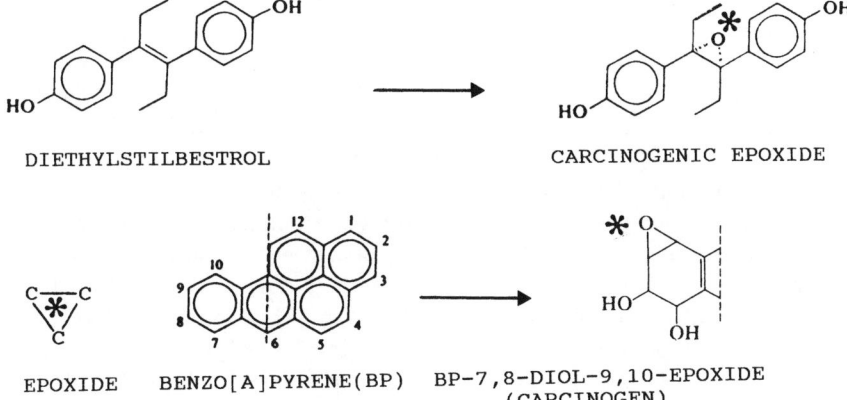

Figure 4 Examples of epoxide formation to potentially carcinogenic metabolites.

Phase II reactions are conjugation (synthetic) reactions which render the agent not only more water-soluble, but biologically inactive, with very few exceptions. A common conjugation reaction is with glucuronic acid. Conjugation also occurs with sulfuric acid, acetic acid, glycine and glutathione. Many Phase I metabolites are still too lipophilic (fat-soluble) to be excreted by the kidneys and are subjected to Phase II conjugation. All chemicals need not be subjected first to Phase I transformations. Many, if they possess the necessary functional groups (e.g., –OH, –NH$_2$) are conjugated directly.

An important concept for understanding toxication and detoxication of xenobiotics is *enzyme induction*. Hepatic enzymes of the smooth endoplasmic reticulum can be stimulated to a higher level of activity by many highly lipophilic agents. Because these enzymes are nonspecific, this has consequences for many other agents transformed by the same enzymes. Induction is accomplished by the increased synthesis of more enzyme, so the SER actually increases in density. The result may be increased detoxication of a chemical or the increased synthesis of a toxic metabolite. Cigarette smoke contains many inducers and may increase the breakdown of many drugs (theophylline, phenacetin, etc.), but conversely it may act through this mechanism as an inducer or cocarcinogen.

Elimination

Every secretory or excretory site in the body is potentially a route of elimination for xenobiotics. They may be excreted in saliva, sweat, milk, tears, bile, mucus, feces and urine. Of these, the most significant site is urine, followed by feces and bile.

The kidney is the principal organ for the elimination of natural waste metabolites, most of which are toxic if they exceed normal levels, as well as the main organ for maintaining fluid and electrolyte balance. It is therefore not surprising that the kidney also is the main site of elimination of xenobiotics, including drugs. Although it constitutes only 0.4% of total body weight, it takes 24% of the cardiac output. It is a highly efficient filter of blood.

The basic physiological unit of the kidney is the nephron, which is composed of the glomerulus (a tightly wound bundle of blood vessels) and the tubule, which is closed at the glomerular end to provide a semipermeable membrane. The tubule is composed of several segments with different functions. These are noted in Figure 5. Substances smaller than 66,000 Daltons are passed through the glomerulus. They may be reabsorbed further down the tubule and even resecreted. This occurs with uric acid, which is completely passed through the filter, 98% reabsorbed and further secreted. The pH of urine will determine the degree of dissociation of acids and bases and hence influence their movement across the reabsorption sites.

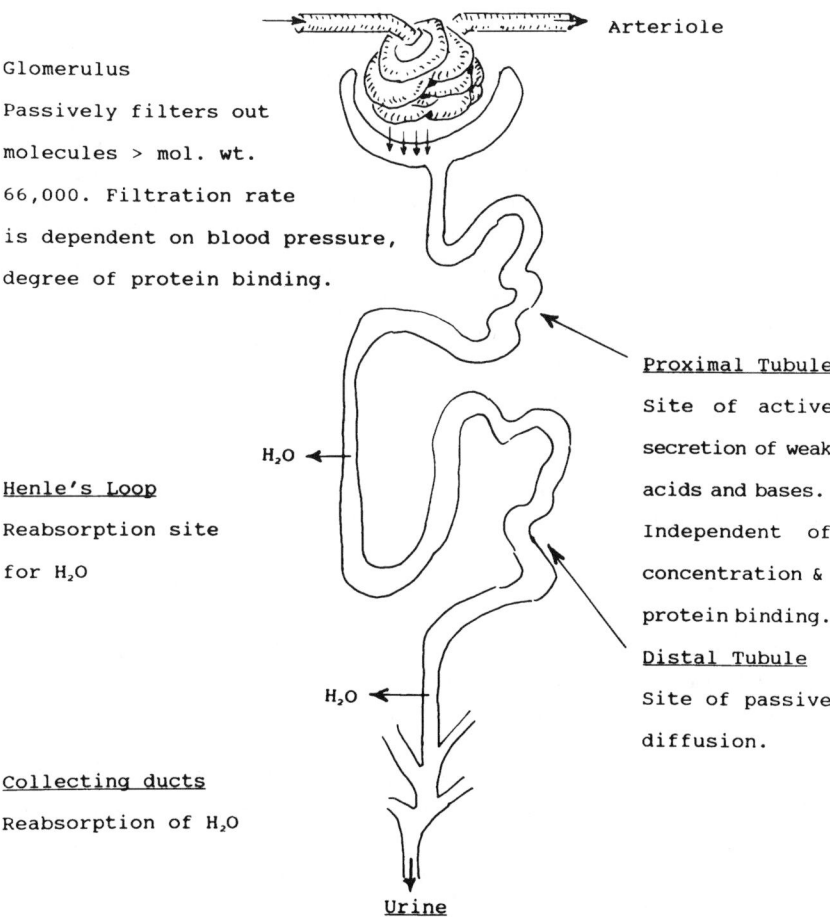

Figure 5 The nephron is the basic renal unit.

Passive diffusion across the distal tubule depends on the degree of ionization in the plasma and extracellular fluid as only the lipid-soluble form will be diffused. The concentration gradient also is an important rate-limiting factor. Very water-soluble agents are passed through the glomerulus if they are small enough, and this is the reason why most biotransformations result in increased water solubility. Other substances are actively secreted (an energy-consuming process) at tubular sites.

It should be noted that the lungs are a very important site of elimination for volatile substances including solvents, alcohols and volatile and gaseous anesthetics. These can, in fact, be smelled on the breath, which can be an important first aid procedure to determine the cause of unconsciousness or stupor. Ketoacidosis in diabetics also can be detected by the acetone-like odor on the breath. Young diabetics have been suspected of glue-sniffing when brought to an emergency department in a stupor or coma because of this fact.

Many drugs and chemicals are excreted into the bile. These tend to be polar agents, both cationic and anionic, the latter including glucuronide conjugates. Nonselective active transport systems, similar to those in the kidneys, are involved in the excretory processes. Once they enter the small intestine, these chemical metabolites may be excreted in the feces or reabsorbed back into the bloodstream. Enzymatic hydrolysis of glucuronide conjugates favors a return to the more lipid-soluble state and hence, reabsorption.

The excretion of xenobiotics in mother's milk may not be an important route of elimination, but it can have significance for toxicity in the infant. The chloracne rash associated with the now-obsolete bromide sedatives appears to be related to the secretion of this halogen in sweat. It is distributed in the body like chloride ion.

Extensive batteries of enzymes in the body may render the chemical non-toxic (detoxication), more water-soluble and hence, more easily excreted, or they may activate it to a toxic form (toxication). The liver is the primary site of xenobiotic biotransformation in the mammalian body, but it is by no means the only one. Indeed, significant biotransformation can occur at the portal of entry. The chemical pathways are often the same. The response of the body to chemical insult also depends on the mitotic activity of the target tissue. Rapidly dividing tissues allow little time for repair to occur before cell division, so that the chance of a mutation is increased. Moreover, tissues which regenerate poorly are vulnerable to permanent damage by toxicants.

■ PHARMACODYNAMICS

Ligand Binding and Receptors

Since only the molecules that are free in solution contribute to the concentration gradient, their binding to tissue components or their chemical alteration by tissue enzymes will contribute to the maintenance of the gradient. The nature and strength of the chemical bond determines how easily the xenobiotic will dissociate when the concentration gradient is reversed. Drugs interact with specific sites (receptors) on proteins such as plasma membrane proteins, cytosolic enzymes, membranes on cell organelles and in some cases, nucleic acids (e.g., certain antineoplastic drugs). Membrane receptors and enzymes have molecular configurations that will react only with certain molecules in a kind of "lock-and-key" manner. Ease of reversibility is an important characteristic for most drugs, so that as concentration of the free substance falls, the drug comes off the receptor and its effect is terminated. This is often expressed by the equation:

$$drug(D) + receptor(R) \leftrightarrow DR\ complex \rightarrow Response$$

The magnitude of the response is determined by the number (percentage) of receptors occupied at any given time. Neither the drug nor the receptor is altered by the reaction, which is defined as "pharmacodynamic".

In many cases, drugs and toxicants interact with receptors that normally accept physiological ligands such as neurotransmitters, hormones, ions and nutritional elements. The proteins of cell surface receptors may penetrate to the interior of the cell in the case of ion channels and exchangers, or they may connect with other proteins in the membrane to transduce signals. Many neurotransmitters operate through a family of receptors that share the property of connecting to a protein having seven, membrane-spanning peptide chains. These "G" proteins ("G" for guanosine triphosphate or GTP) are transducers that interact with enzymes such as adenlycyclase or phospholipase C to initiate intracellular second messengers. G proteins may be inhibitory (G_i), stimulatory (G_s) or operate through other, unidentified mechanisms (G_o). The neurotransmitters noradrenaline, acetylcholine, dopamine, serotonin, histamine, γ-aminobutyric acid (GABA), glycine and glutamic acid have been shown to act through G-protein receptors. Many centrally acting drugs work through these receptors.

Steroid receptors also exist. These are soluble cytosolic receptors that bind to the steroid after it diffuses into the cell and carry it to the nucleus. Opioid receptors in the CNS accept the endogenous peptide endorphins and enkephalins. These receptors are the site of action of the narcotic analgesics.

Any receptor is a potential target for a toxicant interaction. A special case is the aryl hydrocarbon, or Ah, receptor. This cytosolic receptor binds to aromatic hydrocarbons like dioxins, and it is believed that it is involved in their toxicity. No natural ligand for this receptor has yet been identified in mammals. This subject is discussed in detail in the chapter on halogenated hydrocarbons.

The chemical bond with the target receptor can involve covalent bonds, as well as noncovalent bonds including ionic, hydrogen, and van der Waal's forces. If the chemical binds irreversibly with a component of a cell the effect may be long-lasting. Indeed, irreversibility of effect is an important characteristic of many toxicants (organophosphate insecticides are examples of irreversible inhibitors of the enzyme acetylcholinesterase). If a chemical reacts irreversibly with DNA, a mutation may result in carcinogenesis or teratogenesis. This effect is sometimes described as "hit-and-run" because it is

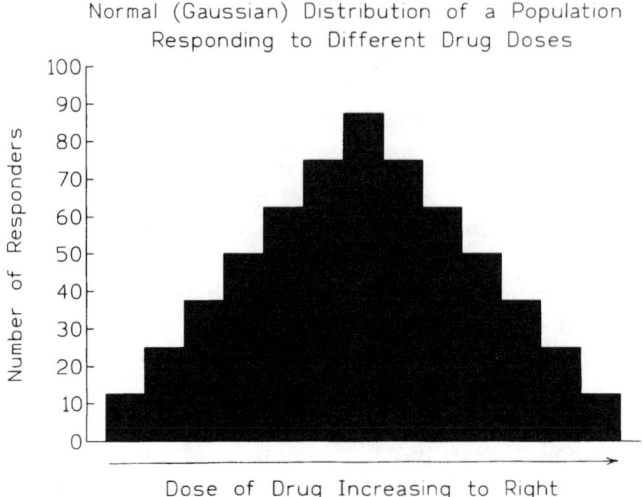

Figure 6 A theoretical normal, or "Gaussian" distribution curve.

unrelated to any measurable concentration of the agent in the serum (see below).

Irreversibility of binding does not always mean irreversibility of effect. The drug acetylsalicylic acid (aspirin) is an irreversible inhibitor of the enzyme cyclo-oxygenase, which accounts for many of its pharmacological actions. Provided that exposure to aspirin is terminated, the effect declines as new enzyme is synthesized.

Biological Variation and Data Manipulation

Within any given population of organisms, there will be some that will respond to a drug or toxin at the lowest concentration, and others that only respond at the very highest concentration, while most subjects will be grouped around the mean response. This is true of all organisms, including human beings and single-cell ones. It is even true of populations of like cells (liver cells, kidney cells, blood cells) within the body and may partly explain why some cells may become malignant while others do not. It is the existence of biological variation that necessitates the use of large populations of test subjects and the development of mathematical treatments of data to permit the comparison of different populations of test subjects. If the responses of the species in question are grouped symmetrically about the mean response, a "normal" or Gaussian distribution curve is obtained (Figure 6). In this case, 68.3% of the population will fall within ±1 standard deviation (SD) of the arithmetic mean, 95.5% between ±2 SDs, and 99.7% between ±3 SDs. Anything lying outside these limits is assumed not to belong to the test population. Sometimes the population is skewed, however, with more subjects falling

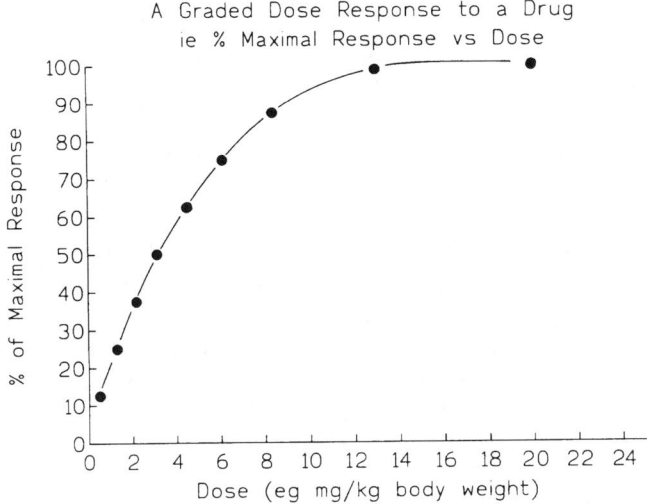

Figure 7 A typical dose-response curve plotted arithmetically.

on one side of the mean than on the other. Factors accounting for variability could include differences in the rate and degree of uptake, distribution, biotransformation, excretion and even the nature and number of binding sites and receptors for the agent. These factors may be under genetic control, or they could be due to environmental differences in such things as temperature, nutrition, disease, the presence of other xenobiotics including medications, and so on. They also tend to vary with age and sex.

Dose Response

A distribution in response to a drug or chemical applies even to a population of like cells within the same organism, or even within a cell culture, because some cells will be more aged than others or may be defective in some way. Therefore, it is impossible, from a single dose of a drug or toxicant, to draw any conclusions about its potency, since it is not known whether one is recruiting only the most sensitive cells or nearly all of them. For this reason, it is necessary to construct a dose response curve whether testing a new drug for its effective dose or a chemical for its toxicity. Typically, the response rises rapidly once the threshold is exceeded and then flattens out as fewer and fewer cells remain to be recruited (Figure 7). This type of response is a graded or *dose-dependent response*. There is another type of response that can be described as yes/no or all-or-nothing, and this is a *quantal response*. Lethality is an example. For graded responses, it is important to establish standard points of comparison, since comparing a dose at the low end of the response for one chemical with one at the top end for another are not statistically reliable. The point usually chosen is that dose which produces 50% of the maximum effect, the effective dose (or concentration) 50% (ED_{50} or EC_{50}). "Dose" is used when the test substance is administered individually to the test subjects, and "concentration" when it is added to the surrounding medium, such as the water in an aquarium or the fluid bathing an isolated tissue in an organ bath.

A quantal response can be converted to a graded one by using several test groups, each receiving a different dose of the agent being tested. The percent of animals showing the expected response can then be plotted against dose. Thus, one can calculate the dose that, on average, will kill 50% of the test animals (lethal dose 50%, LD_{50}). If the response is a toxic one (liver necrosis, kidney failure), the value is the toxic dose 50% (TD_{50}). Variations on this approach include the LD_{10}, TD_{10}, etc. The LD_1 is also called the *minimum lethal dose*. Values such as the LD_1 and the LD_{10} are replacing the LD_{50} in many jurisdictions.

In attempting to compare responses to two different chemicals, it is useful to perform a mathematical manipulation on the data so that differences or similarities in the shapes of the dose response curves are more obvious. This involves plotting the logarithm of the dose against the response, and this converts the exponential curve shown in Figure 7 to the sigmoidal one (S-shaped, Figure 8). It is now much easier to interpolate to the EC_{50} (since the doses selected might not have included it) and to compare these points. Using the log of the dose tends to overcome the fact that large increases in dose result in small increases in response on the right side of the curve, whereas small increases in dose result in large increases in response on the left side of the curve. Thus, the midpoint, the EC_{50} (or TD_{50}), provides the greatest statistical reliability. Parallel slopes of curves suggest similar mechanisms of action, and comparisons based on molar concentrations provide information on relative potencies. Toxicity comparisons may be done by calculating the Therapeutic Index (TI) if the substance is a therapeutic agent. This is the LD_{50}/ED_{50}. The higher the number, the safer the agent. Other estimates of safety, more appropriate to toxicity studies of nontherapeutic agents, involve comparisons of the TD_1 and the EC_{99}.

It is important to note that all toxicity tests contain a temporal factor in that the determination of toxic effects is conducted at a specific time after exposure. Acute toxicity studies generally involve determinations made 72 hr after a single high dose, whereas long-term toxicity requires multiple exposures with measurements made at least 28 days later. These studies are defined by government regulations in jurisdictions where there is a legal requirement for testing new chemicals.

Another value that is frequently used is the *NOEL or NOAEL*, the No Observable (Adverse) Effect Level. The NOEL includes effects, such

PRINCIPLES OF PHARMACOLOGY AND TOXICOLOGY 17

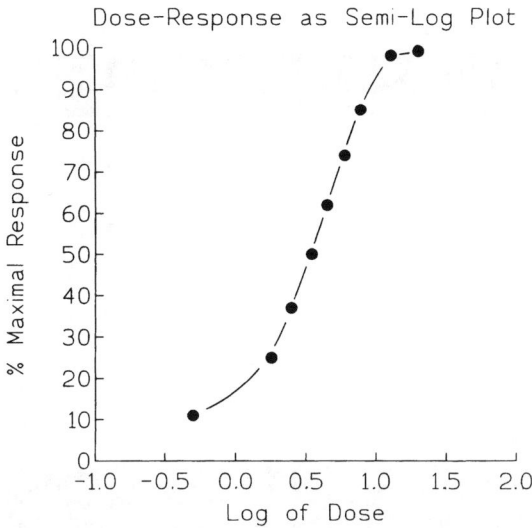

Figure 8 A semilogarithmic plot (response vs. log-dose) of the dose-response curve shown in Figure 7.

as minor weight loss, that are not considered to be adverse. These values are applicable only to that species in which the test was conducted. Extrapolation to other species will require dosage adjustment.

Probit Analysis

It is also often desirable to compare the toxicity of one xenobiotic to that of another, as this information may help to determine whether a substance used commercially or industrially can be replaced with a safer one, or whether a metabolite of a parent compound is more or less toxic than the compound itself. For this purpose, the probit analysis is often used.

When a toxic reaction is expressed as the number of experimental animals in a group displaying that reaction (e.g., kidney failure), the percent of a group responding to a given dose or exposure can be expressed as units of deviation from the mean, and these are called normal equivalent deviations (NEDs). The NED for the group in which there were 50% responders would be zero, since it lies right on the mean. An NED of +1 corresponds to 84.1% responders. NEDs are positive or negative relative to the mean, so a value of 5 is added to each to make them all positive. The result is called a probit (for probability unit, Table 1).

When quantal data are plotted as probit units against the log of the dose, a straight line results, regardless of whether the original data were distributed normally or were skewed. The method, in fact, assumes that the data were distributed normally. It is now easier to compare the quantal data for two different xenobiotics exhibiting

TABLE 1
CONVERSION OF PERCENT RESPONDERS TO PROBIT UNITS

% Responding	NED	Probit
0.1	−3	2
2.3	−2	3
15.9	−1	4
50.0	0	5
84.1	+1	6
97.7	+2	7
99.9	+3	8

TABLE 2
LETHALITY DATA IN FATHEAD MINNOWS FOR TWO TOXIC CHEMICALS

	mg/L	
Lethality	Fluorene	Naphthalene
10%	25.0	0.5
20%	50.0	1.0
60%	100.0	2.0
80%	200.0	4.0

the same toxic manifestation (or their lethality). These concepts apply equally to toxicological studies in mammals and in nonmammalian species. Table 2 illustrates these concepts using hypothetical toxicity data for two toxicants tested in fathead minnows. Each test group consisted of 100 fish. Values listed are milligram per liter concentration in water. Tables are available for conversion to probits.

When using aquatic or marine organisms for toxicity studies, it is important to remember that they are exposed to a given concentration of the test substance continuously, but they may not take it up instantly or even rapidly. A consistent time of exposure must therefore be incorporated into the experimental design. These data are shown as arithmetic (Figure 9), semilogarithmic (Figure 10) and probit (Figure 11) plots.

Cumulative Effects

It may be that the manifestation of toxicity does not occur until the individual has been exposed continuously or repetitively (as with repeated, daily injections) for a prolonged period, perhaps days or weeks. This generally occurs with agents that are metabolized or eliminated very slowly, so that the rate of intake exceeds slightly

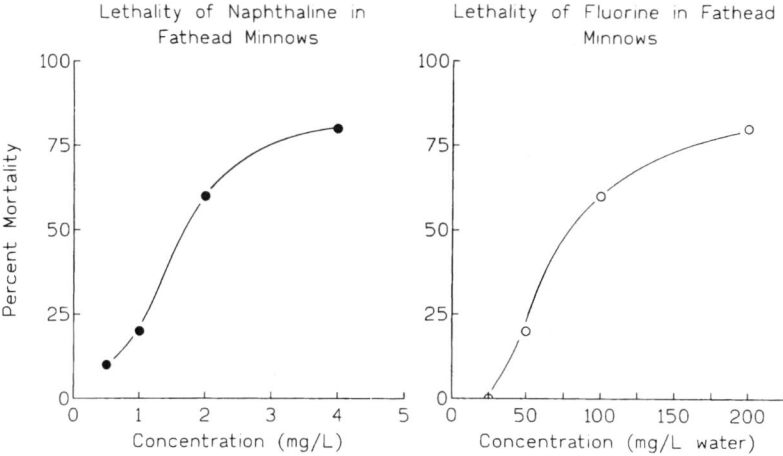

Figure 9 Arithmetic plot of data shown in Table 2.

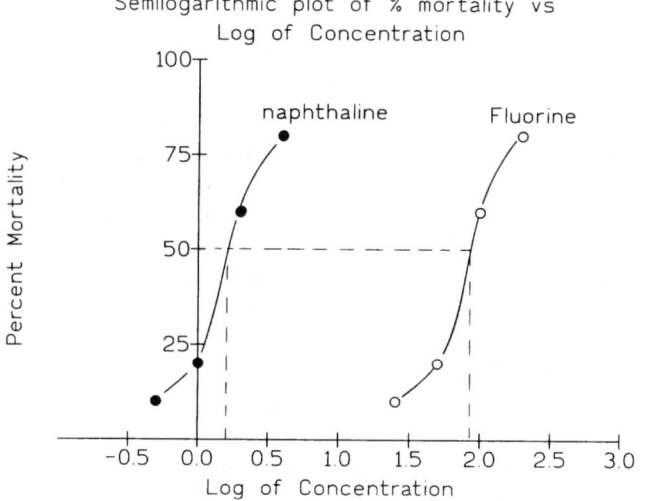

Figure 10 Semi-log plot of data shown in Table 2 and Figure 9.

the rate of detoxification and the drug slowly accumulates until a toxic level is reached. It is analogous to filling a bathtub with a faulty drain. The tub will fill, but only slowly because the water is running out almost as fast as it comes in. This involves the concept of *biological half-life or T1/2*. The plasma T1/2 of a drug or chemical is the time required for the plasma concentration to fall 50%. It is important to note that in most cases, this value is a constant for a given xenobiotic, i.e., it remains the same regardless of the initial

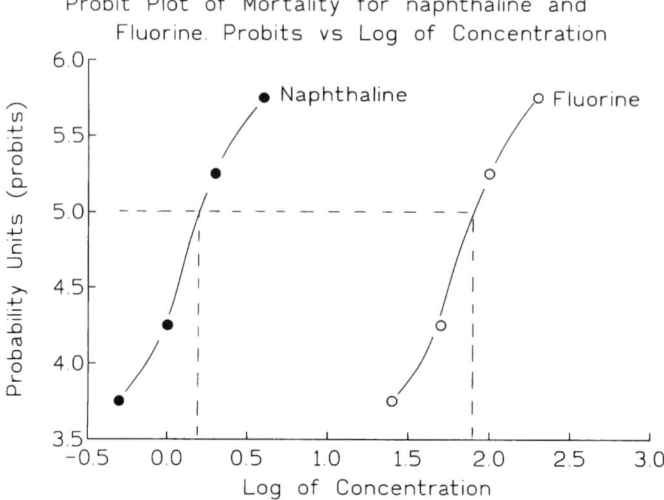

Figure 11 Probit plot of data shown in Table 2 and Figures 9 and 10.

level of the chemical. This is because the rate of biotransformation and excretion are usually concentration-driven, increasing or decreasing according to the plasma level. Biotransformations are enzymatic processes that generally obey first-order kinetics, i.e., the conversion rate is dependent upon the initial concentration of substrate. T1/2 values may also be established for other tissue compartments in the body. If the exposure interval greatly exceeds the T1/2 of a substance, it may be virtually eliminated between exposures. It requires about 5× the plasma T1/2 to achieve virtual elimination. If the exposure interval is equal to, or less than, the T1/2, and the dose is repeated frequently, the xenobiotic may be sequestered in organs and tissues. If no detoxification or elimination occurs, of course, the chemical will accumulate significantly with each exposure. Cumulative effects can occur as the result of repeated exposures even though no detectable levels of the toxicant accumulate, as when genetic damage is induced by carcinogens. Figure 12 illustrates the relationship of frequency of exposure and T1/2 with respect to clearance from tissues. These factors influence whether a toxic reaction is defined as *acute* (within 48 hr), *subacute* (in 7–9 days), *subchronic* (±90 days) or *chronic* (>90 days).

■ FACTORS INFLUENCING RESPONSES TO XENOBIOTICS

It should be evident that anything that influences the absorption, distribution, metabolism or excretion of a xenobiotic will affect its toxicity. Many such factors exist.

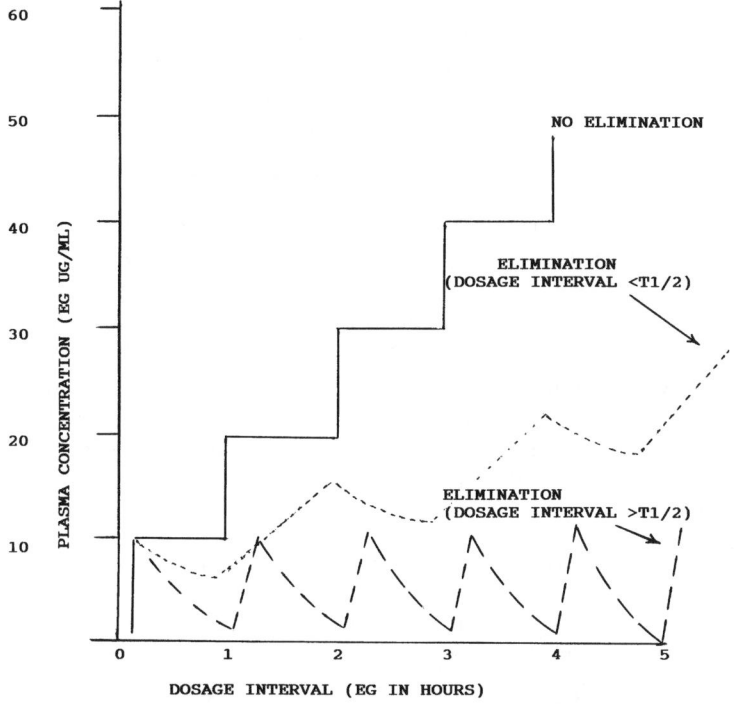

Figure 12 The influence of frequency of exposure or dose on tissue concentration of a chemical.

Age

In general, biotransformation and excretion are less efficient at the extremes of life. Although drug metabolizing enzymes are detectable at mid-gestation, they do not become fully developed until 6–12 months of age. Thus, the toxicity of many substances is higher in neonates. An example is the antibiotic chloramphenicol, which can accumulate to toxic levels because the enzyme glucuronide-conjugase is lacking. Renal function also is underdeveloped so that excretion of drugs is impaired. The T1/2 inulin clearance is 100 min in infants >6 months of age vs. 67 min in adults.

Body composition also differs with age. Total body water is 70–75% of body weight in newborns vs. 50–55% in adults. Extracellular fluid is 40% of body weight in newborns vs. 20% in adults. There is thus a greater fluid volume for dilution of water-soluble drugs in these infants. Their basal metabolic rate is higher than in adults. Other differences include greater permeability of skin which has resulted in toxicity due to absorption of the hospital germicide hexachlorophene. Gastric pH is higher, and gastric emptying is prolonged so that heavy metal absorption is increased. The intestinal flora will differ, and biotransformation by microbes will be different as well. In later childhood these situations may be reversed because systems are functioning at peak efficiency. After age 75, these same systems may have slowed down significantly compared to a 30-year-old. Renal function and respiratory tidal volume are down, drug metabolizing enzymes are less efficient, and even the number of drug receptors may be lower. Body composition also changes. The ratio of fat to lean body mass increases with age, and total body water is down. Cardiac output is reduced, and perfusion is lower. Plasma albumin content is down. The toxicity of water-soluble toxicants like alcohol will be increased because it will be more concentrated. The concentration of the free component of albumin-bound agents will be increased because fewer albumin-binding sites are available. Biotransformation and excretion of xenobiotics will be impaired. The following are some examples of function at age 75 expressed as a percent of function at age 30:

Nerve conduction velocity	90%
Basal metabolic rate	84%
Cardiac output	70%
Glomerular filtration	69%
Respiratory function	43%

Body Composition

Body composition also varies considerably among normal individuals independent of age, and the factors discussed above with respect to fat and water content will be in play here as well.

Sex

Men and women may differ in response to xenobiotics. In part, this is due to differences in body size, fat content and basal metabolic rate. In one study, the T1/2 of antipyrine (an old and toxic analgesic) was 30% longer in young men than in young women. Differences in response to sex steroids obviously occur. Pregnancy is a special situation involving great changes in the metabolism, body composition and fluid content of the mother. Transplacental transfer of many agents occurs, and this may put the fetus at risk (see below). Sex-related differences also occur in experimental animals. Cessation of respiration has been shown to occur more frequently in female than in male rats after barbiturate anesthesia.

Genetic Factors

For the majority of the population, biotransformations are controlled by multigenetic determinants. This is the basis of the continuous variation in response as reflected in the characteristic normal population distribution curve. In some cases, however, a single gene locus may be responsible for altering the metabolism of a substance. This occurs in a subset of the population and affects a particular

PRINCIPLES OF PHARMACOLOGY AND TOXICOLOGY 23

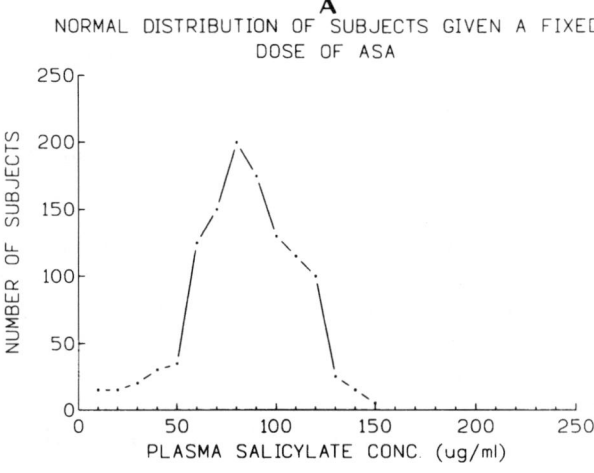

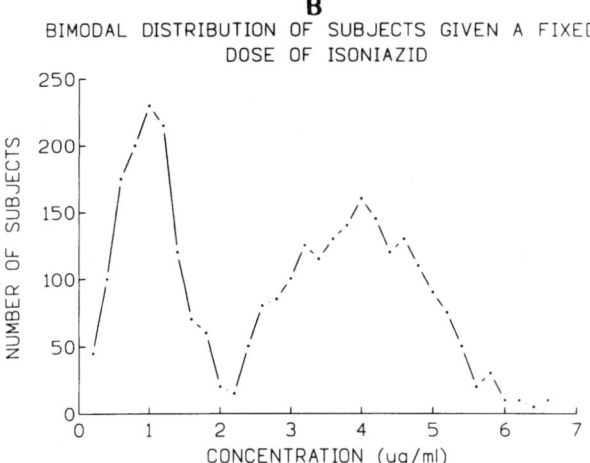

Figure 13 Hypothetical construct of multigenetic vs. single gene influence on drug metabolism.

enzyme. The result is a discontinuous, bimodal distribution which reflects two, overlapping normal distribution curves.

In Figure 13, representative dose-response distributions are shown for acetylsalicylic acid (ASA, aspirin) as an example of the multigene type of control and isoniazid as the single gene type, in this case for an acetylating enzyme.

Pharmacogenetics

The subdiscipline of pharmacology called pharmacogenetics deals with this phenomenon. Over 100 examples have been identified of genetic differences in the biotransformation of drugs and chemicals.

One of special concern for toxicology is the enzyme N-acetyltransferase. It acetylates and detoxifies many drugs and chemicals including the aryl amines which are potential carcinogens, as well as many drugs including isoniazid, an anti-tubercular agent, and sulfa drugs. Slow acetylation is the dominant pattern in most Scandinavians, Jews and North African Caucasians. Fast acetylation predominates in Inuit and Japanese. Similar patterns have been found in other species, including rabbits. Slow and fast oxidative metabolism has also been reported. About 9% of North Americans are slow metabolizers. In one study, a group of dye workers who were exposed to N-substituted aryl compounds were surveyed for the occurrence of *in situ* carcinoma of the bladder. Those with the disease were predominantly slow acetylators, suggesting that slow acetylators probably accumulated a carcinogenic agent normally detoxified by N-acetylation.

Thiopurine S-methyltransferase (TPMT) also is subject to genetic polymorphism. This enzyme is responsible for S-methylation of the antineoplastic drugs azathioprine and 6-mercaptopurine. About 88.6% of humans have high, 11.1% intermediate and 0.3% low or undetectable levels of enzyme activity. Individuals with high levels may be poor responders to cancer chemotherapy unless dosage is adequate, while those with low or nonexistent levels are in danger of developing complete suppression of bone marrow function unless, again, dosage is adjusted downward. Other aromatic and heterocyclic sulfhydryl compounds are methylated by this class of enzyme, and there is little doubt that toxicity can be affected by genetic differences in enzyme activity. Similar genetic polymorphism has been demonstrated for the drugs debrisoquine and metoprolol, which are detoxified by hydroxylation.

It is now technically possible through the use of modern techniques in molecular biology, notably DNA genotyping by the use of appropriate primers and the polymerase chain reaction (PCR) method, to identify individuals in advance who may pose therapeutic problems because of a lack or excess of detoxifying enzyme activity. Such technology may also be applied to the workplace to identify persons who are at excessive risk of a toxic reaction from a chemical in the work environment, or perhaps individuals who have already incurred DNA damage because of past exposure to a mutagen. These people could be excluded from high risk areas (for them) or denied employment. Paradoxically, this possibility has already raised concerns in some union quarters that individuals might be denied their right to earn a living on biochemical grounds. In the future, employers might have to balance union concerns against the possibility of future litigation by workers made ill by their jobs. There is certainly no doubt that new ethical dilemmas will arise out of this technology.

PRINCIPLES OF PHARMACOLOGY AND TOXICOLOGY 25

Genetic factors also determine the emergence of strains of organisms resistant to normally toxic agents. There are many examples, including resistance of mosquitoes to DDT, rats to warfarin, malarial parasites to many drugs, cancer cells to anticancer drugs and bacteria to antibiotics. In all cases, susceptible cells are killed off, leaving resistant mutants to proliferate. Mutations are occurring all the time. In bacterial populations, it has been estimated that a mutation imparting a degree of resistance to a drug occurs once in every 10^9 cell divisions. Unless that particular drug is present, however, the mutant strain will have no selection advantage, and it will be overwhelmed by nonmutant cells. If the drug is present, it will have a distinct survival advantage, and it will become the dominant form.

Genetic factors also may influence the response to a toxicant at the target site as well as at the site of biotransformation. Inherited disorders that render individuals more susceptible to drug-induced hemolytic anemia include glucose-6 phosphate dehydrogenase (G-6-PD) deficiency and sickle cell anemia. A host of drugs, including antimalarials and sulfonamides, will induce a hemolytic attack in these people. There are also several inherited disorders of hemoglobin synthesis that act similarly. Many of the drugs involved are oxidizing, aromatic nitro compounds and nitrates. As noted above, altered gene coding plays a major role in carcinogenesis.

Presence of Pathology

Given that the liver and the kidneys are the major organs of detoxication, it follows that any serious impairment of their function will have a significant impact on the toxicity of xenobiotics. This has been observed in fatty necrosis of the liver, hepatitis and cirrhosis. These will impact most on highly lipid-soluble agents requiring biotransformation to more water-soluble forms. Pharmacologically, this is seen with CNS depressants including the tranquilizers diazepam and chlordiazepoxide. Kidney dysfunction is reflected mainly on water-soluble agents, and their elimination may be greatly impaired by renal disease. Water-soluble antibiotics such as gentamicin have a greatly prolonged T1/2 in the presence of renal disease. Cardiovascular disease may affect tissue perfusion and the delivery of the xenobiotic to, or conversely, its removal from, its target site. Pulmonary disease will also affect the transfer of volatile agents across the alveolar membrane. In meningitis, the presence of inflammation compromises the integrity of the blood-brain barrier, and substances that would normally be excluded may reach significant concentrations in the spinal fluid. This fact makes some antibiotics (e.g., penicillin) useful to treat meningococcal meningitis even though they normally do not penetrate the barrier.

One of the most serious consequences of preexisting pathology may simply be the fact that if an organ has already lost much of its function, further damage by a toxicant may destroy it completely and create a life-threatening situation. This is one of the main reasons why the elderly are more vulnerable to toxic effects of drugs and chemicals.

Xenobiotic Interactions

The effect that one drug may have on the action of another, collectively known as drug interactions (or drug-drug interactions), is an important aspect of clinical pharmacology. The effects of two drugs given together may be *additive* if they induce the same response (even through different mechanisms), *synergistic* if the total response of their combined effect is greater than the predicted sum of their individual effects, or *antagonistic* if one drug diminishes or prevents the effects of the other. Mechanisms involved in drug interactions include altered absorption from the gastrointestinal tract, altered excretion (renal, biliary, respiratory), competition for receptors (antagonism), summation of pharmacological effects and altered biotransformation. Drug interactions are usually associated with adverse drug reactions, but they may also be exploited. All antidotal remedies are based on drug or chemical interactions. Examples include the use of atropine to treat organophosphate poisoning, metal chelators for treating lead and other heavy metal poisoning, the use of activated charcoal to bind toxicants in the gastrointestinal tract and prevent their absorption, emetics to induce vomiting and the removal of toxicants, naloxone to reverse the respiratory depression of opiates and N-acetylcysteine to treat acetaminophen poisoning.

When more than two substances are present, the possibility for a drug interaction is much greater. This is of special concern in the area of environmental toxicology because of the multiplicity of xenobiotics that may be present in water, food and air. Most of the attention has centered on the possibility that the presence of one substance may increase the concentration of carcinogens from other sources by affecting their metabolism through enzyme induction. We have already discussed the effect of cigarette smoke on hepatic enzymes. Many volatile solvents also are enzyme inducers, and chronic exposure to low levels of these, as in the industrial setting, can lead to increased enzyme activity. It has been shown experimentally that bedding rodents on soft-wood shavings induces hepatic microsomal enzyme activity because of the volatile terpenes given off. Workers in soft-wood sawmills could experience similar effects. Literally hundreds of industrial chemicals have been shown to be enzyme inducers, including benzo[a]pyrene and 3-methylcholanthrene. Others include insecticides, (DDT, aldrin, dieldrin, lindane, chlordane), polychlorinated biphenyls (PCBs), polybrominated biphenyls (PBBs), dioxin, and drugs such as phenobarbital and other barbiturates, steroids and others. Thus, even though a substance is itself not toxic at a particular exposure level, it may influence the toxicity of other agents. The enzymes usually involved are the cytochrome P-450 monooxygenases, but conjugating enzymes also may be induced. Food may contain natural enzyme inducers. The potato contains "a-solanin", the tomato "tomatin", both of which are steroidal alkaloids. The bioflavonoids are fairly potent inducers. They are found in species of *Brassica* including Brussels sprouts, where they have been shown to affect the metabolism of some drugs. Rutin is a bioflavonoid found in buckwheat in fairly large amounts. While it is difficult to identify situations where these various inducers have influenced toxicity (with the exception of cigarette smoke), their ubiquity illustrates how different individuals or groups may display different sensitivities to the same toxic exposure on different occasions.

Organic solvents have been shown to influence ethanol toxicity. Toluene depresses alcohol dehydrogenase and prolongs the ethanol T1/2. Ethanol itself may increase the liver and CNS toxicity of CCl_4, trichloethylene and others.

Diet can also affect the toxicity of substances in other ways. A toxicant may be adsorbed to dietary components which reduce its absorption. The ability of dietary calcium to reduce lead toxicity is well documented. Lead and calcium appear to compete for the same absorption site on the intestinal mucosa so that a diet high in calcium will reduce lead absorption. A diet high in fiber will shorten the transit time of the gastrointestinal tract so there is less time available for absorption. This may be one reason why high fiber diet is associated with a lower incidence of colon cancer.

■ SOME TOXICOLOGICAL CONSIDERATIONS

Acute vs. Chronic Toxicity

Acute and chronic toxicity for a single agent may be quite different, and one is not a reliable predictor of the other. For example, acute benzene intoxication involves central nervous system (CNS) disturbances such as excitation, confusion, stupor and convulsions. Chronic toxicity includes depression of the bone marrow and a reduction of all circulating blood cells (pancytopenia), and benzene is carcinogenic in experimental animals. Chronic carbon monoxide (CO) poisoning is experienced by heavy smokers who may suffer from headache, dizziness and shortness of breath. Acute CO poisoning affects fire fighters and fire victims who may become comatose. Numerous other examples exist. A major area illustrating this is that of tumor formation. A short-term exposure to a substance may elicit acute toxicity without significant risk of carcinogenesis, whereas long-term exposure to very low levels may not result in any toxic manifestation but could induce tumor formation. Dioxin (TCDD) is a prime example of this. Acute exposure causes the skin rash known as chloracne, whereas long-term exposure may be carcinogenic.

A detailed discussion of toxic mechanisms is beyond the scope of this chapter. More detailed descriptions are given with the discussions of specific chemicals and groups of chemicals. What follows is a brief overview to illustrate the role of target organs and systems in toxic reactions.

Acute Toxicity

Acute toxicity generally refers to effects that occur following a 24–72 hr exposure to a single or multiple doses of a toxicant. Effects are usually observed within a few days. If the agent is rapidly absorbed, the effect may be immediate. The CNS is very vulnerable to acute toxicity from very lipid-soluble agents.

1. *Peripheral neurotoxins*: Organophosphate and carbamate pesticides inhibit acetylcholinesterase (AChE) so that acetylcholine (ACh) accumulates and overstimulation of receptors occurs. ACh is a neurotransmitter in both the central and peripheral nervous systems. There are many naturally occurring neurotoxins, including tubocurarine, which blocks nerve transmission to voluntary muscle, botulinum toxin (from *Clostridium botulinum*), which prevents the release of ACh from nerve endings, and tetrodotoxin (from the puffer fish), which paralyses nerves by blocking sodium channels. Belladonna alkaloids (atropine and scopolamine) from nightshade are muscarinic blockers with peripheral and central effects, and muscarine from mushrooms is a muscarinic stimulant. Nerve gases are also irreversible AchE inhibitors (see Chapter 11).
2. *Central neurotoxins:* The inhalation of many volatile organic solvents and petroleum distillates can act like anesthetics and may cause unconsciousness. This has occurred in industrial accidents and in substance abuse (e.g., gasoline sniffing). Ethyl, methyl and isopropyl alcohols are CNS depressants.
3. *Inhibitors of oxidative phosphorylation*: Cyanide (CN) in the form of cyanogenic glycosides (e.g., amygdalin) is present in many plant components including almonds, the pits of cherries, apples and peaches, plums, apricots and wild chokecherries. Human poisonings have occurred from consuming too many of these seeds, and livestock are often poisoned from eating chokecherry bushes. Cyanide binds to heme to prevent electron transfer. Cyanide may be present in metal ores, in some pesticides and in metal polishes. The tragic accident at Bhopal, India, which killed over 2,000 people, was due to the release of 40 tons of methyl isocyanate from an American Cyanamid plant. Azide and hydrogen sulfide act like CN.

Carbon monoxide (CO) combines with hemoglobin and cellular cytochromes and prevents association with O_2. It thus causes cell hypoxia and also interferes with O_2 transport by red blood cells.

4. *Uncoupling agents:* Many agents act as uncouplers, preventing the phosphorylation of ADP to ATP, the high energy phosphate. Uncouplers include the herbicide 2,4-D, halogenated phenols, nitrophenols and arsenate. Most contain an aromatic ring structure. O_2 consumption and heat production are increased without an increase in available energy.
5. *Inhibitors of intermediary metabolism:* Certain fluoroacetate compounds of natural origin are used professionally as rat poisons. They inhibit the citric acid cycle to deplete available energy stores. The heart and the CNS are the organs of toxicity.

Chronic Toxicity

Exposure to some toxicants must occur over days, weeks or months before signs of toxicity appear. Heavy metals tend to act in this manner. Several outbreaks of methylmercury poisoning have followed this pattern. Mercury (Hg) from industrial discharges may be converted to methylmercury by microorganisms in the water. Monomethylmercury is CH_3Hg and dimethylmercury is $(CH_3)_2Hg$. These accumulate up the food chain to concentrate in fish and shellfish which may be consumed as food. As the toxicant accumulates in the tissues, severe neurological disorders occur. This happened at Minamata Bay in Japan and on the Grassy Narrows reserve in Northern Ontario. Infants born to exposed mothers may suffer from a cerebral palsy-like syndrome.

Cadmium (Cd), used in nickel-cadmium batteries, in electroplating and in pigments, may accumulate in workers and cause kidney damage. Similar cases in the general population have been reported in Japan, from eating rice and other grains grown in soil contaminated with industrial wastes. Carbon tetrachloride (CCl_4) was used extensively in the drycleaning industry before it was discovered that it caused hepatic necrosis because it was activated to a free radical by cytochrome P-450-dependent monooxygenase.

■ MUTAGENESIS AND CARCINOGENESIS

Some toxic effects cannot be related to the level or frequency of exposure because the xenobiotic or its metabolite attaches irreversibly by binding to an essential component of the living cell to disrupt normal function. The binding sites are usually on macromolecules such as nucleotides, nucleosides, regulatory proteins, RNA and DNA. The effects may be cumulative, and they cannot be related to blood or tissue levels at the time of their manifestation. There may be a considerable delay from the initial exposure to the emergence of toxicity. This kind of reaction is involved in mutagenic and carcinogenic effects and also causes a type of anemia, called aplastic anemia, in which the capacity of the bone marrow to produce blood cells is permanently destroyed. Some older drugs such as chloramphenicol and phenylbutazone have caused aplastic anemia in a very small number of patients, perhaps 1/40,000 exposed persons. These low frequency toxicities are difficult to identify and necessitate careful risk/benefit analysis when therapeutic interventions are contemplated.

A toxic effect of great concern in society today is, of course, the propensity of some xenobiotics to induce cancer. A wide range of natural and synthetic chemicals may induce alterations in DNA that, depending on the nature of the defect and the timing of its occurrence, can cause a neoplasm, a heritable change (*mutation*) or a developmental birth defect. It should be noted that birth defects (*teratogenesis*) may result from chemical interference with many other cell processes not involving altered DNA, such as interference with essential substrates and precursors, impaired mitosis, enzyme inhibition, altered membrane characteristics, etc. Most anticancer drugs are teratogenic for the same reason they are antineoplastic. Mutations are not always harmful, and they provide the necessary genetic diversity for natural selection to occur, but they can also be responsible for fertility disorders, hereditary diseases, cancers and malformations. Three types of genetic abnormalities may be induced. Point mutation, also called gene-locus mutation, involves the alteration in some way of a small number of base pairs. This may involve deletion, addition or the substitution of an incorrect base pair. Strictly speaking, the term *mutagenesis* refers exclusively to this type of DNA alteration. When the total number of chromosomes is altered, this is termed *aneuploidization*. Chromosomal aberrations involving gaps, breaks and translocations are referred to as *clastogenesis*. Genetic diseases of eukaryotes can arise from all three types of chromosomal abnormalities. A characteristic of chemically induced carcinogenesis is that it is believed to involve decades-long exposure to very low levels of carcinogens, making predictions from animal studies exceedingly difficult. The induction of a cancer is thought to be a complex, multistage process involving interactions among the carcinogen and environmental and endogenous factors. Three stages are generally recognized: (1) initiation, (2) promotion and (3) progression.

1. *Initiation*: The induction of a mutation by an electrophilic chemical or metabolite (a genotoxic carcinogen) that binds to DNA is believed to be the initial step. The heritable characteristic is not expressed.
2. *Promotion*: Promoters are agents that increase the number of tumors, increase their growth rate, or decrease the latency period. They do not bind to DNA and are therefore called epigenetic carcinogens. Exposure to the initiator must occur first. Promoters do not themselves generally cause tumors, although there is evidence that some promoters are merely weak carcinogens that act

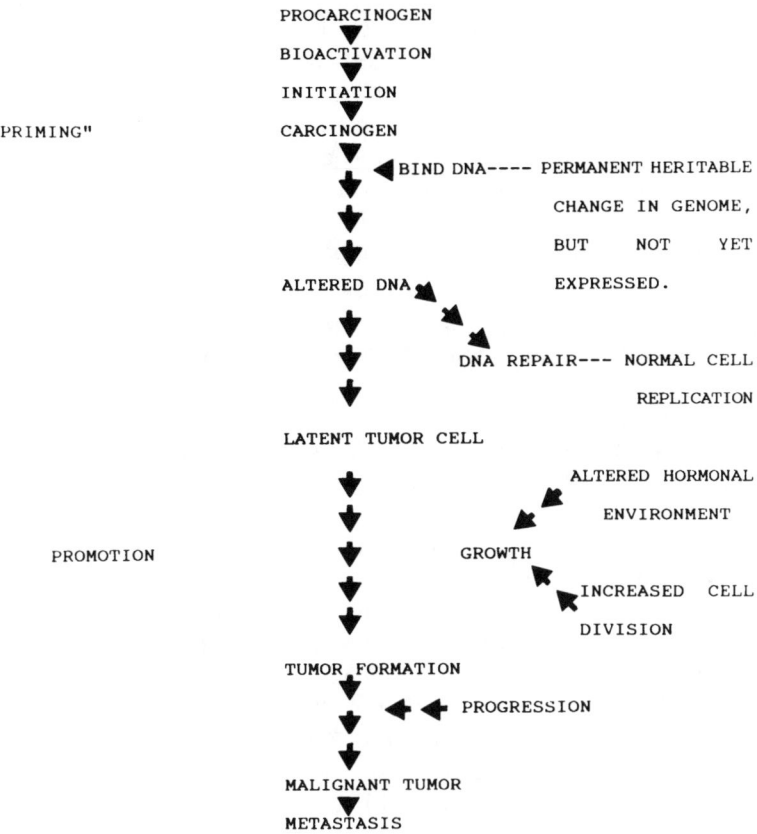

Figure 14 The steps in the progression of carcinogenesis.

synergistically with other carcinogens. Cocarcinogens are agents that, when present just before, or together with, a carcinogen, result in significantly higher tumor yields in experimental animals. Some substances may be both promoters and cocarcinogens (e.g., phorbol esters). Indeed, these categorizations may be largely artificial, as nearly all promoters have been shown to be carcinogenic in animal studies. Nutritional factors (e.g., saturated fat intake), hormones, trauma and viruses may act as cocarcinogens.

3. *Progression* refers to the natural history of the disease. Figure 14 summarizes the steps involved in carcinogenesis.

An important advance in understanding carcinogenesis was the discovery of the existence of oncogenes. These are genes present in some cells that, when activated, convert them to malignant forms. They may preexist in some cells, or they may be introduced by oncogenic viruses, usually RNA retroviruses. Over 30 human oncogenes have been identified to date. An area of research that is currently attracting much attention relates to the existence of tumor suppressor genes. These have been called the "flip side" of the oncogene story and appear to modulate normal cell growth to prevent the emergence of neoplastic cells. Inactivation of these suppressor genes may be an essential step in tumorogenesis, at least in some cases, and prevention of this inactivation by inserting normal copies of tumor suppressor genes, or by mimicking their function pharmacologically, may lead to new therapies.

THE ROLE OF CELL REPAIR AND REGENERATION IN TOXIC REACTIONS

The ability of a tissue to repair itself after chemical insult is dependent on the proportion of cells damaged, the nature of the damage and the rate of cell division. Cells of the epithelium (skin, hair, nails) and mucous membranes (gastrointestinal tract, lung, urogenital tract) are capable of rapid cell division and tissue regeneration. Sloughing of these tissues is protective because it carries away any toxicants accumulated in the cells from low-level exposures. Bone marrow also has a rapid rate of cell division as do the germ cells of the developing fetus. Damaged bone marrow may regenerate if enough cells survive. Conversely, such rapidly dividing tissues are vulnerable to destruction when exposed to higher levels of cytotoxic agents. It is for this reason that hair loss, anemia, diarrhea and gastrointestinal hemorrhage may occur with the use of antineoplastic drugs and following exposure to radiation. These adverse reactions occur because the interval between cell damage and cell division is too short for repair to occur. Cell division cannot proceed or is so impaired that second generation cells cannot themselves reproduce.

Cells of the liver, kidney, exocrine and endocrine glands and connective tissue have less rapid rates of cell division, but are capable of proliferation and repair.

Heart cells, nerve cells, and voluntary muscle cells regenerate poorly, but recent evidence suggests that these may be replaced to a greater extent than previously thought. Increased muscle mass from conditioning is due to increased actin and myosin content of the cells.

Response of Tissues to Chemical Insult

Tissue Regeneration

The destruction of some cells may lead to the release of growth factors which stimulate cell division and proliferation, followed by maturation of the cells to a functional state. Function is thus restored.

Cell Regeneration
If a portion of a cell is destroyed but the nucleus remains intact, the cell may regenerate. Thus, the axon of a neuron may eventually regenerate if the cell body is undamaged.

Hyperplasia
This refers to an increase in the size of an organ or tissue in response to increased demands upon it. This is a normal adaptive phenomenon, and it involves an increased number of normal cells. It tends to occur in organs of intermediate specialization such as the liver, pancreas, thyroid, adrenal cortex and ovary. Hyperplasia of the liver may occur in response to an increased demand to detoxify a xenobiotic presented in high concentration or for a long period.

■ FETAL TOXICOLOGY

Teratogenesis
The existence of malformed infants and animals has been recorded for thousands of years. Early explanations tended to be supernatural. Deformed infants were often regarded as warnings from the gods. The word "monster" comes from the Latin "monstrum", meaning portent. Malformed animals were thought to be the result of copulation with humans. Those suspected of such an act were often put to death painfully.

Although it was recognized as early as 1932 that dietary deficiencies could result in birth defects, it was not until 1962 that the possibility of drug-induced teratogenesis was recognized as a result of the phocomelia observed in the thalidomide infants in Germany, Japan, the U.K. and other countries, including Canada. This tragedy resulted in the introduction of regulatory requirements for tests for teratogenic effects of new drugs in all industrialized countries.

The placenta is a semipermeable membrane which will pass many xenobiotics and their metabolites. Whereas most nutrients cross the placental membrane by energy-consuming active transport systems, toxicants cross mainly by passive diffusion, so that lipid solubility is an important determinant of toxicity to the fetus. The exceptions are antimetabolites (anticancer drugs) which are analogs of natural substrates and which may utilize their transport pathways. Highly lipid-soluble agents will establish equilibrium between the maternal plasma and the fetus very quickly. More polar agents will take longer to accumulate, but *there is no such thing as a "placental barrier"*. The placenta possesses biotransforming enzymes so that some detoxication may occur, but it is not enough to protect against any but very low exposures. The human placenta is capable of Phase I and II

TABLE 3

CAUSES OF TERATOGENESIS

Known genetic transmission	20%
Chromosomal aberration	3–5%
Environmental causes	
Radiation (therapeutic, nuclear)	<1%
Infections (rubella, herpes, toxoplasma, cytomegalovirus, syphilis)	2–3%
Maternal metabolic imbalance (endemic cretinism, diabetes, phenylketonuria, virilizing tumors)	1–2%
Drugs and environmental chemicals	2–3%

biotransformations, but the enzymes are not inducible. Biological functions in the fetus, even near term, are poorly developed. The blood-brain barrier is imperfect; biotransforming enzymes are undeveloped, as is the excretory function of the kidney. The earlier in gestation, the more this is so. There are several consequences of fetal exposure to toxicants:

1. If the exposure is high enough, embryonic death will occur and possibly resorption. This tends to occur early in gestation.
2. If the exposure occurs during structural development, teratogenesis may occur and an anatomical defect may result, the nature of which will reflect the stage of development at which the exposure occurred. There is thus a critical period during which a target site exists which does not exist before or afterward. This is what occurred when pregnant women took the drug thalidomide to control morning sickness.
3. If structural development is more or less complete, a functional defect may occur that may not become evident until later in life when the affected function would normally come into play. The newborn infant may thus be vulnerable to developmental toxicity for several weeks after birth until organ systems become fully matured.

Teratogenesis may occur from causes other than drugs and environmental chemicals. In fact, in 65–70% of cases, the cause is never established. Developmental toxicity can occur in the absence of any maternal toxicity. Of the remainder, the breakdown is as shown in Table 3. Samples of known teratogenicity resulting from fetal exposure to toxicants include the following (others no doubt exist):

1. Folic acid antagonists (aminopterin) used as anticancer agents.
2. Androgenic hormones (natural and synthetic progesterones) used to treat breast cancers (and by athletes to increase muscle mass) will masculinize female offspring.
3. Thalidomide, a very potent teratogen, used for a single day during day 20–50 of pregnancy was associated with phocomelia (flipper-like limbs). Most animal species are much less sensitive than humans.

4. Fetal alcohol syndrome (FAS) involves impaired growth, impaired mentation and distinct facial characteristics (small head, small eye openings, thin upper lip). Fetal alcohol effects (FAE) involve only one or two of these criteria.
5. Infants exposed to methylmercury *in utero* may develop severe neurological disturbances similar to cerebral palsy. This occurred in the Minamata exposure in Japan.

Transplacental Carcinogenesis

The development of cancer as a result of fetal exposure is a possibility if the reproductive system is involved. The best-known example is the occurrence of carcinoma of the vagina and cervix in young women exposed *in utero* to diethylstilbestrol (DES). DES was given to prevent impending abortion, mostly between 1950–1970.

Carcinogens abound as natural substances in the environment. Table 4 compares some relative risks of these with synthetic ones. The problems associated with attempting to predict the carcinogenic potential of new chemicals from animal studies will be discussed in the next chapter. The problem is critical, due to the high cost and protracted period it takes to evaluate a new chemical. Table 5 lists some potential human carcinogens that occur in industry and that may contaminate the environment.

TABLE 4

RELATIVE CARCINOGENIC HAZARDS IN THE ENVIRONMENT

Source	Cancers/100,000 population
U.S. average lifetime risk (all types)	25–30,000
Foods	
Four tablespoons peanut butter/day (aflatoxin)	60
One pint milk/day (aflatoxin)	14
8 oz. broiled steak/week (nitrosamines, polycyclic aromatic hydrocarbons, PAHs)	3
One diet soda/day (saccharin, U.S.)	70
Average U.S. fish consumption/day	33
Lake Michigan sport fish (based on median consumption levels and EPA potency values)	480–3300
U.S. sport fish (based on Kim and Stone potency values)	77–340
Drinking water	
Average U.S. groundwater 2 L/day	1
Niagara River water 2 L/day (EPA potency value)	0.3
Air	
Various estimates, U.S. urban	10–560

TABLE 5
SOME POTENTIAL CARCINOGENS ENCOUNTERED IN INDUSTRY

Asbestos from mining, in insulation.
Benzene (32 million tons/year produced); chemical and petrochemical industries, solvents (paint industry), as an impurity, etc.
Carbon tetrachloride (CCl_4); solvent in industrial processes.
Chlorinated dibenzofurans.
Formaldehyde (embalmers, pathologists).
Halogenated hydrocarbons (PCBs, PBBs, TCDD); in lubricants, transformer insulation, pesticides, etc.
N-nitrosodimethylamine (NDMA, softener for copolymers, lubricant additive).
PAHs (e.g., benzo[a]pyrene); produced by any combustion process, refining and distilling of petroleum, breakdown of lubricants, in welding and foundry processes.
Vinyl chloride in plastics industry.

It is generally accepted that the root of most anthropogenic environmental problems is overpopulation. The doubling time of the Earth's population is down to about 30 years, and this has created incredible pressures on our ability to provide food, living space, energy and manufactured goods, and particularly on our ability to deal with the waste products of our society. Yet there has been little official recognition of this in the West, nor specific actions to control the population explosion. This is changing, but fear of reaction from religious groups and right-to-life activists, who tend to link population control to such issues as abortion and birth control, has made the issue unattractive to politicians. The lessons of nature should convince us of the need to address this issue. When a population exceeds the ability of its ecosystem to support it, it dies out.

■ REVIEW QUESTIONS

1. Match the following terms with the appropriate definition:
   a. Pharmacodynamic
   b. Probit
   c. Economic toxicology
   d. Xenobiotic
   e. Anthropogenic

   ___. i. A substance foreign to living systems.

   ___. ii. A chemical process in biological systems in which neither of two agents which interact are permanently altered.

   ___. iii. The application of toxicology to achieve an advantage for humankind.

___. iv. Resulting from human activity.

___. v. A probability unit, used for making comparisons of potency or toxicity.

2. Select the correct statement.
   a. A quantal response is one that is "all or nothing".
   b. The response to increasing doses of a drug or toxin continues to increase indefinitely.
   c. The minimum lethal dose value is applicable to all species.
   d. A quantal response can never be converted to a graded one.
   e. Semilogarithmic dose-response plots are used to clean up bad data.

Answer the following questions True or False.

3. If drug A has a Therapeutic Index (TI) of 500 and drug B has a TI of 10,000, drug A is safer than drug B._____

4. An acute toxicity reaction is defined as one which occurs within 48 hr of exposure._____

5. Parallel dose-response curves suggest that the two agents in question probably work through the same mechanism._____

6. If the T1/2 of an agent exceeds the dosage/exposure interval, it will never accumulate in the body._____

7. Insecticides which inhibit the enzyme acetylcholinesterase cause rapid loss of consciousness._____

8. Nerve toxins may work by blocking axonal conduction, blocking the enzymatic destruction of a neurotransmitter, blocking the attachment of a neurotransmitter to its receptor or blocking its release from the nerve terminal._____

9. Methylmercuries are less toxic than elemental mercury._____

10. Cadmium is a nontoxic heavy metal._____

For the following questions, answer A if statements 1, 2 and 3 are correct; B if 1 and 3 are correct; C if 2 and 4 are correct; D if 4 only is correct and E if all (1,2,3,4) are correct.

11. 1. Most anticancer drugs are teratogenic.
    2. Genetic abnormalities may involve point mutations or chromosomal breaks.
    3. Cancer induction is likely a multistage process.
    4. Viruses have nothing to do with cancer.

12. 1. Promoters increase tumorogenesis in response to other agents.
    2. Exposure to the promoter must occur before exposure to the initiator.

3. Promoters do not bind to DNA.
4. Promoters may cause cancer in their own right.

13.
1. Selective toxicity is usually absolute; i.e., a substance is completely toxic for the target species and completely harmless for other ones.
2. The portal of entry may significantly affect the toxicity and carcinogenicity of a xenobiotic.
3. Cancer studies in animals are highly predictable for cancer risk in humans.
4. Oncogenes may be activated to convert a cell to a malignant form.

14. Highly water-soluble chemicals are
1. Excreted by the kidney without having to be biotransformed.
2. Poorly absorbed from the gastrointestinal tract.
3. Very polar.
4. Unable to easily cross the blood-brain barrier.

15. Highly lipid-soluble chemicals
1. Are well absorbed from the gastrointestinal tract.
2. Do not enter the cerebrospinal fluid.
3. Require biotransformation by the liver before being eliminated by the kidneys.
4. Generally have a very short T1/2.

16. Which of the following statements is/are true?
1. Serum albumin has numerous binding sites for weakly acidic xenobiotics.
2. Substances with epoxide bonds tend to bind irreversibly to macromolecules within cells.
3. Mixed function oxidase enzymes in the smooth endoplasmic reticulum of the liver cells are responsible for much drug metabolism.
4. Conjugation of a xenobiotic with glucuronide is defined as a Phase II reaction.

Answer the following questions True or False.
17. The kidney is the sole route of elimination of xenobiotics._____

18. Water-soluble agents of small molecular size pass through the glomerulus._____

19. Body composition has no influence on the fate of chemicals in the body._____

20. Some agents may be secreted by an active process across the wall of the renal tubule._____

21. The presence of one chemical in the body may influence the metabolism of another._____

22. List five chemicals that can induce hepatic microsomal enzymes.

23. List five known industrial carcinogens.

24. Define the following:
    a. Teratogenesis
    b. Carcinogenesis
    c. Mutagenesis

25. List three biological variables that can affect the body's response to xenobiotics.

■ ANSWERS

1. i. d, ii. a, iii. c, iv. e, v. b; 2. a; 3. F; 4. T; 5. T; 6. F; 7. F; 8. T; 9. F; 10. F; 11. A; 12. B; 13. C; 14. E; 15. B; 16. E; 17. F; 18. T; 19. F; 20. T; 21. T.

For Q. 22–Q. 25, find the answers in the text.

■ ESSAY QUESTION

Discuss, with examples, the various ways in which one xenobiotic can alter the response of the organism to another xenobiotic. Refer to other chapters in this text.

■ FURTHER READING

American Medical Association Council on Scientific Affairs, Formaldehyde Report. *JAMA,* 261, 1183, 1989.

Assennato, G., Cervino, D., Emmett, E.A., Longo, G. and Merlo, F., Followup of subjects who developed chloracne following TCDD exposure at Seveso. *Am. J. Ind. Med.*, 16, 119, 1989.

Benet, L.Z., Mitchell, J.R. and Sheiner, L.B., Pharmacokinetics: the dynamics of drug absorption, distribution and elimination, in *Goodman and Gilman's Pharmacological Basis of Therapeutics.* Gilman, A.G., Goodman, L.S., Rall, T.W., Nies, A.S.,Taylor, P. (Eds.), 8th Edit. Macmillan, London, 1990, chap. 1.

Cartwright, R.A., Glashan, R.W. et al., Role of N-acetyltransferase phenotypes in bladder carcinogenesis: a pharmacogenetic epidemiological to bladder cancer. *Lancet,* 2, 842, 1982.

Covello, T., Flamm, W.G., Rodericks, J.V. and Tardiff, R.G. (Eds.), *The Analysis of Actual vs. Perceived Risks.* Plenum Press, New York, 1981.

Hayes, A.W. (Ed.), *Principles and Methods of Toxicology.* 2nd Edit. Raven Press, New York, 1988.

Hodgson, E. and Levi, P.E., *A Textbook of Modern Toxicology*. Elsevier, New York, 1987.

Kalant, H. and Roschlau, W.H.E. (Eds.), *Principles of Medical Pharmacology*. 5th Edit. BC Decker, Toronto, 1989.

Klaassen, C.D., Amdur, M. and Doull, J., *Casarett and Doull's Toxicology: The Basic Science of Poisons*. 3rd Edit. Macmillan, New York, 1986.

Lennard, M.S., Silas, J.H. et al. Oxidation phenotype — a major determinant of metoprolol metabolism and response. *New Eng. J. Med.*, 307, 1558, 1982.

Marx, J., Research news: learning how to suppress cancer. *Science*, 261, 1385, 1993.

News and Comment, Animal carcinogen testing challenged. *Science*, 250, 743, 1990.

News and Comment, Experts clash over cancer data. *Science*, 250, 900, 1990.

Research News, Dioxin revisited. *Science*, 251, 624, 1990.

Richardson, L. (Ed.), *Risk Assessment in the Environment*. Royal Society of Chemistry, London, 1988.

Ross, E.M., Pharmacodynamics: mechanisms of drug action and the relationship between drug concentration and effect. *Goodman and Gilman's The Pharmacological Basis of Therapeutics*. Gilman, A.G., Rall, T.W., Nies, A.S., Taylor, P. (Eds.), 8th Edit. Macmillan, London, 1990, chap. 2.

Sutter, G.W. II (Ed.), *Ecological Risk Assessment*. Lewis Publishers, Chelsea, MI, 1993.

Van Loon, J. and Weinshilboum, R.M., Thiopurine methyltransferase isoenzymes in human renal tissue. *Drug Metab. Dispos.*, 18, 632, 1990.

Zeckhauser, R.J. and Viscusi, W.K., Risk without reason. *Science*, 248, 559, 1990.

2 RISK ANALYSIS AND PUBLIC PERCEPTIONS OF RISK

"Risky Business"

INTRODUCTION

It is estimated that between 60 and 70 thousand industrial and commercial chemicals are currently in use in North America with the possibility of more coming onstream every day. Only about 3,500 of these have been studied sufficiently to conduct any sort of risk assessment regarding human health, and such studies use, characteristically, only one route of administration (portal of entry). About 600 chemicals are currently judged to constitute a significant potential risk to human health either because of their toxicity or because they are manufactured in such quantities that there is likely to be a high level in the environment. The public seems unwilling to give up the advantages accruing from such chemicals (plastics, pesticides, petroleum fuels, etc.), but also it is increasingly vociferous in its demands to be protected from any adverse effects arising from their use. The environmental damage caused by some of these agents is becoming more and more evident, and indeed this may be the real danger facing humankind.

Nevertheless, legislators and regulators are faced with the task of making decisions regarding safe limits for thousands of chemicals, often on the basis of very limited data and in the face of pressure from consumer groups, environmental activists, and industry lobbies.

ASSESSMENT OF TOXICITY VS. RISK

Toxicity assessment is the determination of the potential of a substance to act as a poison, the conditions under which this potential will be realized, and the characterization of its action. Conversely, the *assessment of risk* involves the quantitative assessment of the likelihood of these deleterious effects occurring in a given set of conditions. *Hazard* is similar to toxicity, but includes things like explosiveness and carcinogenicity. These subtle differences are not always appreciated by the public and especially not by the news media. Thus, statements frequently appear to the effect that dioxin is the "most potent poison known to man". In fact, botulinum toxin is 100× more potent in mice, and the toxicity of dioxins in man has not been fully established. Moreover, the real question of risk must consider the following factors:

1. The biological half-life of the substance (dioxins are very stable).
2. The partition coefficient (dioxins are very lipid-soluble; therefore, they are sequestered in the body).
3. Does the toxin concentrate up the food chain? Yes, because of the partition coefficient.
4. What are the long-term effects? Is the substance carcinogenic? Yes, in experimental animals. In humans, the evidence is much less conclusive.
5. What are the predicted risks to humans and the environment based on known levels of contamination? This is the area that causes most controversy because it is highly speculative.
6. What are the costs of avoiding these risks? This is very difficult to estimate and therefore controversial.

While risk to the general public is difficult to assess and usually of a very minor nature, risks encountered by industrial workers may be much greater because of the higher exposures and because of the risk of accidental contamination. Populations in some regions, however, may be exposed to similar risks from industrial accidents or from uncontained dump sites.

PREDICTING RISK: WORKPLACE VS. THE ENVIRONMENT

Acute Exposures

Information from industrial accidents and from preregulation exposures is very valuable because it eliminates the need to make extrapolations from test animals. Prediction of risk following defined exposures is thus fairly accurate as, for example, in the case of cholinesterase-inhibiting insecticides. Animal data is still useful, however, because it too deals with acute exposure.

Chronic Exposures

Predictions are less reliable due to biological variations in susceptibility to chronic, lower levels of exposure. Individual susceptibility to lung damage from paraquat, for example, may vary considerably.

Very Low-Level, Long-Term Exposures

It is more difficult to predict organ toxicity from animal studies with this type of exposure, but they are still useful. Epidemiological data from human exposures are most useful if available. For example, extensive data have accumulated over many decades regarding pneumoconiosis (black lung disease in miners).

Carcinogenesis

At best, predictions from animal data can only be a rough approximation due to the need to extrapolate from very high to extremely low exposures and the possibility of species differences. Differences in the nature of the exposure may further complicate extrapolations from animal data to the human situation. Moreover, predictions of risk due to low-level exposures are complicated by the presence of other risk factors, many of them from natural sources. For example, volcanic eruptions can pour huge volumes of gases and particulates into the atmosphere, equal to years of industrial pollution. After the Mt. St. Helen volcanic explosion, "pneumoultramicroscopicsilicovolcanopneumoconiosis" was coined as the longest word in the English language. It refers to pneumoconiosis from inhaling volcanic ash. Smoking would be an example of an "anthropogenic" risk factor (i.e., of human origin).

■ RISK ASSESSMENT AND CARCINOGENESIS

As already noted, this is the most complicated and least reliable area regarding the prediction of risk to human health in the general population from exposure to very low levels of environmental pollutants. There are several mathematical models for predicting carcinogenic risk, by extrapolation either from animal data or from human industrial exposures. Regarding animal studies, there is general agreement among these models for extrapolation to human exposures at high doses, but at very low exposure levels, predictions of cancer risk can vary by several orders of magnitude, and this is the very type of exposure which creates the greatest concern in the public's mind. These differences arise because of the application of different theories of carcinogenesis to the development of models for calculating risk. Here are some examples of these models:

1. Distribution models (log probit, logit) assume that every individual has a threshold below which no adverse effect will occur (a "No Observable Adverse Effect Level" or NOAEL).
2. Mechanistic models are based on presumed mechanisms of tumorogenesis and assume that a cancer can arise from a single mutated cell. The single-hit model assumes that the exposure of DNA to a single molecule of a carcinogen is sufficient to induce carcinogenesis. The gamma multihit model assumes that more than one "hit" is required. Multistage models assume that carcinogenesis is a process requiring several stages (a series of mutations, biotransformations) involving carcinogens, cocarcinogens and promoters that can best be modeled by a series of multiplicative mathematical functions. Predicted dose-responses are linear at very low exposure levels and assume that there is not a NOAEL.

All of these methods differ in the nature and shape of the dose-response curve at the low-exposure end. Figure 15 illustrates how these differences affect predictions at the low end of the exposure curve.

The U.S. Environmental Protection Agency (EPA) uses the "Linearized, Multistage Assessment Technique" which assumes that there is no NOAEL and which involves the following steps (see also Figure 16):

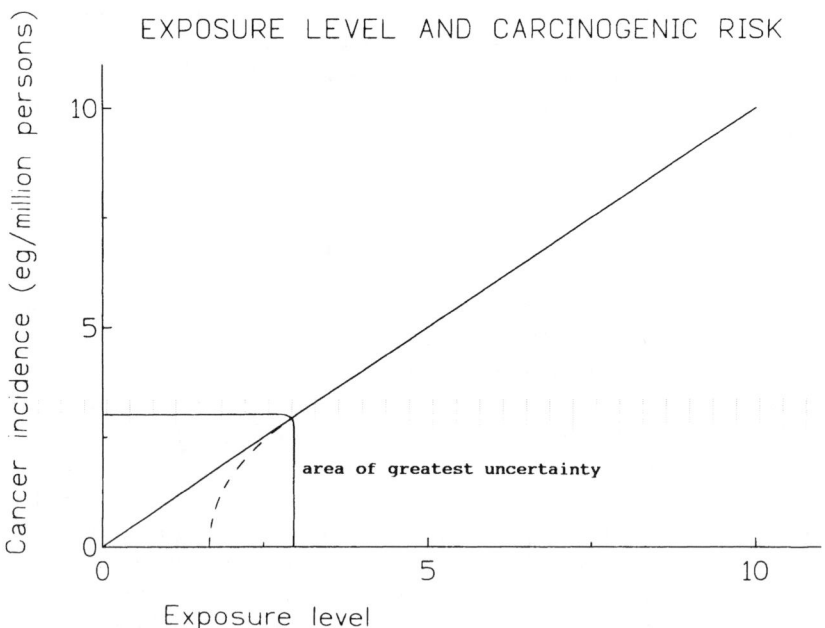

Figure 15 Area of greatest inaccuracy (threshold vs. no threshold) in predicting cancer risk.

RISK ANALYSIS AND PUBLIC PERCEPTIONS OF RISK 45

EVIDENCE OF TUMOR FORMATION
(usually from a single portal of entry in a single animal model)

+

POSSIBLY EVIDENCE OF MUTAGENESIS (Ames Test)

▼

CALCULATION OF EXPOSURE TO CAUSE ONE CANCER IN ONE MILLION ANIMALS
(EPA "red-line, assumes no threshold)

▼

CALCULATION OF EQUIVALENT HUMAN EXPOSURE
(known species differences in absorption, metabolism etc. may be employed in the calculation)

▼

KNOWLEDGE OF AVERAGE DAILY INTAKE OF FOOD, WATER, PLUS A SAFETY FACTOR OF x 1/100 USED TO CALCULATE MAXIMUM ALLOWABLE INTAKE OF TOXICANT

Figure 16 Stages in the process of cancer risk prediction. There are several points of uncertainty.

1. Evidence of carcinogenesis is obtained from animal studies in rabbits, rats and mice, with dose-response data for oral, inhalation or dermal portals of entry (routes of administration).
2. From this dose-response data, the dose is calculated that would theoretically cause one cancer per million animals. The assumption is made that the dose-response curve is linear all the way to zero; i.e., that there is not a "no effect" level for the carcinogen.
3. An equivalent human dose is calculated that would cause the same incidence of cancer. This stage employs arbitrary factors to adjust for differences in absorption, metabolism and excretion based on what data are available for humans, or simply uses a safety margin if no data are available. The 1/1,000,000 risk level is the "red line" that EPA has set for acceptable risk, and it is used to determine safe limits in the environment.
4. Using knowledge of the average human intake of air and water, maximum allowable limits are set for the toxicant that would keep daily intake below the level that would induce one additional cancer per million people. An additional safety margin may be introduced, based on the lowest levels that can be achieved at an acceptable cost. In Canada, the Canada Environmental Protection Act (CEPA) defines the tolerable daily intake (TDI) as the maximum to be permitted. It uses a safety factor of 100× the threshold obtained from animal studies. It also uses the Exposure/Potency Index (EPI), a value that takes into account the level of environmental exposure as well as the known toxicity of a substance, to rank chemicals as to degree of risk. So far, Canada has identified some 44 "priority" chemicals that are felt to be significant risks; the U.S. has 128 on a similar list.

The linearized, multistage model assumes that there is no threshold for carcinogenesis, a reasonable assumption for electrophilic carcinogens affecting DNA, but this may not be true for epigenetic carcinogens such as dioxin. Canada and some European countries set dioxin limits 170–1,700 times higher than EPA limits because they do not apply the linear approach to dioxin risk analysis. The CEPA defines such "threshold" chemicals where possible and treats them separately from those where no threshold exists or where none has been demonstrated.

Sources of Error in Predicting Cancer Risks

Obviously, there are several points in this method that require estimations, and therefore, there may be wide variations in resulting predictions. This is the greatest source of contention between governments and various special interest groups. Environmentalists generally press for reductions in allowable levels, whereas industry may lobby for higher levels if lower ones involve significant cost factors. Some specific sources of contention in risk analysis are discussed below.

Portal-of-Entry Effects

1. The method may not be reliable when exposure of humans involves multiple portals of entry. Volatile chemicals, for example, may be inhaled, ingested and absorbed through the skin.
2. Toxicity may be affected by differences in absorption or biotransformation that occur at the portal of entry, so that data obtained from one type of exposure may not be applicable to others. As an extreme example of portal-of-entry effects, the purest air can be fatal if injected intravenously, as can the purest water if inhaled. Ethyl acrylate produces a 77% incidence of tumors in rats at 200 mg/day orally. The same dose applied to the skin causes no tumors. Cadmium (Cd) is carcinogenic by inhalation but not orally or dermally. Conversely epichlorhydrin will cause tumors at the point of contact with any epithelium. It has been stated that of the >500 risk assessments that have been completed, nearly all involve a single route. This applies both to carcinogenic and noncarcinogenic effects. Numerous examples of route-specific effects exist; e.g., trichlorethylene causes CNS depression at 7 ppm if inhaled, but the same concentration taken orally has no effect because of incomplete absorption.
3. The area of contact may affect uptake, even for the same portal of entry. Thus, if a large area of skin is exposed to a toxicant, more will be absorbed. Moreover, the skin of the forehead absorbs 20×, and that of the scrotum 40×, more effectively than the skin of the forearm. Transit time for ingested material in the intestinal tract may vary from 10 to 80 hr depending on age, diet and other factors, so the time available for absorption will vary as well. The relatively rapid transit time through the small bowel may partly explain the rarity of cancer in this area.

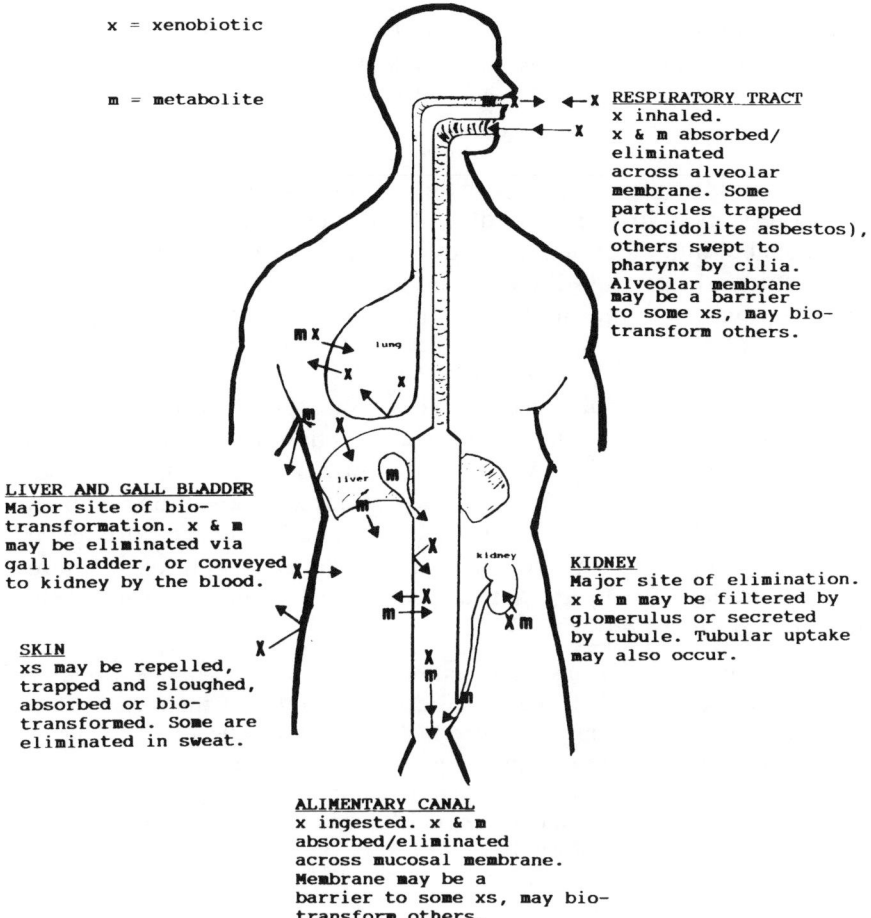

Figure 17 The possible fate of xenobiotics in the body.

Figure 17 summarizes the possible fate of xenobiotics (literally, "foreign to life") that can occur at various portals of entry and thereafter. The mammalian body can be visualized as a thick-walled tube with the outer surface (skin) and inner surface (gastrointestinal tract) in contact with the environment.

Excretion of toxicants and waste products back to the environment takes place through sweat, expired air, feces, urine and the sloughing of cells in contact with the environment.

Age Effects

The age of the population at risk may affect the degree of risk. Infants, especially premature ones, absorb chemicals through the skin much more efficiently than adults. Infants have died from absorbing pentachlorophenol used as an antibacterial agent in hospital bedding before the practice was abandoned. Even data acquired from human industrial exposures usually deal with adult males and may not be applicable to the elderly or to females.

Presence of Cocarcinogens and Promoters

It is often difficult to control for the presence of cocarcinogens and promoters, even in animal studies. Regarding human data, such factors as smoking, alcohol consumption, intake of nitrites and nitrates and saturated fats may differ considerably from an exposed, industrial population to the public at large.

Species Differences

These are the focus of considerable attention both from the scientific community and from animal rights activists who use them to trivialize the value of animal data. It must be stated at the outset that pharmacokinetic differences can be far greater among human beings than between them and experimental animals. Biological variation is a governing force in all living things. Nevertheless, there are important differences that are known and others that are only now being identified.

1. Human skin, for example, is much more impervious than that of laboratory animals, being more similar to that of the pig.
2. The rat forestomach is devoid of secretory cells and is a better model of squamous epithelium than of secretory tissue.
3. Moreover, it contains an active microflora that can alter chemicals, whereas the stomach and upper bowel of the human are virtually sterile because of the acidity.
4. This same acid medium can serve to denature and detoxify potentially harmful chemicals.
5. Anatomical differences in the branching patterns of bronchi exist in the lungs of rodents vs. primates. This can result in vastly different deliveries of inhaled volatile toxins. The pattern in humans is described as "dichotomous-asymmetric", whereas that in the rat is "monopodial-symmetric". In the latter case, the primary bronchi penetrate deeply into the lungs and have secondary bronchi branching off their length. The distance to the terminal bronchiole may vary greatly and hence too, will the target cell exposure (see Figure 18).
6. The rat has no gall bladder, so bile flow tends to be continuous and unaffected by food. Stasis of the bile, which can affect contact time, is rare.
7. There are numerous differences in the nature and location of biotransforming enzymes. Knowledge of these differences, e.g., for cytochrome P450, can be exploited to select the most appropriate model for study.

Despite the problems with extrapolation from animals to humans, it should be remembered that DNA varies from the human array by only 5% in mice, by less than 2% in most primates and by less than

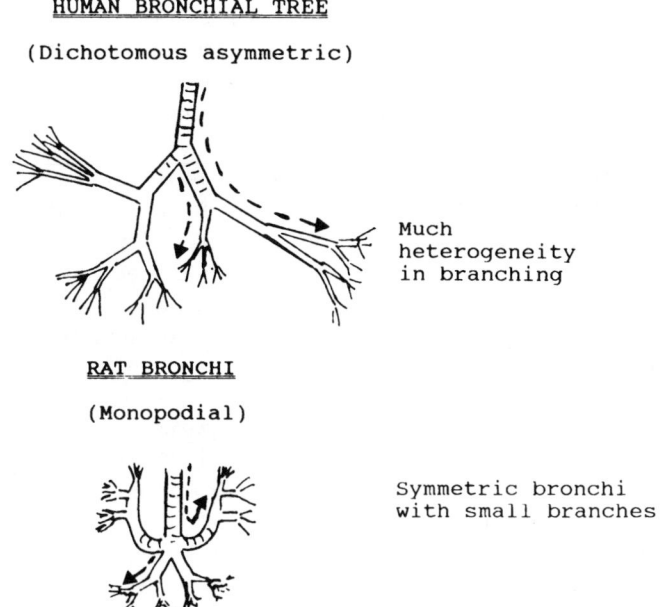

Figure 18 Comparative anatomy of human and rat bronchial trees. Dotted line represents the differences in distances to terminal bronchioles.

1% in chimpanzees. The similarities are far greater than the differences. Moreover, the extrapolation of risk to the general public from data acquired from industrial exposures, including accidents, has its own problems. Numerous differences usually exist between workers and the populace. The former tend to be predominantly males 18–65 years old, from the lower end of the socioeconomic scale, and possibly with different habits regarding such health factors as smoking, alcohol consumption, and diet. There is also the need to extrapolate from moderate to high exposures (and possibly from very high single exposures as in an industrial accident) in the workplace to very low ones in the environment.

Extrapolation of Animal Data to Humans

One source of continuing dispute is the reliability of animal data in extrapolating cancer risk to humans. One critic of the current system is Bruce Ames, inventor of the Ames test for mutagenicity. He now feels that it is too sensitive, and thus it is predicting cancer risks that are artificial for many chemicals. One basis for his argument is that many chemicals are cytotoxic at the high concentrations in standard tests for carcinogenicity, and therefore they induce a high rate of cell proliferation for repair. This in turn increases the likelihood of mutations that could lead to malignancy. Critics claim that any substance, at high enough doses, can be carcinogenic. The debate revolves around the use of the "estimated maximum tolerated dose" or EMTD, as the high dose level in cancer bioassays. This is defined as the highest dose in chronic studies that can be predicted not to alter the animals' longevity from effects other than cancer. According to the proliferation-mutagenicity theory, lower doses should not be carcinogenic if they do not induce cell proliferation. Defenders of the current (U.S.) National Toxicology Program, however, point out that about 90% of chemicals defined as carcinogens induced tumors at doses well below the EMTD, and that, of 33 proven human carcinogens, 91% were shown to be carcinogenic in the animal tests.

Natural vs. Anthropogenic Carcinogens
A more valid criticism perhaps, also raised by Bruce Ames, is the fact that there are hundreds of natural carcinogens in foods to which we are exposed daily, and of 77 which have been tested by the standard methods, about half (37) were carcinogenic. We are thus likely exposed to many more natural carcinogens than synthetic ones. Our natural defenses probably take care of many of these. We slough our epithelial layer regularly (skin, gastrointestinal tract and lungs) and with it, its accumulation of toxins. We have killer lymphocytes which destroy abnormal cells, and we have detoxifying mechanisms which render many toxins harmless unless these defenses are overwhelmed by high doses. The natural decline in these defenses with age is a major factor in the increasing incidence of cancer in the elderly. In view of the abundance of natural carcinogens, elimination of all synthetic ones may not reduce cancer incidences as much as one might expect. The counterargument is that we have had several million years to evolve defenses against natural carcinogens, and these may not work as efficiently against synthetic ones. One indisputable statistic, however, is that life expectancy has been steadily increasing for several decades.

Reliability of Tests of Carcinogenesis
Another concern has been raised recently about the reliability of tests of carcinogenicity. A particular strain of mice, $B_6C_3F_1$, has been widely used in such tests because of its known tendency to readily develop tumors in response to a wide variety of chemicals. Information has been accumulating concerning significant differences in the metabolic processes of rodents and people, and it is likely that using this cancer-sensitive strain of mice may yield too many false positive findings of carcinogenicity. For example, the volatile solvent butadiene, used in the production of synthetic rubber, is exhaled unchanged in most animals and thus does not display carcinogenicity. In mice,

however, much more is retained in the lungs and absorbed (33× as much as in monkeys). It is then oxidized to a mutagenic epoxide. The mouse has a much lower activity of a detoxifying enzyme "epoxide hydrolase", than do humans, so that the carcinogenic risk is many times greater in mice. This strain of mice also harbors a murine leukemia virus that has been shown to enhance carcinogenicity. The (U.S.) National Institute of Occupational Safety and Health (NIOSH) based its estimate of butadiene cancer risks entirely on studies of this strain of mice. THe NIOSH model predicted that exposure to 2 ppm butadiene for 45 years would cause 597 excess cancers in 10,000 workers. In fact, 1,066 workers who had been exposed to levels as high as 1,000 ppm since the industry began in the 1940s, had only 75% of the cancer incidence in the overall population.

The cost of overestimating cancer risks can be enormous, as unnecessary and expensive protective measures are legislated. This can drive industry to seek homes in countries with less restrictive legislation, which may then lead to loss of control over other, truly hazardous, industrial chemicals.

■ ENVIRONMENTAL MONITORING

Environmental monitoring occurs in two ways. Ambient monitoring refers to measurements in water or air downstream or downwind from the source and is primarily a measure of the state of the environment. So-called "end-of-pipe" or point of emission monitoring refers to the measure of effluent levels from drains and stacks and is used to ensure compliance with legislative regulations.

Bioassays are used to look for effects rather than to identify specific chemicals. For water, the water flea, salmonid fingerlings and the opossum shrimp are used. Earthworms and germinating plants are used for testing soil. Sensitive bacteria are used to detect mutagens. The Ames test can detect a few mutations in several million cells, and a newer test, the Microtox assay, measures reduced bacterial luminescence resulting from inhibited cell division. Genetically engineered species such as nematodes are being developed to detect contamination in indoor air — the modern version of the canary in the coal mine. EC_{50}, LD_{50} and TLV values (see below) can be calculated for these.

■ SETTING SAFE LIMITS IN THE WORKPLACE

As noted above, because of data collected after industrial accidents and from preregulation exposures, setting acceptable limits for toxic substances in the workplace can be done with somewhat greater confidence. Exposures are generally higher but of relatively short duration as compared to those in the environment. It should be noted that tolerance limits apply only to the particular type of exposure stated (inhalation, skin contact, etc.). Jurisdictions over worker safety vary widely from country to country. Most western, industrialized countries have similar legislation. In Canada, provincial ministries are responsible for occupational health and safety. In Ontario, this comes under the Occupational Health and Safety Act, Revised Statutes of Ontario, 1980. Regulations made under this act deal specifically with biological, chemical and physical agents in the industrial, construction and mining settings and, most recently, regulations governing the Workplace Hazardous Information System (WHMIS). Under federal guidelines, each province has enacted comparable legislation.

Historically, the development of safety legislation in Ontario dates to the Ontario Factories Act of 1884. The formation of the Royal Commission on the Health and Safety of Workers in Mines in 1976 led to significant additions to legislation dealing with workplace safety. In 1979, the Occupational Health and Safety Act was proclaimed, together with regulations for the industrial, mining and construction settings. In 1984, a Royal Commission on Asbestos resulted in a further regulation under the Act in 1985. In 1981, a list of "designated substances" was begun. These are chemicals that are considered to be especially hazardous and therefore require special controls, restrictions or even prohibition. The regulations apply to all workplaces and other projects (except construction sites) where designated substances are likely to be inhaled, ingested or absorbed. The 1990 list contained the following:

> acrylonitrile
> arsenic
> asbestos
> benzene
> coke oven emissions
> ethylene oxide
> isocyanates
> lead
> mercury
> silica
> vinyl chloride

In 1988, the WHMIS regulations were brought into effect. These define the information that must be on labels of chemical containers in the workplace and the information that must be readily available to the worker in the form of "Material Safety Data Sheets". Ontario Bill 208 expands on the Act. Some important features of Bill 208 are as follows:

1. It is designed to work on the "Internal Responsibility System", the key aspect of which is that workers have
   a. the right to be informed about the nature of the hazards they might be exposed to in the workplace,
   b. the right to participate in the decision process concerning job safety, and
   c. the right to refuse to work in conditions they consider to be unsafe without fear of reprisal from the employer.
2. Workplaces (as defined in the Act) must form a Joint Health and Safety Committee with representation from both the employer and the employees (union) who must have at least numerically equal membership. This committee may make recommendations concerning health and safety, but the employer is not bound to accept them. This is a weakness in the current system.
3. Responsibilities for safety in the workplace must be shared by the employer and the employees.
4. The definition of a worker is "a person who performs work or provides services for monetary compensation", but does not include inmates of a correctional institution, owners or occupants of a private residence or their servants, farmers, or hospital patients. Persons performing work in their homes for monetary compensation are considered to be workers.
5. The Act contains a blanket clause to the effect that "the employer must take every reasonable precaution to ensure the health and safety of the worker" where specific regulations do not exist (they do for such things as protective clothing and equipment).
6. Some important definitions are as follows:

   TWAEV: Time-weighted average exposure value; the average concentration in air of a biological or chemical agent to which a worker may be exposed in a work day (8 hr) or work week (40 hr).

   STEV: Short-term exposure value; the maximum concentration in air of a biological or chemical agent to which a worker may be exposed in any 15 min period. If not specifically defined in the regulations, this is taken as 3× the TWAEV for up to 30 min.

   CEV: Ceiling exposure value; The maximum concentration in air of a biological or chemical agent to which a worker may be exposed at any time. If not specifically defined in the regulations, this is taken as 5× the TWAEV. Levels in air are expressed as ppm or mg/m^3 of air (sometimes called an excursion limit).

   The NOEL and the ADI are values that are employed in animal tests. NOEL: No Observable Effect Level; the highest level at which no toxic effect is observed in experimental animals. ADI: Acceptable Daily Intake. This is the NOEL divided by an arbitrary number, at least 100. Reliability depends on the use of a large number of animals to determine the NOEL. The ADI is not a predictor for carcinogenic risk. ADIs are also calculated for human exposures to determine allowable levels in air, water and food.

The Act also provides regulations governing noise exposures. Standards are different for the mining and industrial sectors.

In the U.S., somewhat different abbreviations are employed, but with similar definitions, for example, threshold limit values (TLVs). These apply mostly to vapors, e.g.:

> TLV-C (ceiling); that concentration that should never be exceeded (= CEV).
>
> TLV-STEL (short-term exposure limit); the exposure that can be tolerated for up to 15 min without irritation, tissue damage or sedation (= STEV).
>
> TLV-TWA (time-weighted average); that level to which a worker may be exposed continuously for up to 40 hr/week without adverse effects (= TWAEV).

▬ ENVIRONMENTAL RISKS: PROBLEMS WITH ASSESSMENT AND PUBLIC PERCEPTIONS

Establishing acceptable levels of toxic substances in the environment is difficult and inexact for the following reasons:

1. Exposure may be for a lifetime, and no data usually exist regarding long-term, very low-level exposures.
2. Toxic effects may be difficult to identify unless they are unusual. Angiosarcoma of the liver was readily associated with vinyl chloride because it is otherwise very rare. For the same reason, Kaposi's sarcoma was associated with AIDS. If hare lip were the teratogenic effect of a chemical pollutant, however, the association would be difficult because it is a fairly common congenital defect.
3. Public perceptions of risk may be very exaggerated, forcing the legislation of much lower maximum levels than are necessary.
4. The concept of risk-benefit analysis may seem callous and calculating to segments of the populace. In their view, there may be no such thing as an acceptable level of risk, and they frequently have an imperfect understanding of the meaning of statistical probability. Moreover, the possibility that avoiding one risk may increase another is often overlooked. The debate over the expansion of nuclear vs. fossil fuel power plants is a good example of this. The elimination of nuclear power plants would force greater reliance on coal-fired generators, with increases in acid rain production, release of greenhouse gases and a greater risk of fatal mining accidents (see Chapter 12).
5. Risk factors may be additive or even synergistic, making analysis even more difficult.

The Psychological Impact of Potential Environmental Risks

One of the greatest adverse effects of environmental pollutants is undoubtedly psychological in nature. Human beings have a natural fear of the unknown and incomprehensible, and the field of environmental toxicology brings both of these into play. The human mind is a highly impressionable instrument. Many years ago, when one city's drinking water was first fluoridated, it was announced that the fluoride would be added on a certain date. On that date, the switchboard of the municipal offices was flooded with calls complaining about the taste of the water. It was then announced that the fluoride had actually been added a week earlier! Only then did the phone calls subside. It is well established that after every major environmental accident, there is a rash of vague medical problems such as headache, nausea, dizziness, etc. Cancerphobia (fear of cancer) leads to attributing every new case of cancer in the area of exposure to the accident. Such beliefs may persist even after extensive studies have failed to reveal any difference in incidence between exposed and nonexposed populations. A sheep farmer living near a nuclear generating station may attribute every abnormal birth and fetal deformity (not uncommon in lambs) to radiation from the power plant despite evidence from extensive testing indicating that there were no radiation abnormalities on the farm.

Voluntary Risk Acceptance vs. Imposed Risks

Public pressure may force the expenditure of vast sums of money to avoid risks that are practically nonexistent. Conversely, a large segment of the public may steadfastly refuse to take steps that are inexpensive and proven to reduce premature deaths. The use of seat belts in cars is a prime example of this. In the U.S., motor vehicle accidents affect 1 of every 50–60 persons each year. The statistical probability of incurring such an injury during one's lifetime is thus quite high, but the risk of being injured on any given trip is very low, and this is what influences the public's perception of risk. Recent surveys indicate that only about 15% of American drivers routinely use passenger restraints despite extensive efforts at educating the public. This pattern has not changed much in the last decade. Users differed from nonusers in a number of ways. Nonusers tended to be less well educated, were more often smokers, rated seat belts as uncomfortable or inconvenient more often, often considered seat belt legislation to be an infringement of personal liberty and less frequently knew someone who had been injured in a car accident. This pattern appears to be repeated when the mandatory use of helmets by bicyclists is newly introduced to a jurisdiction, with indignant letters to newspaper editors protesting this latest infringement on personal freedom of choice. A recent report evaluated similar measures for horseback riders and found that the rate of serious injury per number of riding hours was greater than for either motorcyclists or automobile racers. In a recent 2-year period, there were nearly

93,000 emergency room visits in the U.S. for riding-related injuries with over 17,000 head and neck injuries.

Competition riders are required to wear safety equipment. In North America, the National Hockey League recently chose to make helmets for players optional, illustrating the degree of risk elements of the public will accept in some circumstances.

Costs of Risk Avoidance

Because it is not possible to truly "save" a life, but only to postpone death, it is common in this field to refer to "premature deaths avoided". This requires an arbitrary decision regarding normal life expectancy and what constitutes a premature death, and this definition will keep changing as life expectancy is extended. About one in three people will develop cancer by age 70 as a result of the gradual deterioration in immune and cellular defenses. If a carcinogen in the environment significantly increases the incidence of cancer but mostly in the >70 segment of the population, are these premature deaths?

The following are some calculated costs, in U.S. dollars, of avoiding a premature death by instituting some simple safety and health measures:

1. Screening and education for cervical cancer — $25,000.
2. Installation of smoke alarms in homes — $40,000.
3. Installation of seat belts in autos at time of manufacture — $30,000.
4. Stopping smoking — Actually a saving of $1,000/year plus medical costs avoided. Ironically, a Canadian group calling itself the "Smokers Freedom Society", using data from 1986, recently claimed that smokers actually save society money by dying prematurely and costing less in pension benefits and custodial care. They failed to take into account such factors as the damage done to the health of others by sidestream (secondhand) smoke, the cost in work days lost because of generally poorer health (a higher incidence of respiratory infections, for example) and the cost in lives and property from fires started accidentally by smokers.
5. Installing fire detection and control systems in commercial aircraft cabins — $200,000.
6. In contrast, banning of diethylstilbestrol, a synthetic estrogen used as a weight-gain promoter in cattle and suspected of being carcinogenic — $132 million.

A recent example of this cost-effectiveness problem comes from the experience of some American states which introduced compulsory AIDS testing for couples applying for a marriage license. While this approach may seem somewhat draconian in the current social climate, it is by no means a new concept. Not many decades ago, most jurisdictions required a Wasserman test for syphilis before a marriage license would be granted. As a result of the mandatory AIDS tests, 160,000 tests were performed at a cost of $5.5 million.

23 subjects were identified as HIV-positive, for a cost of $239,130 each.

Cost-benefit analyses frequently involve conclusions that may be repugnant to a segment of the public. The cost of detecting one breast cancer through annual mammography in 40–50-year-old women has been estimated at $144,000. In the 55–65-year-old group, it drops to $90,000. Legislators are required to make such difficult choices because of fiscal restraints, often in the face of severe criticism.

■ SOME EXAMPLES OF MAJOR INDUSTRIAL ACCIDENTS AND ENVIRONMENTAL CHEMICAL EXPOSURES WITH HUMAN HEALTH IMPLICATIONS

Radiation

In 1979, a serious accident occurred at the Three Mile Island Nuclear Generating Station near Middletown, Pennsylvania. This was a disaster without noise, smoke or visible evidence of damage. The information the public received came from the press which, early on, labelled the event a manifestation of the "nuclear disease". Over 10 years later, the cleanup is still in progress. It has produced 2.3 million U.S. gallons of weakly radioactive water, mostly from tritium. This could have been discharged into the river without exceeding federal standards, but public pressure forced the installation of a special evaporator at a cost of $5.5 million. The calculated cost of avoiding one premature death is as follows: the maximum exposure from the river discharge would have been 2 microrems (rems). This is equivalent to a 4 min exposure to natural source radiation such as radon or cosmic radiation. The collective dose (mean exposure × number of persons exposed) would have been about 1 person-rem (prem). The calculated incidence of cancer for radiation exposure is 1/5,000 prem. The computed cost of avoiding one premature death is about $25 billion. The Chernobyl disaster will be discussed in Chapter 12.

Formaldehyde

In the late 1970s, a report was published indicating that formaldehyde fumes caused nasal cancer (squamous cell carcinoma of the mucosa) in rats and in one strain of mice. No evidence of carcinogenesis in hamsters could be shown, and there were no human studies suggesting a carcinogenic potential for formaldehyde. This type of tumor is rare in people, and no evidence of increased incidence in workers exposed to high levels of formaldehyde could be found. The EPA used the linear extrapolation multistage method to evaluate risk. Their initial evaluation (1981) was overturned by a federal court on the grounds that there were not enough data to determine risk. Legal wrangles between the EPA and the chemical industry continued for the next few years. Urea formaldehyde foam insulation was banned in the U.S. and Canada, and many homes were ripped apart and the insulation removed. Homeowners agitated for government subsidies to finance this. Over the next few years, additional epidemiological data accumulated on populations deemed to be exposed to higher than average levels of formaldehyde. Whereas people exposed to 0.4 ppm (medical, dental, nursing and science students) had cancer incidences little different from the general population, pathologists and funeral service workers exposed to 3.0 ppm had almost 2,000 additional cases of cancer per 100,000 over the expected frequency. In 1987, the EPA defined formaldehyde as a probable human carcinogen. The EPA set the time-weighted average exposure value (TWAEV) at 1 ppm and the short-term exposure level at 2 ppm. An action level of 0.5 ppm was set. Above this, regulations come into force governing such things as medical surveillance, protective equipment and training. Most people can detect formaldehyde by odor at 0.5 ppm. For example, the characteristic smell of new carpeting is due to formaldehyde used in the adhesive. Particle board also contains it in the binding adhesive. The smell of formaldehyde is rarely if ever detectable in homes insulated with urea formaldehyde insulation, which releases it as the foam slowly breaks down. Levels are unlikely to reach the "action level" set by the EPA for industry, especially if ventilation is adequate, except in new homes where levels up to 1 ppm have been measured due to its release from building materials and carpeting. The Canadian government, however, subsidized removal of such insulation to the tune of $10,000 per home. An important source of toxic aldehydes as a potential health hazard is cigarette smoke. Acrolein is much more irritating than formaldehyde, and the industrial TLV is set at 0.1 ppm. It is a major contributor to the irritant properties of cigarette smoke and photochemical smog.

Dioxin (TCDD)

On July 10, 1976, a serious explosion and fire occurred at a chemical plant near Milan, Italy. As a result, over 1 kg of TCDD (tetrachlorodibenzo-p-dioxin, the most toxic of the dioxins) was spewed over the adjacent countryside. The chemical fallout was heaviest in the town of Seveso, where concentrations in some parts reached 20,000 $\mu g/m^2$ of surface area. By the end of July, 753 people were evacuated from the area. 3,300 animals died, and 77,000 were killed. 500 people were treated for acute skin irritation, and 192 eventually developed chloracne. There also was evidence of liver damage as indicated by increases in serum enzyme levels. Several followup studies were reported, the most recent in 1985. The chloracne was completely reversible; all but one had recovered by 1983. Enzyme levels also returned to normal, and no other abnormalities were reported. To date, no conclusive evidence of excess cancer incidence has been detected. Although recent findings suggested increases in the incidence of leukemia and lymphoma in contaminated areas, the overall incidence of cancer was not elevated. Moreover, it was not possible to relate cancer incidences to exposure levels. In the area thought to be most heavily exposed, none of the cancer rates was significantly elevated. One suggestion is that other carcinogens, such as 4-aminobiphenyl, may have played a role in cancer generation (for a fuller discussion of TCDD carcinogenicity, see Chapter 5).

▬ SOME LEGAL ASPECTS OF RISK

De minimis Concept

Recently, courts in the U.S. and Great Britain have begun to apply an old legal tradition to the question of risk. This is the concept of "de minimis non curat lex" which means "the law does not concern itself with trifles". Applied to risk analysis, the "de minimis concept" means that in some cases the computed risk is so small that it does not justify regulation. For example, the U.S. Supreme Court ruled in 1980 that the Occupational Safety and Health Administration (OSHA) could not further limit levels of benzene fumes in the workplace unless it could demonstrate significant risk to workers from existing exposure levels. Similarly, the appeals court of Washington, D.C., ruled that regulations governing plastic containers could not be introduced on the purely theoretical grounds that toxic substances might leach into the contents.

Delaney Amendment

The Delaney Amendment to the U.S. Food and Drug Act has had a significant impact on public attitudes worldwide about cancer risks and on the responses of politicians to those attitudes. It states that no substance that has been shown to induce cancer in animal experiments can be permitted in any foodstuff. Strictly interpreted, it would prohibit the addition of even one molecule of a suspected carcinogen to any food. The U.S. Food and Drug Administration sought to apply the de minimis principle to two food colors that had been shown, in very high doses, to induce cancer in animal tests. A review court ruled against the FDA on the grounds that Congress intended the Delaney clause as an absolute prohibition against carcinogenic dyes in foods. Other authorities, however, are stating that it is time to reassess the Delaney clause in light of the extreme sensitivity of current methods of detecting impurities in food. In all areas of toxicology, the sensitivity of test methods has outstripped our knowledge of the significance of such low exposure levels for human health.

Cancerphobia, an unreasonable fear of developing cancer, has spawned a new phenomenon: suing out of fear of developing the disease in the absence of any signs or symptoms. Some litigants are winning. In 1989, residents of Toone, Tennessee, accepted an out-of-court settlement from the Velsicol Chemical Company of $9.8 million because they had been exposed to contaminated drinking water as a result of corroding barrels in a toxic dump site owned by the company.

A recent controversy concerning a chemical hazard in the environment centered on the use of daminozide sprayed on red apples to prevent premature windfall. Early in 1989, the Natural Resources Defense Council, an American environmental activist group, published a report claiming that children who consume large amounts of apples and apple products could be at increased risk of cancer because a breakdown product of daminozide, unsymmetrical dimethylhydrazine (UDMH), has been shown to be carcinogenic in animals. Maximum levels of Alar in apples is set by the FDA at 20 ppm. In Canada, it is 30 ppm. The NRDC claims that these limits were set before information regarding the cancer risk was available and that the increased risk of cancer could be 45/1,000,000, which exceeds the "socially acceptable" level of 1/1,000,000 that the EPA uses as a guideline. Moreover, this group claims that the acceptable level (1/1,000,000) is exceeded when levels of Alar exceed 0.01 ppm. This contention has been hotly disputed by the FDA and by many scientists, including Bruce Ames, developer of the Ames test for mutagenesis. Thomas Jukes, a prominent microbiologist at Berkeley, points out that "organic" apple juice, recommended by the NRDC, may contain up to 45 ppm of patulin, a carcinogen produced by a penicillium mold. A study undertaken by Health and Welfare Canada in response to this debate found that 13% of 30 samples of apples tested had daminozide levels of from 0.06–3.0 ppm, well below the legal limit but above the acceptable level claimed by NRDC. A dilution effect could well lower it further, even to their limit. How real is the risk? The question is now academic, as the manufacturer has withdrawn the product in the face of adverse publicity.

▪ STATISTICAL PROBLEMS WITH RISK ASSESSMENT

In toxicity studies, to reach the 95% confidence limit, one must have 12% responders in a group of 50 subjects, 30% in 20 subjects and 50% in 10 subjects. To have a chance of seeing one single case in a population at risk, one must test a population three times as large.

In other words, to detect an incidence of 1/100, one must test 300 subjects, or to detect 1/1,000, one must test 3,000, etc. This is an extremely important concept when dealing with toxic reactions that are not dose-dependent. Aplastic anemia from bone marrow depression occurs very rarely in response to some chemicals. For example, the antibiotic chloramphenicol is estimated to cause this in one of 35,000–50,000 treated patients. Even assuming that the same genetic predisposition existed in a test animal as it appears to do in humans, one would have to test 150,000 animals in order to see one case. The effects of high doses, moreover, cannot always be extrapolated to low-dose situations, even if the effect is dose-dependent. The time available for the repair of chemically damaged DNA may be adequate at very low doses, so that the defect is not expressed.

Risk assessment is thus a very inexact process when applied to low-risk situations relating to environmental pollutants. The public tends to overestimate risk and underestimate or ignore avoidance costs. Vested interests tend to do the reverse, and politicians sometimes are influenced by the most powerful lobby, which may be either an industry or an environmental organization. Before the Challenger disaster, NASA estimated the risk of a shuttle accident to be 1/100,000. Empirical data now suggest that the real risk was 1/25. Unconscious bias, small sample size, lack of human data, failure to consider interactive factors such as food chain biomagnification, public pressure resulting from unjustified fears, all may affect the decision process. It is a characteristic of human nature that people will accept significant levels of risk if they can exert some personal choice in the situation, but they will almost universally reject even the slightest degree of a risk which appears to be imposed upon them by government or industry. Examples of voluntary risks discussed above include smoking, not wearing seat belts, not wearing protective helmets and, of course, there is a whole range of hazardous sports which involve risk. The question of relative risk assumes great significance with respect to various sources of energy. Is nuclear energy inherently more dangerous than that from coal-fired generators? This subject will be considered in Chapter 12, Radiation Hazards.

▬ REVIEW QUESTIONS

For the following questions, answer A if statements 1, 2 and 3 are correct; B if statements 1 and 3 are correct; C if statements 2 and 4 are correct; D if 4 only is correct and E if all (1,2,3,4) are correct.

1. Which of the following factors contributes to the degree of risk of a toxicant in the environment?
   1. The biological half-life of the substance.

2. The partition coefficient of the substance.
3. Its toxicity.
4. The level of exposure likely to occur.

2. Which of the following statements is/are true?
   1. Individual susceptibility to a toxicant may vary considerably.
   2. The prediction of risk associated with very low levels of exposure can be done with reasonable accuracy.
   3. Mechanistic models assume that cancer can arise from a single mutated cell.
   4. Distribution models assume that there is no threshold below which a cancer-causing agent will induce tumor formation.

3. Which of the following statements is/are true?
   1. The carcinogenicity of a substance is not affected by the portal of entry.
   2. Toxicity studies are generally conducted using a single portal of entry.
   3. The age of the population likely to be exposed to a cancer-causing agent does not affect the degree of risk.
   4. Cancer incidence data acquired from accidental or industrial human exposures are of greater use for predicting risk in the population at large, if exposed to the same chemical, than animal data.

4. Which of the following statements is/are true?
   1. A population of workers may not be representative of the population at large.
   2. The DNA of other mammals differs by 50% compared to human beings.
   3. Infants generally absorb chemicals through the skin much more efficiently than adults.
   4. Smoking habits have no bearing on cancer incidence due to other carcinogens.

5. Safety in the workplace requires
   1. A means of measuring levels of potentially hazardous substances in the work environment.
   2. A knowledge of the toxicity of the substance.
   3. A knowledge of the influence of level and duration of exposure on risk.
   4. A knowledge of the effectiveness of appropriate safety measures such as respirators.

6. Match the appropriate acronym to the definitions given below.
   a. STEV   b. TWAEV   c. NOAEL or NOEL   d. CEV

   _____.i. The average concentration, in air, of a toxicant to which a worker may be exposed during an 8-hr day or a 40-hr week.

____.ii. The level of exposure at which no (adverse) effect is observed.

____.iii. The maximum concentration, in air, of a toxicant to which a worker may be exposed at any time.

____.iv. The maximum level of a toxicant, in air, to which a worker may be exposed in any 15-min period.

Answer the following questions True or False.

7. Industrial exposure to vinyl chloride has been associated with an increased incidence of angiosarcoma.____

8. About one in three North Americans will develop some form of cancer by age 70.____

9. The "de minimis" concept means that no level of risk from industrial pollutants is acceptable.____

10. The cost of one "premature death avoided" from installing smoke alarms in homes is about $1,000.____

11. Most people can detect formaldehyde by odor at a concentration in air of 0.5 ppm.____

12. The main toxic effect related to the dioxin accident at Seveso has been chloracne.____

13. The rat stomach is a good model of the human one.____

14. The skin of the forehead absorbs chemicals much more efficiently than the skin of the forearm.____

15. The acceptable level of risk from environmental anthropogenic chemicals is defined by the EPA as an one additional cancer per 1,000,000 population.____

16. There are no natural carcinogens of great concern.____

17. Patulin is a carcinogenic mycotoxin.____

18. Highly lipid-soluble toxicants with long T1/2 values tend to concentrate up the food chain.____

19. Skin does not have any biotransforming properties.____

20. By sloughing our skin cells we may also eliminate some carcinogens which they have accumulated.____

■ ANSWERS

1. E; 2. B; 3. C; 4. B; 5. E; 6. i. b, ii. c, iii. d, iv. a; 7. T; 8. T; 9. F; 10. F; 11. T; 12. T; 13. F; 14. T; 15. T; 16. F; 17. T; 18. T; 19. F; 20. T.

ENVIRONMENTAL HAZARDS AND HUMAN HEALTH

■ CASE STUDY #1

A population of miners has been identified as having a high incidence of a rare form of cancer. It accounted for 18% of all deaths in workers who had been exposed 20–30 years previously. Those living close to the mine, and family members of miners, regardless of the location of their home, also had an elevated incidence of this same tumor (2–3% of all deaths). The incidence of death from this cancer in the general population is only 0.0066% of all deaths. The incidence of lung cancer in miners who smoke is 60× the incidence in miners who do not smoke. In the general population, smoking increases the risk of lung cancer by about 20×.

Question 1: What factors, including biological and social, could account for this distribution of the rare cancer?

Question 2: Why is the incidence of lung cancer so much higher in the smoking miners?

Question 3: Why should family members of miners who live some distance from the mine have an elevated incidence of the rare cancer?

■ CASE STUDY #2

Five men were employed in plugging leaks in, and waterproofing, an underground storage tank using epoxy resin paint. The tank contained several inches of water, so the ventilating fans that were available were not used for fear of electrocution. The tank measures 20 yards long by 6 yards wide and was 3 yards deep. It was divided into three connecting compartments, and a single ladder and hatch provided the only means of entry and egress. At 10:00 a.m. one worker left the tank because he was drowsy and nauseated. On reaching the surface, he vomited. At 11:30 another man left the tank to get a coffee. When both men returned, they found the remaining three men dead in the tank.

Question 1: What violations of safe working procedures occurred here?

Question 2: What safety measures could have been instituted to prevent these deaths?

Question 3: What can you say about the source and nature of the toxic substance involved in these deaths?

For both of the above case studies, apply common sense and some of the information provided in this chapter.

■ FURTHER READING

Abelson, P.H., Testing for carcinogenicity with rodents (Editorial). *Science*, 249, 1357, 1990.
Abelson, P.H., Exaggerated carcinogenicity of chemicals (Editorial). *Science*, 256, 1609, 1992.
Ames, B.N. and Gold, L.S., Pesticides, risk, and applesauce. *Science*, 244, 755, 1988.
Ames, B.N. and Gold, L.S., Too many rodent carcinogens: mitogenesis increases mutagenesis. *Science*, 249, 970, 1990.
Ames, B.N. and Gold, L.S., Response to Perera. *Science*, 250, 1645, 1990.
Ames, B. and Swirski. L., Response to Cogliano et al. *Science*, 251, 607, 1991.
Assennato, G., Cervino, D. et al., Followup of subjects who developed chloracne following TCDD exposure at Seveso. *Amer. J. Indust. Med.*, 16, 119, 1989.
Cogliano, V.J., Farland, W.H. et al., Carcinogens and human health: part 3 (letter). *Science*, 251, 607, 1991.
Covello, T., Flamm, W.G., Rodericks, W.V. and Tardiff, R.G. (Eds.), *The Analysis of Actual vs. Perceived Risks.* Plenum Press, New York, 1981.
Klaasen, C.D., Amdur, M.O. and Doull, J. (Eds.), *Casarett and Doull's Toxicology: The Basic Science of Poisons.* 3rd Edit. Collier Macmillan, Toronto, 1989.
Flamm, W.G., Pros and cons of quantitative risk analysis. *Food Toxicology: A Perspective on the Relative Risks.* Taylor, S.L., Scanlon, R.A. (Eds.). Marcel Dekker, New York, 1989, chap. 15.
Formaldehyde — council on scientific affairs (AMA) report. *JAMA*, 261, 1183, 1989.
Goldfarb, B., Beyond reasonable fear. *Health Watch*, Sept./Oct., 14, 1991.
Infante, P.F., Prevention versus chemophobia: a defence of rodent carcinogenicity tests. *Lancet*, 337, 538, 1991.
Marshall, E., A is for apple, alar, and ... alarmist? News and comment. *Science*, 254, 20, 1991.
Marx, J., Animal carcinogen testing challenged. *Science*, 250, 743, 1990.
Perera, F.P., Carcinogens and human health: part 1 (letter). *Science*, 250, 1644, 1990.
Rall, D.P., Carcinogens and human health: part 2 (letter). *Science*, 251, 10, 1991.
Richardson, L.(Ed.), *Risk Assessment in the Environment.* Royal Society of Chemistry, London, 1988.
Samoiloff, M., Environmental contamination: testing, monitoring and detoxification. *Future Health,* Winter, 13, 1990.
Stone, R., New Seveso findings point to cancer. *Science*, 261, 1383, 1993.
Weinstein, I.B., Mitogenesis is only one factor in carcinogenesis. *Science*, 251, 387, 1991.
Weinstein, N.D., Optimistic biases about personal risks. *Science*, 246, 1232, 1989.
Zeckhauser, R.J. and Viscusi, W.K., Risk without reason. *Science*, 248, 559, 1990.

3 WATER POLLUTION

Hang your clothes on a hickory limb, but don't go near the water.

INTRODUCTION

Three components of the biosphere can serve as toxicological sinks: soil, air and water. These are often considered separately, but it should be obvious that they function as an integrated system. Thus, rain will transfer toxicants to soil and water, and evaporated surface water, and soil as airborne dust, can move them back into the air where they may be transported over great distances by wind. Moreover, runoff from the soil, as well as sewage and industrial discharge, are the main source of water contamination. Seepage into deep aquifers from soil and surface water also may occur, and fresh water reservoirs are connected to the sea by rivers and estuaries. Thus, while the separate consideration of water is convenient, it should not detract from an understanding of the integrated nature of the biosphere.

Water pollution is of considerable importance for several reasons. The most obvious is the possibility that xenobiotics may enter drinking water supplies and constitute a direct threat to human health. The contamination of fish and shellfish obtained from both the sea (marine organisms) and fresh-water lakes and rivers (aquatic organisms) may further threaten human health when these foods are consumed. Larger (and older) fish often have higher levels of lipid-soluble toxicants.

Many toxicants are taken up initially by unicellular organisms that serve as a food source for larger (but still microscopic) ones, which in turn are food for bigger ones and so on. This process can lead to increasingly higher concentrations progressing up the food chain, and this is called *biomagnification*. Fresh-water and marine organisms are themselves vulnerable to toxicants, and their survival may be threatened by them. Toxicants can shift the selection advantage for species, so that hardier ones may proliferate to the detriment of others. A classical example of this is the process known as "eutrophication", which results from excessive phosphate levels arising from fertilizer runoff from farmlands and from sewage effluent containing detergents. The high phosphate levels favor the growth of certain algae which bloom extensively and consume available oxygen until there is not enough to support other life forms. Sunlight also will be blocked out, further altering the nature of the ecosystem.

■ FACTORS AFFECTING TOXICANTS IN WATER

All natural water contains soil and all soil contains water, but there is considerable variation in the mix. In fact, it is necessary to distinguish among various types because the behavior of pollutants differs in them. Moreover, the nature of the water itself may vary with regard to hardness, pH, temperature and light penetration with consequences for the fate of pollutants. These modifying factors will be considered in more detail below.

Exchange of Toxicants in an Ecosystem

Figure 19 is a schematic representation of a body of water showing sources of contamination (rain, runoff, effluent discharge, percolation through soil) and some of the means of transferring toxicants to aquatic organisms. Of particular note is the layer of soil/water mix at the bottom interface. This is described as the "active sediment", and it contains, at the surface, a layer of colloidal particles suspended in pore water. The sediment contains organic carbon that tends to take up lipophilic substances. An equilibrium state is thus established with the pore water, which is in equilibrium with the body of water itself. The active sediment is a rich environment for many forms of aquatic life, and particle feeders may concentrate toxicants from the suspended particles, whereas filter feeders will do so from the pore water. Dilution of the toxicant in the principal body of water will shift the equilibrium and release more from the bound state. Thus, removing a source of contamination may not be reflected in improved water quality for some time. The active sediment can thus be both a sink and a source of toxicants (see below).

Limnologists who work primarily in small bodies of water may think of the floccular, low-density active sediment as being only a few centimeters thick, but commercial divers on the Great Lakes describe the phenomenon of sinking up to their helmet top in soft, bottom sediment, a disturbing sensation when first experienced.

Factors (Modifiers) Affecting Uptake of Toxicants from the Environment

Modifiers are classified as abiotic (not related to the activity of life forms) or biotic (related to the activity of life forms).

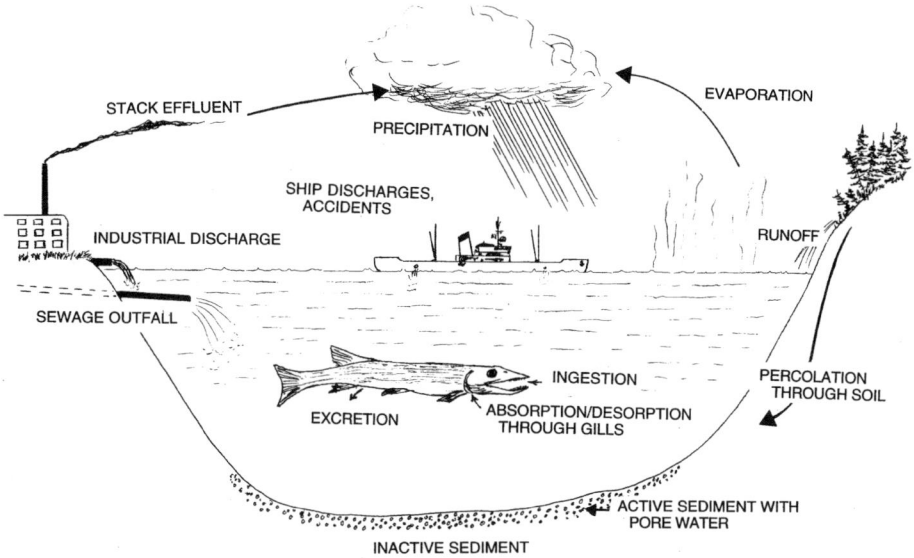

Figure 19 The dissemination of toxicants within an ecosystem.

Abiotic Modifiers

pH. As is the case in any solvent/solute interaction, the pH of the solvent will affect the degree of ionization (dissociation) of the solute. Since the nondissociated form is the more lipophilic one, this will influence uptake by organisms. The wood preservative pentachlorophenol, for example, dissociates in an alkaline medium so that, in theory at least, acid rain would increase the bioavailability of the toxicant by favoring a shift to the lipophilic form. Copper, which is very toxic to fish and other aquatic life forms, exists in the elemental cuprous (Cu^{2+}) form at more acidic pH, but as less toxic carbonates at about pH 7. Toxicity to rainbow trout decreases around neutral pH. An important aspect of pH concerns the methylation of mercury by sediment microorganisms. This occurs over a narrow pH range and is a detoxication mechanism that allows the microorganisms to eliminate the mercury as a small complexed molecule. About 1.5%/month is thought to be converted under optimal conditions (pH 7) (see also Chapter 6).

Water Hardness. Carbonates can bind metals such as cadmium (Cd), zinc (Zn) and chromium (Cr), rendering them unavailable to aquatic organisms. Of course, there will be an equilibrium established between the bound and the free forms, so that removal of the dissolved copper will cause the carbonate to give up some of its copper. There also is an intimate interaction between hardness and pH, so that the lethality curve for rainbow trout will be bimodal at a given degree of hardness, with dramatic increases in the LC_{50}

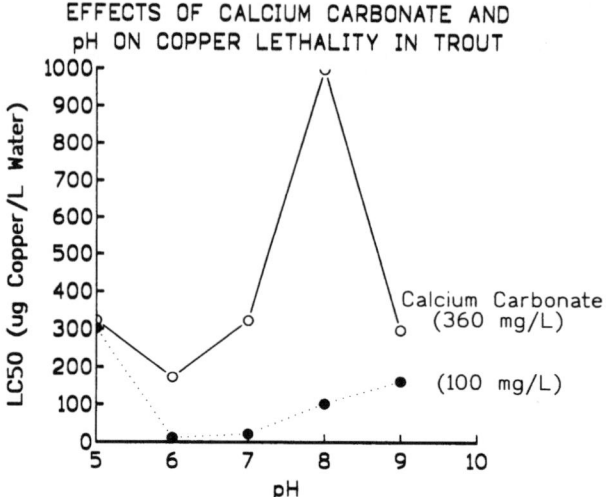

Figure 20 A hypothetical bimodal lethality curve for copper in rainbow trout showing the influence of pH and water hardness.

(Lethal Concentration 50%) at pHs 5 and 8. It should be noted that Canadian Shield lakes tend to be soft because they do not receive drainage from limestone. Figure 20 illustrates this relationship for copper toxicity at a single degree of hardness.

Temperature. Apart from a few mammals, aquatic and marine species are poikilotherms, so that water temperature greatly affects their metabolic rate, which in turn will be reflected in the circulation time of blood through gills, the activity of transport processes and hence, the rate of uptake of xenobiotics. Rates of biotransformation and excretion also may be affected. Temperature also will affect the rate of conversion of mercury to methylmercury.

Dissolved Organic Carbon. These will complex with a variety of lipophilic toxicants and serve as a contaminant sink in sediment and suspended particles. Again, an equilibrium state will exist and if dissolved toxicants are removed or diluted, more will be released from the sink. Sediment typically consists of inorganic material (silt, sand, clay) coated and admixed with organic matter, both animal and vegetable, living and dead.

Oxygen. As noted above, oxygen depletion by algae blooms will compromise other life forms that may be involved in processes of toxification or detoxification including the microbes that form methylmercury.

Light Stress (Photochemical Transformations). Ultraviolet radiation may induce chemical changes in contaminants that may result in more toxic forms of a chemical. Thus, photo-oxidation can increase the toxicity of polycyclic aromatic hydrocarbons (PAHs) through the formation of highly reactive free radicals. In clear water, this effect can be significant at a depth of 6 meters, and it can have a marked impact on levels of toxicants.

Biotic Modifiers

These are essentially the same factors that may affect a patient's response to a drug.

Age. Old trout are less sensitive than fry to some toxicants; larval forms of aquatic organisms usually differ metabolically from adult forms and may concentrate or metabolize toxicants differently.

Species. Many differences exist regarding species sensitivity to toxicants. Salmonid species are generally more vulnerable than carp, which can exist under a wider variety of environmental conditions. Disturbances of the natural ecosystem by the introduction of foreign species can have drastic consequences. The Great Lakes are especially vulnerable to this type of disturbance since the introduction of the system of locks connecting them with the sea. Periodic invasions of marine species have included the lamprey eel, still a major problem for sport and commercial fisheries, the alewife, a small coarse fish which died by the thousands and fouled the beaches until the introduction of the coho salmon controlled them, and the recent plague of zebra mussels which clog water intakes and foul ships' hulls. These are an inhabitant of fresh-water rivers in Europe. Natural transfer of mollusks from marine to aquatic environments is rare because the larvae are not strong enough to swim against river currents. When adults or larvae hitchhike in the hold of a ship, however, it is quite another matter. The current invasion of zebra mussels is thought to have occurred as a result of a ship emptying, in the Great Lakes, its hold of ballast water taken on in a European port. This practice is, in fact, prohibited by law.

Overcrowding. This may constitute an additional stress factor that can influence responses to toxicants.

Acclimation. This refers to the process of adaptation to a single environmental factor under laboratory conditions. Acclimation to heavy metals such as cadmium occurs because of an increase in the levels of metalothionein, a metal-binding protein.

Nutrition. The level of nutrition will affect such factors as depot fat (an important storage site for lipophilic toxicants) and the efficiency of detoxifying mechanisms. The nutritional state in turn may be affected by abiotic factors.

Genetic Variables. Unidentified genetic variables are undoubtedly at work, influencing the response of individuals to xenobiotics.

SOME IMPORTANT DEFINITIONS

Acclimation
This refers to the adaptation of an organism to multiple environmental factors under field conditions.

Bioconcentration
This refers to the uptake of the dissolved phase of a toxicant to achieve total body concentrations which exceed that of the dissolved phase in the water, i.e., against a concentration gradient.

Bioaccumulation
This refers to the uptake of the dissolved plus the ingested phases of a toxicant; e.g., gill breathers absorb lipophilic substances through the gills and consume them in food.

Biomagnification
This refers to the concentration of a toxicant up the food chain, so that the higher, predatory species contain the highest levels; e.g., polycyclic aromatic hydrocarbons (PAHs) such as benzo[a]pyrene (BaP) (complex ring structures, implicated as carcinogens, formed from incomplete combustion during forest fires or coming from oil spills) are concentrated but not metabolized by mollusks. These may bioaccumulate, and BaP has been detected in the brains of beluga whales taken from a polluted area of the St. Lawrence River.

Anthropogenic
This refers to anything that arises out of human activity.

TOXICITY TESTING IN MARINE AND AQUATIC SPECIES

A wide variety of marine and aquatic organisms is employed for toxicity testing. This is important because of the biomagnification factor discussed above. Testing species only at the top of the food chain would not provide any information regarding the likelihood that those species might bioconcentrate and bioaccumulate the toxicant. Nor would it give any indication of how the toxicant might distribute in the aquatic or marine environments. Species commonly employed include the organisms *Daphnia magna* (water flea, an aquatic crustacean 2–3 mm in length) and *Selenastrum* (duckweed), rainbow trout and fathead minnows. Fish species are important because the gills are an important mechanism for uptake of toxicants. The gills will pass molecules <500 Daltons. Large molecules may clog the gills and suffocate the fish. A marine species gaining importance is the opossum shrimp, *Mysidopsis bahia*. This is a tiny, live-bearing estuarine species with a rapid life cycle and adaptability to laboratory culture conditions. It is being used as a bioassay for sewage effluent and petroleum spill toxicity.

■ WATER QUALITY

Liquid fresh water (as opposed to water vapor) exists on earth either as surface water (lakes, rivers, streams, ponds, etc.) or as ground water which, in turn, may be in the form of a shallow water table that rather quickly reflects changing levels of xenobiotics at the surface or as much deeper aquifers which acquire surface contaminants more slowly, but just as surely nonetheless. An aquifer is a layer of rock or soil capable of holding great amounts of water. Subterranean streams and pools also exist. A significant difference between surface water and ground water is the accumulation of sediments by the former. It is estimated that 50% of croplands in the U.S. lose 3–8 tons of topsoil/acre/year and another 20% lose >8 tons/acre/year. Soil erosion contributes >700× as much sedimentary material to surface water as does sewage discharge. Both surface and ground water can serve as a source for drinking, household and industrial use. Ground water, however, provides a supply for 50% of all of North America, 97% of all rural populations, 35% of all municipalities, and 40% of all agricultural irrigation.

Sources of Pollution

Agricultural Runoff

Drainage systems conduct any soil contaminants to surface water and, by seepage, to ground water. This includes agricultural chemicals (pesticides, chemical fertilizers), heavy metals leached from the soil by acid rain, atmospheric pollutants carried to the soil in rainfall, bacteria from organic fertilizers, seepage from farm dump sites (old batteries, used engine oil), etc.

Rain

Rain will transfer atmospheric pollutants directly to surface water. Gases may be dissolved directly in water.

Drainage

Drainage from municipal and industrial waste disposal sites and effluent from industrial discharge is an important potential source of contamination if not controlled.

Runoff

Runoff from mine tailings which may be rich in heavy metals can contaminate both surface and ground water.

Municipal Sewage Discharge

Even if treated, this may carry phosphate detergents, chlorine and other dissolved xenobiotics into rivers, etc.

Municipal Storm Drains

These constitute another source of pollution through runoff. In the Great Lakes basin, salt is used extensively on roads to melt ice and improve traction for vehicles. The salinity of rivers and lakes is increasing as a result. Used engine oil from home oil changes in automobiles may be dumped down storm drains. In Canada, an estimated 30,000,000 liters from such usage is not recycled annually. Calcium chloride also may be conducted to lakes, along with residues from vehicle exhaust.

Natural Sources

Although the primary concern of many people is toxicants of anthropogenic origin, it must be remembered that natural toxicants such as methylmercury can form as a result of bacterial action on mercury leached from rock, and of special concern is the level of natural nitrates in drinking water. Nitrates form from nitrogenous organic materials derived from decaying vegetation. Natural levels are not usually a source of concern, but the addition of nitrates from agricultural activity (nitrate fertilizers, animal wastes) may increase the content to dangerous levels. Nitrates are converted by intestinal flora to nitrites which oxidize ferrous hemoglobin to ferric methemoglobin, which cannot transport oxygen. Infants are especially sensitive and cases of poisoning numbering in the thousands have been reported, with a significant mortality. Adults and older children possess an enzyme, methemoglobin reductase, that can reform normal ferrous hemoglobin. Normal nitrate levels in water are about 1.3 mg/liter, contributing about 2 mg/day to the total intake of 75 mg per person per day. Levels as high as 160 mg/liter have been reported in some rural areas where wells serve as the source of water (see also section on food additives). Both the United States Environmental Protection Agency (EPA) and the Environmental Health Directorate of Health and Welfare Canada set maximum acceptable limits for toxicants in drinking water. For example, the EPA limit for nitrates is 10 mg/liter measured as nitrogen.

Water pollutants may be described as oxygen-depleting (contributing to eutrophication), synthetic organic chemicals (detergents, paints, plastics, petroleum products, solvents) which may be very persistent in the environment, inorganic chemicals (salts, heavy metals, acids) and radioactive wastes from nuclear generating plants. Low-level radioactive liquid wastes are produced in the primary coolant.

Some Major Water Pollutants
Specific classes of xenobiotics will be dealt with in detail later in this text, as they may serve to contaminate soil, water or air. Here we will briefly review the more important groups in water.

Detergents
A wide variety of substances is employed as wetting agents, solubilizers, emulsifiers and antifoaming agents in industry and in the home. They have in common the ability to lower the surface tension of water (surfactant effect) and, as cleaning agents, this increases the interaction of water with soil, solubilizing the latter. Chemically, they possess discrete polar and nonpolar regions in the same molecule. The nonpolar region is usually a long aliphatic chain. Sodium dodecylbenzenesulfonate (an anionic detergent) and polyphosphates such as sodium tripolyphosphate are in this group. The latter, $Na_5O_{10}P_3$, is used as an engine oil additive. In sewage, it is readily hydrolyzed to form orthophosphate. Removal efficiencies for sewage treatment are typically 50–60%, so that considerable amounts can enter surface water to contribute to the process of eutrophication (discussed above). Despite a ban by most Great Lakes states and provinces on phosphate detergents, water phosphate levels have not dropped significantly. The ban has apparently been offset by the use of phosphate fertilizers. The average North American uses about 23 kg of soaps and detergents yearly. The biochemical, or biological, oxygen demand (BOD) is a measure of the organic material dissolved in the water column and hence of the oxygen requirement for its decomposition. It includes natural sources such as phytoplankton, zooplankton and organic material from vegetation as well as nitrates. Pure water has a defined BOD of 1 ppm. BODs >5 ppm suggest significant pollution. Pulp mill effluents may have levels >200, and agricultural animal wastes may approach 2,000.

Pesticides
This class of chemicals has generated great public concern, sometimes in the absence of any hard evidence of toxicity for humans at the level of exposure likely to be encountered. For example, the European Economic Community, in its Drinking Water Directive of 1980, set limits of 0.1 µg/L for any single pesticide and 0.5 µg/L for all pesticides combined without regard for their toxicity or their economic importance to agriculture. Included in this group are insecticides, herbicides, fungicides, rodenticides and specific agents to kill snails (molluscicides) and nematodes (nematocides or nemacides). Nematodes (roundworms) from the Greek "nema" meaning thread, are a huge class of parasites that infect humans and animals as well

as many plants. The galls that one sees sometimes on leaves of trees are usually due to nematode infestation. Although not strictly pesticides, the public tends to include other agricultural chemicals used to improve growth or ripening in this category. Alar, for example, holds red apples on the tree to allow for even color development. It was recently withdrawn voluntarily by the manufacturer because of concern over carcinogenicity.

Chemical Classification of Pesticides. The following are classifications of common pesticides:

1. Chlorinated hydrocarbons such as DDT, lindane, aldrin, dieldrin, heptachlor (also called organochlorine insecticides). PCBs are also chlorinated hydrocarbons, but are not insecticides.
2. Chlorphenoxy acids include the herbicides 2,4-D and 2,4,5-T which contain dioxins as impurities.
3. Organophosphate insecticides such as parathion, malathion, DDVP and TEPP.
4. Carbamate insecticides, which act like organophosphates (cholinesterase inhibitors) but which are derivatives of carbamic acid. There are also carbamate herbicides which lack significant anticholinesterase activity.
5. Paraquat is a bipyridyl herbicidal agent that is not considered to be an important environmental contaminant but which is extremely toxic to handlers if used incautiously.

■ HEALTH HAZARDS OF CHEMICAL WATER POLLUTANTS

Chlorinated Hydrocarbons

These are very persistent in the environment. They are slowly degraded by bacteria and other microbes. In addition, they are very lipid-soluble and thus have very long biological half-lives. Although this group is considered to have low acute toxicity, the combination of lipophilicity and long T1/2 leads to biomagnification up the food chain and greater potential for chronic toxicity. This is not easily demonstrable in humans, but in nature, DDT (dichlorodiphenyltrichloroethane) and its metabolites have been shown to interfere with calcium metabolism, causing softening of the egg shell in many species of birds (gulls, peregrine falcon, bald eagle, brown pelican, etc.) with consequent loss of reproductive efficiency. Human fat may contain up to 10 ppm, with a clearance of about 1% of content/day. Acutely, DDT is a neurotoxin, causing tremors and convulsions. The oral LD_{50} for humans is estimated at 400 mg/kg. Polychlorinated biphenyls have been used for many years for their insulating properties and the fact that they can withstand temperatures up to 800°C. These properties make them ideal for use in electrical transformers, hydraulic fluids, brake fluids, etc. Their stability means that they are very persistent in the environment when contamination occurs through accident or improper disposal. In the U.S. the EPA has set a maximum allowable level of 0.01 ppb in streams. In the Baltic Sea, PCBs have been incriminated in reproductive failure in seals (pinnipeds).

Considerable concern has been generated over seepage of PCBs into ground water and streams from deteriorating containers in dump sites. Levels of 5–20 ppm have been detected in Lake Ontario fish. Toxic effects in the environment include reproductive defects in phytoplankton and, in mammals and birds, microsomal enzyme induction, tumor promotion, estrogenic effects and immunosuppression. The potential for human toxicity is therefore high, but existing data are somewhat controversial (see below).

Chlorphenoxy Acid Herbicides

The chlorphenoxy acid herbicides 2,4-D and 2,4,5-T have been widely used on lawns and along road and railway rights-of-way. They mimic plant growth hormones, so that accelerated growth exceeds the energy supply. 2,4,5-trichlorophenoxyacetic acid (2,4,5-T) is weakly teratogenic, but the main concern is the presence of the contaminant 2,3,7,8-tetrachlorodibenzo-p-dioxin (TCDD, dioxin) a by-product of synthesis. It is teratogenic and very toxic to some animals. The LD_{50} for rats is 0.6–115 µg/kg. It causes degenerative changes in the liver and thymus, weight loss, changes in serum enzymes, porphyria, chloracne and cancer. In humans, the only confirmed toxic effect is chloracne (see previous chapter on the Seveso accident). Although Vietnam veterans have been very concerned about the use of "Agent Orange" (contains equal parts 2,4-D and 2,4,5-T) as a defoliant, several epidemiological studies have failed to confirm long-term effects. Some recent evidence suggests that industrial workers with prolonged exposure to high levels may have a slightly increased incidence of cancer (see also Chapter 5).

Organophosphates

Organophosphates irreversibly inhibit acetylcholinesterase, and the symptoms of acute toxicity are those of massive cholinergic discharge and include profuse sweating and diarrhea, tremors, mental disturbances, convulsions and death. Although parathion is fairly toxic for humans, it does not persist in the environment, and thus, is not a significant environmental hazard. These agents are water-soluble, but they are hydrolyzed to nontoxic by-products.

Carbamates

Carbamates act generally like the organophosphates. The exception is aldicarb, which is not hydrolyzed and which has entered ground water in several locations including Long Island, New York, where it is predicted to exceed maximum allowable levels of 7 ppb for up to 20 years. It is highly toxic, but it is a reversible inhibitor and is rapidly degraded and excreted. Pesticides will be considered further in Chapter 9.

Others

Another hazardous chemical introduced as a result of water treatment is chloroform. It is introduced as a contaminant in the process of chlorination, and it is a known carcinogen. Others, such as benzene and carbon tetrachloride, may enter ground water from industrial sources.

■ ACIDITY AND TOXIC METALS

Acidity, largely from acid rain, the causes of which will be discussed in Chapter 4, can have several deleterious effects on water quality. This subject has already been introduced (see "Abiotic Factors", page 71). Acidity can leach toxic metals such as lead (Pb), cadmium (Cd) and aluminum (Al) from soil into ground water. It can contribute to the formation of more toxic methylmercury from mercury. It may also leach lead from solder in the plumbing of houses and cottages as has been shown in a study by Health and Welfare Canada. Water at pH 4.5–5.2 was allowed to stand in plumbing systems and reached maximum leaching rates after 2 hr. After 10 days, the water showed levels of 4,560 µg/L for copper (Cu), 3,610 µg/L for zinc (Zn), 478 µg/L for Pb, and 1.2 µg/L for Cd. The upper limits recommended for Canadian drinking water are 100 µg/L for Cu and 50 µg/L for Pb. Arsenic (As) and selenium (Se) have also been detected. It is highly advisable to flush plumbing systems in houses and cottages that have been standing vacant for any length of time.

At pH 5 or less, aluminum can be leached from soil and transported as complexes with bicarbonate, organic materials and in the ionic form. Acid surface water may contain 4–8 µmol/L which can be toxic to fish. In humans, high concentrations of Al may be deposited in bones and brain tissue to cause osteomalacia and symptoms of dementia. Microcytic-hypochromic (i.e., small, pale cells) anemia can also occur. These problems have been encountered in hemodialysis patients due to the leaching of Al from the dialyzer into the blood of the patient and from oral aluminum hydroxide given them in antacids. Bauxite miners suffer respiratory problems from inhaling ore dust. Al also appears in drinking water because of the use of alum [$Al_2(SO_4)_3 \cdot 12H_2O$] to precipitate suspended organic material in the third (tertiary) stage of water treatment. The first stage involves the removal of large solids by screens, and the second stage removes most of the organic material with filters.

In 1980, the U.S. Congress commissioned the National Precipitation Assessment Panel (NAPAP), consisting of over 2,000 scientists from virtually all of the major universities, to study the acid rain problem. In 1987, it issued a highly controversial interim report which concluded that the situation was not as bad as previously suspected. Of several thousand lakes studied in upper New York State, only 75 were found to be seriously affected. Sulfur emissions were found to have declined by 25% in the last decade. The final 6,000-page report was released in 1990 (total cost over $570 million) and concluded that 10% of eastern lakes and streams were adversely affected, that acid rain had contributed to the decline of red spruce at higher altitudes and to the corrosion of buildings and materials. A more controversial statement was that there was no evidence of widespread decline of forests in the United States or Canada. Acid precipitation, however, can be deposited hundreds of miles from its site of formation. Moreover, lakes which drain limestone bedrock areas are much more resistant to acidification because of their buffering capacity. Lakes which drain granite, however, are very susceptible because they have virtually no buffering capacity. This includes all of the lakes in the Canadian Shield.

Again, aluminum plays an important role. Scientists have discovered that fish in a laboratory setting can withstand a pH of 4.5 or less, while in the natural setting, such a low pH is frequently fatal. Aluminum silicates are a major soil component, and soft-water lakes which drain soil areas acquire significant levels of these. A suspension of fine aluminum precipitate forms in water, blocking the sodium and oxygen exchange systems in the gills of fish, which then expire. Fresh-water fish must take up sodium across the gills to compensate for that lost in urine. Fresh-water fairy shrimp have "chloride cells" which also regulate sodium levels and which accumulate toxic levels of aluminum. Some success has been achieved in selectively breeding aluminum-resistant strains of aquatic plants such as duckweed, which may be used to revitalize dead lakes. Selective breeding of plant species was developed early in the century to combat the effects of acidic soils which poison plants by interfering, through metal solubilization, with calcium and phosphorus. Phosphorus, especially, binds to aluminum. Since sodium regulation in nearly all cells involves sodium/potassium ATPase (the sodium pump), the aluminum-bound phosphate can attach to, and disable, the sodium pump. This phosphorus link may be involved in the massive diebacks of forests exposed to high levels of acid rain, and selective breeding of resistant species may provide a partial solution.

CHEMICAL HAZARDS FROM WASTE DISPOSAL

In addition to the types of hazardous contaminants discussed above, numerous substances may enter water from industrial, agricultural, institutional and domestic sources. They may be solids, liquids, sludges or gases and may be corrosive, flammable, explosive, radioactive or biologically toxic. Risks range from minimal to extreme, and there may be short- or long-term effects on human health. Usual disposal methods for these substances include surface impoundments used in industry (48%), landfill sites (domestic and other, 30%), burning (10%) and other means, e.g., disposal at sea (2%).

An idea of the extent of the problem of buried toxic substances may be gleaned from the experiences of workers in tunnel construction projects. Traditionally, compressed air has been used in tunnel construction for the purpose of excluding water from the tunnel and also for stabilizing the surrounding ground. A new use is emerging, however. Increasingly, tunnel projects in urban areas (for sewer mains, rapid transit systems, auto tunnels under rivers, etc.) are encountering pockets of toxic materials such as gasoline, probably leaked from old service station storage tanks, chlorinated hydrocarbons and other dump-site toxins. The use of compressed air in the tunnel prevents the seepage of toxic fumes and liquids into the tunnel and provides a safer working environment.

Water from drinking wells continues to be a source of concern regarding chemical contamination. In 1987 a study of the Tutu well fields in St. Thomas, U.S. Virgin Islands, showed that 22 wells were contaminated with benzene, trans-1,2-dichloroethylene, trichlorethylene, and tetrachlorethylene originating from several sources. Although levels were low, an estimated 11,000 people were exposed for about 20 years. In Minnesota in 1981–1982, 41 of 137 private and commercial wells located downhill from an industrial complex were found to be contaminated with trichlorethylene and trichlorethane. Such wells generally should be sealed with concrete or clay and abandoned.

THE LOVE CANAL STORY

The Hooker Chemical Company, between 1942 and 1953, disposed of about 420,000 metric tonnes of around 300 organic chemicals by burying them in steel drums in the abandoned Love Canal near Niagara Falls, New York. The site was sold to the Niagara Falls Board of Education for $1.00 in 1953 and subsequently, a subdivision was built over it. As the barrels rusted out, chemicals seeped into the ground water. There were some early warning signs such as chemical odors in people's basements, sinking areas over collapsing barrels and some exposed pools of waste, and some minor health problems such as rashes and eye irritation after contact with exposed chemicals, but overall, the residents of Love Canal seemed unaware of any unusual health problems or circumstances. In 1976, however, the International Joint Commission on the Great Lakes undertook a study to find the source of the banned pesticide mirex in Great Lakes fish, and the chemical contamination was discovered. The New York State Department of Environmental Conservation studied the situation over the next 2 years amid great controversy and in the face of emerging anecdotal claims of serious health problems.

In August of 1978, a series of dramatic events occurred. The state health commissioner declared a health emergency and recommended that pregnant women and children under age 2 be evacuated. President Jimmy Carter declared the area a federal disaster zone, thereby creating a mechanism for federal assistance to be provided. The governor of New York State announced that 239 families would be relocated at state expense. The immediate consequence was that Love Canal became a ghost town, and considerable anxiety was created about long-term health effects, not only in the evacuees, but in those who lived near the edge of the arbitrary demarcation line. In 1988, a federal district court found Hooker Chemical (by then Occidental Petroleum) to be liable for the cost of the cleanup estimated at $250 million. The State Commissioner of Health declared that some areas were safe to resettle, but new information challenged this decision (see below).

Numerous health studies had been conducted in the intervening decade. They generally failed to reveal any significant evidence of an increased incidence of illnesses. Cancer registries are relatively new, and some states still do not have one. An analysis of the New York State Cancer Registry found that the incidence of lung cancers was generally higher throughout the Niagara Falls region, but no differences in the occurrence of liver cancer, lymphomas or leukemias were noted in the Love Canal region. The increased incidence of lung cancer is intriguing in light of a study by The University of Toronto and Pollution Probe which found that the mist from Niagara Falls contains PCBs, benzene, chloroform, methylene chloride and toluene. Although the levels were not themselves high enough to pose a risk, they could be additive with other carcinogens in cigarette smoke, auto exhausts, etc., and the mist could settle out on crops to enter the food chain. Statistics compiled in the U.S. and by Health and Welfare Canada showed statistically significant differences in types of cancer mortalities among and within counties bordering the Great Lakes, but did not indicate that these populations suffered substantially higher cancer mortalities than nonbasin counties. Nor were there consistent differences among municipalities within Niagara County.

Currently, attempts are being made to track the some 15,000 persons who lived in the Love Canal area prior to the evacuation in order to identify increased incidences of cancer and other diseases. If this can be accomplished, some answers about the nature of the Love Canal risks may finally be obtained. One fairly convincing bit of datum is that the families living closest to the canal had a higher occurrence of miscarriages, abortions and low birth weight babies. Health and Welfare Canada also attempted to uncover evidence of an increase in birth defects by surveying the occurrence of cleft palate in Ontario. There was definitely a higher incidence in the Niagara peninsula, but the area of higher incidence extended all the way north to Lake Simcoe, and other areas of concentration were discovered. There is a small one at the southern tip of Lake Huron (but upstream from the "chemical valley" of the St. Clair River) and another very large one running from the north shore of Lake Superior north to Hudson's Bay. These areas are highly diverse both demographically and geographically, so it is difficult to identify a common factor to explain this distribution.

There is no question that a major component of the health impact of exposure to such toxic disasters is psychological in nature, which is not to say that it is not real. People tend to accept natural disasters with much greater equanimity. Flood victims do not appear to suffer the same prolonged mental stress as victims of a toxic disaster, nor do they attach blame as readily, even when there may be legitimate questions about the adequacy of flood control measures. Such things tend to be accepted as acts of God, and there may even be some essence of pioneer spirit to be taken from battling the forces of nature. In contrast, victims of toxic disasters tend to feel at the mercy of forces that are man-made and out of control. They are not just victims; they feel victimized by greed and callousness. Their sense of helplessness may be exacerbated by contradictory statements from government officials and by sensational reporting by the news media.

An understanding of this phenomenon may be an important step toward minimizing the human price of such disasters. Following the Love Canal disaster, there was a high frequency of insomnia, severe depression, an increase in suicide attempts, lowered performance in school children and other manifestations of emotional illness, including symptoms of malaise for which no organic cause could be found. The PCB fire at St. Basile-le-Grand, Quebec, and the tire-dump fire at Hagersville, Ontario, are two more recent incidents in Canada with the potential for similar costs in mental anguish.

In April of 1989, a peer review panel consisting of 10 scientists familiar with the Love Canal situation met to examine charges that the area was not safe for resettlement because the site had not been adequately cleaned up and because an area in a zone designated as a control area for purposes of comparison had been shown to contain a "hot spot" with high levels of chlorinated organics under a church parking lot. The charges were brought by the Environmental Defense Fund, which also was responsible for the ALAR controversy (see Chapter 2). It released its report in May through the EPA, concluding that the area comparison method was still valid for the Love Canal and that the habitability decision should not be reconsidered.

Problems with Love Canal Studies

1. Initial studies were incomplete and not well organized. Most of the chemicals involved were very volatile and would not show up in human blood or tissues except for a short time after exposure. Only lindane and dioxins were persistent enough to be detected later, and lindane was not found in the subjects who were monitored. Dioxin assays are expensive and time-consuming.
2. When the Centers for Disease Control was asked to conduct a study in 1980, those people with the highest risk of exposure had been evacuated. Little information was available on previous exposures.
3. The episode was highly emotionalized. These were the homes of working-class people who had invested their life savings in their properties.
4. The dispersion of the evacuated families will make it hard to collect valuable data on health effects.

The story of Love Canal has been repeated in many other communities in smaller ways. In London, Ontario, a playground in St. Julien Park was discovered to have been built over a waste disposal site. There were anecdotal reports of local clusters of brain tumors and other problems, but a study conducted by local health officials failed to confirm increased health risks. Cleanup and preventive measures have been instituted. Chemical spills in the St. Clair River have on several occasions threatened the water serving Wallaceburg, Ontario, and the Wallpole Island Indian Reserve.

■ TOXICANTS IN THE GREAT LAKES AND THE IMPLICATIONS FOR HUMAN HEALTH AND WILDLIFE

While the Great Lakes are not threatened by acid rain because of their large size and buffering capacity, they are distinctly threatened by toxic chemicals. In 1991, the Joint Commission issued a report on toxicants present in Great Lakes water which they felt posed a potential threat to the populations of provinces and states in the Great Lakes basin by virtue of their ability to interfere with reproduction. Table 6 is a partial list of these.

It is important to emphasize that these threats are largely potential, but there is an obvious need to improve the situation dramatically.

TABLE 6
SOME GREAT LAKES CONTAMINANTS WITH POTENTIAL REPRODUCTIVE EFFECTS

PCBs — linked to embryolethality, deformities, development.
DDT — now banned, it disrupts hormone balance.
Dieldrin and aldrin (pesticides) — linked to the death of adult bald eagles.
Chlorinated dibenzo-furans (carcinogenic).
Toxaphene — an insecticide used in the cotton belt, it has been found as far afield as the high arctic.
Dioxin (TCDD) — probably carcinogenic for humans exposed to high levels.
Polycyclic aromatic hydrocarbons (PAHs) — carcinogens.
Hexachlorobenzene — fungicide, causes organ toxicity, is carcinogenic and may cause infertility.
Mirex — carcinogenic insecticide, causes reproductive problems.
Mercury — toxic to the CNS, liver and kidney.
Lead — toxic to the CNS.
Based on information presented in *Human Health Risks from Chemical Exposure*. Flint, R.W. and Vena, J. (Eds.), Lewis Publishers, Chelsea, MI, 1991.

Recent studies of white suckers from Lake Ontario have shown that, despite the fact that these bottom-feeding fish contain large amounts of PAHs, only about 5% showed cancer, and these all had parasitic liver disease. A study at the University of Guelph suggested that glutathione-S-transferase (GST) enzymes in the liver normally protect the fish by detoxifying the PAHs, but the efficiency of these enzymes is destroyed by the liver parasites. This illustrates that there are many factors, including genetic ones, which likely must combine before a tumor will develop. Nevertheless, there is abundant evidence that numerous pollutants are toxic to wildlife. These include lead, mercury, hexachlorobenzene, polychlorinated biphenyls (PCBs), polycyclic aromatic hydrocarbons (PAHs), dioxin (TCDD), mirex, DDT, dieldrin and toxaphene. The embryos and chicks of fish-eating birds (herring gulls) have been shown to suffer from a disease termed "expanded chick edema disease" which is very similar to a disease that occurs in domestic poultry accidentally exposed to high levels of polychlorinated dibenzodioxins (PCDDs) and PCBs. TCDD (2,3,7,8 tetrachlorodibenzodioxin) has been shown to have an LD_{50} in lake trout eggs of 80 parts per trillion, and in hamsters, 58% fetal mortality at the ninth gestation day has been shown at a dose of 18 µg/kg. Neurobehavioral and neurochemical abnormalities have been detected in fish-eating birds that consumed contaminated fish from Lake Michigan. Abnormalities have also been shown in laboratory rats fed a diet containing 30% salmon from Lake Ontario.

One of the difficulties in attempting to estimate the risk to humans of exposure to toxicants in the environment is the fact that little is known about the combined effects of several of these. An attempt has been made to deal with this problem as it relates to the polyhalogenated aromatic hydrocarbons that are structural analogs of TCDD. Recent evidence indicates that all of these are inducers of an hepatic microsomal mono-oxygenase enzyme aryl hydrocarbon hydroxylase (AHH). The potency of these agents in inducing this enzyme in cultured hepatocytes correlates well with their experimental toxicity, and this has led to the development of a set of "toxicity equivalency factors" or TEFs. The EPA has adopted the TEF method as a means of determining the toxicity of mixtures of these agents, as their effects on AHH seem to be additive. This approach, however, has limitations. The use of an *in vitro* culture system cannot take into account differences in pharmacokinetic parameters among different compounds nor the extent to which they bioaccumulate. For example, one of the hexachlorobiphenyls has been shown to be a significant contaminant of mother's milk, even though it is one of the less common environmental contaminants.

Evidence of Adverse Effects on Human Health

It is axiomatic that toxicity to humans resulting from exposure to PAHs is related to total body burden. Portals of entry include the lungs (contaminated air) and the gastrointestinal tract (water and diet). Polychlorinated dibenzodioxins (PCDDs) and polychlorinated dibenzofurans (PCDFs) are known products of municipal incinerators. The Ontario Ministry of the Environment has calculated a worst-case scenario of 126 pg TEQ/day for people living near a municipal incinerator with inadequate pollution controls. In contrast, because of their poor solubility in water, drinking water is estimated to contain less than 2 pg TEQ/L. Diet is probably the most significant source of exposure and may account for 95% of all human intake. Levels of TCDD in human fat have been calculated to average 10 ppt worldwide, with little variation from place to place (evidence of the ubiquity of the substance). EPA has set an acceptable daily intake of TCDD of 1.0 pg/kg/day.

Accidental poisonings in Yusho, Japan, and Yucheng, Taiwan, resulted in total body burdens of PCDDs and PCDFs 200–300 times the North American average. These levels were associated with nausea and anorexia, increased frequency of premature births, low birth weights, impaired growth, impairment of neuromuscular and intellectual development and a higher frequency of health problems. The Michigan State Department of Health has assembled three sizable cohorts of individuals exposed to increased levels of organohalogens and maintains a registry of these individuals.

Included are the farmers who were exposed accidentally to high levels of PBBs when this fire retardant was accidentally labeled as a mineral supplement for cattle, farmers who were exposed to chlorinated biphenyls ("silo farmers") and a population of Lake Michigan anglers and their families who consumed large amounts of fish from the lake and who were exposed to a mixture of contaminants. These cohorts have been followed for over 20 years. Mothers in the "angler" families consumed an average of three meals of contaminated Great Lakes fish per month while their children were *in utero*. Levels of PCBs in the fish were low and believed to be nontoxic, and the degree of impairment of their infants was much lower than in the Yusho and Yucheng incidents, but the differences were nonetheless significant, even when a number of confounding variables such as smoking and alcohol consumption were taken into account. The effects were greatest in infants whose umbilical vein serum PCB levels were greater than 3.5 mg/ml. It must be emphasized that the effects seen in the Michigan infants were largely subclinical, and therefore it is difficult to determine the true degree of risk to the population at large.

In 1989, a workshop was held at which findings from these studies were reviewed and compared to a similar study from North Carolina in which 800 infants believed to have moderate *in utero* exposure to PCBs (estimated from maternal breast milk levels) were followed for over 1 year. Most studies showed a delay in cognitive performance as measured by the Brazelton and Bayley scales. These decrements in performance were detectable in newborns and in children at age 4, as were deficits in body size. The Michigan studies have been criticized for such flaws as using anecdotal reports of the type and amount of fish consumption going back over 6 years, differences between control and test groups (e.g., in alcohol consumption, use of medications, caffeine consumption, and in the frequency of assisted births) and limiting the studies' statistical power by restricting participants to one third of those exposed. If nothing else, this illustrates the difficulties in conducting epidemiological studies looking for subtle, subclinical findings. Nevertheless, the potential for toxicity and the significance of these warning signs cannot be ignored. The large body of evidence from laboratory studies and from the examination of numerous wildlife species provides ample indication that many of the halogenated hydrocarbons (organohalogens) have serious reproductive consequences if sufficient amounts are consumed. The questions of extrapolation to humans and of the effects of much lower exposures are as troublesome here as they are for carcinogenesis (see Chapter 5 for a further discussion of these accidental exposures).

Not all toxicants in water are anthropogenic. Many microorganisms such as algae, diatoms, etc. are capable of producing lethal toxins which can concentrate up the food chain. There is presently concern that algal blooms in the ocean are creating a hazard for marine life and people alike. This subject is discussed in Chapter 11.

■ REVIEW QUESTIONS

For Questions 1 to 10, define each of the following terms (answers can be found in the text).

1. Abiotic
2. Biotic
3. Eutrophication
4. Bioconcentration
5. Biomagnification
6. Acclimatization
7. Acclimation
8. Bioaccumulation
9. Aquatic
10. Marine

For Questions 11 to 20, answer True or False.

11. Species is a biotic modifier.____
12. Pore water is bottom water in which sedimentary particles are suspended.____
13. pH has no effect on the methylation of mercury by microorganisms.____
14. Metals bound to carbonates in water become more toxic.____
15. Concentration equilibriums are established between bound and unbound forms of toxicants in water.____
16. Methylmercury never forms from natural sources.____
17. DDT is an example of a chlorinated hydrocarbon.____
18. Chlorinated hydrocarbons have a short biological half-life.____
19. Human fat may contain up to 10 ppm of DDT.____
20. Organophosphate insecticides are reversible inhibitors of acetylcholinesterase.____

For Questions 21–25, use the following code: answer A if statements 1, 2 and 3 are correct; answer B if statements 1 and 3 are correct; answer C if statements 2 and 4 are correct; answer D if statement 4 only is correct; answer E if all statements are correct.

21. 1. Dioxin (TCDD) is a proven carcinogen for humans at levels widely encountered in the environment.
    2. Chloracne (a skin rash) is the most common toxic manifestation of TCDD.
    3. TCDD is used as a herbicide.
    4. TCDD is a contaminant of the herbicide 2,4,5-trichlorophenoxyacetic acid.

22. Which of the following symptoms is/are *not* characteristic of organophosphate poisoning?
    1. Profuse sweating
    2. Tremors
    3. Confusion and other mental disturbances
    4. Constipation

23. Acidity in water can contribute to
    1. Leaching of lead from solder joints in plumbing.
    2. Acceleration of the transfer of metals from soil to ground water.
    3. A shift of copper to the elemental cuprous form (Cu^{2+}) from the carbonate form.
    4. The existence of pentachlorophenol in the undissociated, more lipid-soluble state.

24. Polychlorinated dibenzodioxins (PCDDs)
    1. Have been associated with accidental poisonings.
    2. Are used industrially as pure chemicals.
    3. Are products of municipal and industrial incinerators.
    4. Are not themselves toxic.

25. Which of the following statements is/are true regarding polycyclic aromatic hydrocarbons (PAHs)?
    1. Liver parasites may increase the toxicity of PAHs in fish by impairing their detoxication.
    2. Behavioral abnormalities have been observed in experimental animals exposed to PAHs.
    3. Herring gull embryos have shown signs of "expanded chick edema disease" which has been associated with exposure to PAHs.
    4. Acid rain is a threat to the Great Lakes by increasing the toxicity of PAHs.

■ ANSWERS

11. T; 12. T; 13. F; 14. F; 15. T; 16. F; 17. T; 18. F; 19. T; 20. F; 21. C; 22. D; 23. E; 24. B; 25. A.

FURTHER READING

Bart, A., They can't do it alone: some carcinogens in pollutants need help to cause cancer. *The (University of Guelph) Crest*, 2, (July), 9, 1991.

Levi, P.E., Chemical pollutants in the soil and water, in *Textbook of Modern Toxicology*. Hodgson, E. and Levi, P.E. (Eds.), Elsevier, New York, 1987, chap. 7.

Colbourn, T.E., Davidson, A., et al., *Great Lakes Great Legacy?*. The Conservation Foundation, Washington, D.C., and The Institute for Research on Public Policy, Ottawa, Canada, 1990.

Continued use of drinking water wells contaminated with hazardous chemical substances — Virgin Islands and Minnesota, 1981–1993. *Morbid. Mortal. Weekly Rep.*, 43, 89, 1994.

Edelstein, M.R., *Contaminated Communities: The Social and Psychological Impacts of Residential Toxic Exposure*. Westview Press, Boulder, 1988.

Gotfryd, A., Aluminum and acid: a sinister synergy. *Can. Res.*, June/July, 10, 1989.

Henderson-Sellers, B. and Markland, H.R., *Decaying Lakes: The Origin and Control of Cultural Eutrophication*. John Wiley and Sons, Toronto, 1987.

Humphrey, H., Environmental contaminants and reproductive outcomes. *Health Environ. Digest.*, 5, 1, 1991.

Kimbrough, R.D., Health impact of toxic wastes: estimation of risk, in *The Analysis of Actual Versus Perceived Risks*. Covello, V.T., Flamm, W.G., Rodricks, J.V. and Tardiff, R.G. (Eds.), Plenum Press, New York, 1981.

Menzer, R.E. and Judd, O.N., Water and soil pollutants, in *Casarett and Doull's Toxicology*. Klassen, C.D., Amdur, M.O. and Doull, J. (Eds.), Macmillan, New York, 1986, chap. 26.

Olsson, M., Contaminants and diseases in seals from Swedish waters. *Ambio*, 21, 561, 1992.

Paneth, N., Human reproduction after eating PCB-contaminated fish. *Health Environ. Digest.*, 5, 4, 1991.

Habitability of Love Canal questioned after new discoveries. *Pesticide and Toxic Chemical News.*, 17, #19, 1, #25, 7, #30, 21, 1990.

Roberts, L., News and comment. Learning from an acid rain program. *Science*, 251, 1302, 1991.

Steward, R., Olson, J., et al. Toxicology and environmental chemistry of exposure to toxic chemicals, in *Human Health Risks from Chemical Exposure: The Great Lakes Ecosystem*. Flint, R.W. and Vena, J. (Eds.), Lewis Publishers, Chelsea, MI, 1991, chap. 3.

Swain, W.R., Human health consequences of consumption of fish contaminated with organochlorine compounds. *Aquatic Toxicol.*, 11, 357, 1988.

4 AIR-BORNE HAZARDS

The work is going well, but it looks like the end of the world.
S. Rowland, co-discoverer of the CFC effect, to his wife

INTRODUCTION

When potentially noxious substances are discharged into the atmosphere at a rate that exceeds its capacity to disperse them by dilution and air currents, the resulting accumulation is *air pollution*. It may take the form of haze, dust, mist (which may be corrosive) or smoke and may contain oxides of sulfur and nitrogen and other gases which may irritate the eyes, respiratory tract or skin, and other substances which may be harmful to the environment or to human health. Absorption may occur in amounts sufficient to cause acute or chronic systemic toxicity.

TYPES OF AIR POLLUTION

Air pollutants may be gaseous or particulate in nature, and particulates may be either solid or liquid.

Gaseous Pollutants
These are derived from materials which have entered into chemical reactions or combustion processes. They include carbon-based compounds like hydrocarbons, oxides and acids, sulfur compounds such as dioxide, trioxide and sulfides, nitrogen compounds (ammonia, amines, oxides) and halogenated substances (organic and inorganic halides).

Particulates
Particle or droplet size may range from 0.01 to 100 μm in diameter. The smaller particles are referred to as aerosols and can remain

suspended, scattering light and behaving much like a gas. Below 10 μm, particles are capable of penetrating to all sites in the respiratory tract. Industrial particulates are usually solid and are carbonaceous, metallic oxides, salts, or acids, and their porosity is such that they will absorb other gases and liquids.

Smog
The word is a combination of smoke and fog and is a popular term for a fairly uniform mixture of gaseous and particulate pollutants that accumulate over urban centers and persist for a prolonged period. Smog is a brown or yellow haze, and it usually occurs during the phenomenon of temperature inversion when a high-level mass of cold air traps warmer air beneath it to prevent mixing and dispersion. An especially bad "killer smog" occurred in London, England, in 1952. It lasted over a week and was responsible for about 4,000 deaths, mostly from respiratory diseases. As a result, the Clean Air Act was passed in 1956, banning the use of soft coal for home heating.

■ SOURCES OF AIR POLLUTION

Air pollution may arise both from natural sources and from human activities. Volcanic eruptions, forest fires and dust storms are natural sources, the importance of which should not be underestimated. The 1980 Mt. St. Helen explosion in Washington State pulverized half of a mountain and released millions of tons of dust. It affected weather patterns as far east as the Great Lakes. In 1912, a similar explosion of a volcano in Alaska released about 30× as much dust as Mt. St. Helen. The recent eruptions of Mount Pinatubo in the Philippines, together with smoke from the Gulf oil fires, have been blamed for unusually cool summers and excessive rainfall throughout most of North America in 1991–1992. Additional major eruptions in the "ring of fire" are predicted for the near future.

Human sources include discharge from coal-fired electrical generating stations, nuclear generating stations, industrial emissions and domestic heating. Transportation sources include passenger autos, trucks, diesel locomotives, etc. Pollution may arise from all sources of combustion, industrial fuming and volatilizations, dust-making processes, photochemical reactions and biological sources (include microorganisms such as viruses, bacteria and fungi), pollen, and chemicals from decaying organic matter. The breakdown of pollution sources in industrial countries is approximately as follows: transportation, 60%; industry, 18%; electricity generating, 13%; heating, 16%; waste disposal, 3%. Considerable concern is arising over the problem of indoor air pollution. The hazards of sidestream cigarette smoke seem firmly established, and this has lead to increased restrictions on smoking in the workplace and in public buildings. Recent studies have shown that 4-aminobiphenyl, a potent human bladder carcinogen present in both mainstream and sidestream cigarette smoke, has been found in fetal hemoglobin, indicating that it crosses the placenta. Other indoor pollutants may include formaldehyde gas (see Chapter 2, 2–12 to 2–13), other toxic chemicals, particulates such as asbestos fibers and fiberglass wool and radon-source ionizing radiation (see Chapter 12). Airtight houses and buildings, constructed during the energy crisis of the 1970s, increase the risk of adverse health effects. Industrial indoor pollution is a special problem. In Ontario, the Ministry of Labour has jurisdiction over levels of air pollutants in the workplace and defines acceptable limits under various conditions (see Chapter 2).

ATMOSPHERIC DISTRIBUTION OF POLLUTANTS

Air pollution generally starts out as a local problem, but it may become global if the pollutants enter the atmospheric circulating system. Pollutants may enter the atmosphere in the form of gases, vapors (from volatile liquids), aerosol droplets or fine dust particles (see Chapter 3 for a discussion of distribution of pollutants in the biosphere).

Movement in the Troposphere

The troposphere is the air mass up to an altitude of about 10 miles. In the upper troposphere, the winds are predominantly westerly and average 35 meters/sec (mps) to disperse pollutants worldwide in about 12 days. Vertical movement circulates air north and south from the equator in systems called Hadley cells. In a band from 30°N Latitude to 30°S Latitude, other cells called Ferrel cells circulate air toward the poles. Speeds may reach 30 mps. Microscopic particles are retained for 1 or 2 months in the upper and mid-troposphere and about 1 week in the lower troposphere (<1 mi).

Movement in the Stratosphere

The stratosphere extends from 10–30 miles above the Earth. Movement occurs very slowly, at the rate of a few centimeters/sec, but particles may stay for 2 or 3 years at an altitude of 20 miles and about 1 year at 11 miles. Certain gaseous pollutants such as freon, chlorofluorocarbons (CFCs) and some rare radioactive isotopes (e.g., krypton-85 from nuclear reactors, T1/2 10.5 yr) are not readily removed by physicochemical means and may persist in the atmosphere for very long periods. Recent studies suggest that fluorinated gases will persist in the atmosphere for 300 to 2,000 years or more, depending on the chemical.

Water and Soil Transport of Air Pollutants

The subject of the exchange of pollutants among various components of the biosphere was introduced in Chapter 3. Gaseous atmospheric pollutants may be dissolved in rainwater and solid particles carried in it mechanically. Precipitation thus carries them into the soil and ground water, and they can reach oceans, lakes and rivers by runoff and soil erosion and deep aquifers by seepage. The oceans are the ultimate repository for pollutants, and surface evaporation may conduct them back into the atmosphere. Several studies have confirmed this biospheric circulation of toxicants. In the 1950s, atmospheric tests of nuclear bombs resulted in widespread dissemination of radioactive fallout. Of particular concern was the presence of strontium 90, which exhibits chemical characteristics similar to calcium, including deposition in bone. Strontium 90 reached significant levels in cow's milk, other dairy products and in fruit and vegetables, and concern about its accumulation in the bones of children was a major factor in the discontinuation of atmospheric nuclear testing. The estimated North American exposure from all anthropogenic radionuclides is estimated now to be <1 mrem/yr. In 1969, contamination of Antarctic snow with DDT was identified. The only way it could have reached there was through precipitation. Presently, the most compelling concern is the problem of acid rain, the pH of which may be <4. Acid rain may be deposited far from its source.

■ TYPES OF POLLUTANTS

Gaseous Pollutants
These include

1. Sulfur dioxide (SO_2), which forms acid rain as sulfurous acid.
2. Sulfur trioxide (SO_3), which forms acid rain as sulfuric acid.
3. Nitrogen monoxide (nitric oxide, NO), oxidized to nitrogen dioxide (NO_2); it is part of photochemical smog and acid rain.
4. Carbon monoxide (CO), a product of incomplete combustion; it forms carboxyhemoglobin which is incapable of transporting oxygen to the tissues.
5. Ozone (O_3), which contributes to photochemical smog.
6. Hydrogen sulfide (H_2S), which is very toxic.
7. Various hydrocarbons (C_xH_y), from automobile emissions.
8. CFCs, freon, vinyl chloride, and radioactive isotopes.
9. Methane from several sources.

Particulate Pollutants

Dusts

Fine particle solids may arise from sawdust, cement, grains, metals, rock (in quarrying operations), incomplete combustion of fossil fuels (producing particles of <1.0 μm), i.e., smoke, and any substance including chemicals (pesticides, etc.) existing in powder form.

Liquids

Any liquid that forms droplets 1.0–2.0 μm in diameter will remain in suspension in air as a "mist" (e.g., sulfuric acid). Droplets <1.0 μm are defined as an aerosol. The term is also applied to solid particles of this size. It is important to note that water vapor is by far the most significant greenhouse "gas", accounting for about 85% of infrared trapping, but its level fluctuates widely.

■ HEALTH EFFECTS OF AIR POLLUTION

Acute Effects

Short-term exposure to hazardous levels of air pollutants may result in irritation to the eyes and the respiratory tract. Populations at high risk include the very young and the elderly, whose respiratory and cardiovascular systems are not fully functional, people with asthma, emphysema, heart disease, and heavy smokers. These groups had the highest mortality rates during the killer smog in London, England. The accidental release of toxic chemicals from industrial plants has caused serious health problems and death, the most tragic being the release of 40 tons of methyl isocyanate from the American Cyanamid plant in Bhopal, India, in 1984. Nearly 3,000 people died.

Chronic Effects

Long-term exposure to lower levels of pollution may result in, or aggravate, chronic bronchitis, pulmonary emphysema, bronchial asthma and lung cancer. Cigarette smoke will cause all of these problems. Excessive secretion of bronchial mucus and a chronic cough are the hallmarks of chronic air pollution effects. Dust and other allergens, including pollen, 1–90 μm in diameter, can induce or trigger allergic reactions in susceptible people.

■ AIR POLLUTION IN THE WORKPLACE

Systemic poisoning has occurred to workers inhaling toxic levels of metals such as lead, arsenic, mercury, manganese, zinc, cadmium, as well as pesticides and drugs. Oxides of all of these metals plus those of copper, tin and nickel and brass dust can cause a febrile reaction (fever, joint and muscle aches) called "metal-fume fever". Cutting with an acetylene torch generates temperatures high enough to vaporize metals, including lead. Workers exposed to vinyl chloride gas have a high incidence of hepatic angiosarcoma, an otherwise rare tumor. Pneumoconiosis or coal miner's lung results from the inhalation of coal dust with the formation of localized lesions with silica crystals, emphysema, fibrosis, loss of vital capacity and, eventually, right heart failure due to increased cardiac output to compensate for inadequate oxygenation of the blood. Organic solvents may be hazardous because of their CNS-depressing action.

Some recent studies have suggested that the offspring of firefighters have a higher incidence of birth defects in locales where firefighters, or their spouses, are responsible for washing their work clothes. This presumably is the result of the absorption of toxic contaminants on the clothing through the skin, although absolute confirmation of this risk source is yet to come.

Asbestos

Asbestos workers are exposed to a variety of health hazards including "white lung syndrome" (asbestosis, a form of fibrotic pneumoconiosis), carcinoma of the lung, mesothelioma (cancer of the pleural and peritoneal membranes) and possibly gastrointestinal cancer, although animal studies have not been able to confirm this. Mesothelioma is a rapidly fatal cancer occurring most often 30–40 years after the first exposure. The linings of the chest (pleura) and abdomen thicken, fluid accumulates and widespread metastasis occurs. This cancer occurs rarely in people not exposed to asbestos. There are several forms of asbestos fiber, and not all of them may cause mesothelioma. There is no doubt that the form known as crocidolite is carcinogenic, but controversy has centered on whether the form known as chrysotile is also carcinogenic. There is an ongoing study of Quebec chrysotile miners born between 1881 and 1920 and employed for at least 1 month. Over 70% of these miners have now died, and an estimated 30 cases of mesothelioma would be expected. Seven of these men were also exposed to crocidolite in a small factory, and contamination with tremolite (another form) could account for additional cases. It now seems that the risk is at least much lower for chrysotile asbestos. In order to cause pleural mesothelioma, asbestos fibers must traverse the lung and appear in the pleura. Chrysotile fibers will do this, and they have been shown to cause mesothelioma-like lesions in experimental animals. The risk associated with chrysotile fibers has not been firmly established in humans. According to some studies, a very large number of fibers must be inhaled for this to occur.

Carcinoma of the lung occurs 60× more often in asbestos workers who smoke than in those who do not. Asbestos becomes a hazard for the general populace when building insulation begins to break down or is disturbed during construction. Wear of brake linings releases asbestos particles in the air. There is increasing concern that glass wool fibers can cause the same type of cancer as asbestos. An excess in cancer incidence has been shown in workers in the glass wool industry, but no direct evidence linking this to the inhalation of fibers has been uncovered. In the Fiberglass Canada plant in Sarnia, an increased incidence was shown in the 2,500 workers, but it was not statistically significant. In the U.S., NIOSH recommended that allowable air levels of glass fibers in plants be reduced.

Silicosis

This results from the inhalation of silica particles or silicates or other mineral fibers. Histiocytes are transformed into fibrocytes, alveoli harden, and loss of elasticity and lung function occur. Emphysema results, as it does from cigarette smoking.

Pyrolysis of Plastics

Prior to about 1980, firefighters did not routinely wear a breathing apparatus unless dealing with a fire involving known toxic fumes. There is some evidence (still largely anecdotal) that fire fighters who attended fires involving plastics are beginning to show increased cancer rates. It is now known that when polyurethane smolders, fine particles of degraded polymers are produced which may have toxic chemicals adsorbed to them. These release lytic enzymes in the lung to cause massive tissue damage and edema.

Dust

Even barn dust can be an environmental hazard in the workplace. It may contain dried fecal material, animal dander, protein from feed grains and hay, skin parasites and microorganisms. A Scandinavian study found a high incidence of respiratory and other health problems in farm workers who spent a lot of time in hog barns. 30% of workers lost work time due to respiratory problems.

CO and NO_2

Chemicals which are involved in atmospheric pollution may sometimes become a problem indoors. There is increasing concern over indoor events that involve the use of internal combustion engines. These include tractor pulls, monster truck rallies and mud races. CO levels have been shown to peak as high as 250 ppm during such events. Peak levels should not exceed 30 ppm. NO_2 levels may also be elevated because of incomplete combustion. CO is colorless, odorless and nonirritating. It can produce headache, nausea and mental impairment. NO_2 is irritating and may cause pulmonary edema. High concentrations may be fatal.

There is growing evidence that particle pollution at levels encountered in the environments of most large urban centers may be more hazardous than previously believed. There are elevated incidences of premature deaths, hospital admissions and a variety of health problems. There is a statistically significant association between acute exposures to particles and increased mortality regardless of the source of the particles or the climatic conditions prevailing at the time of exposure. This seems to suggest that the particles are the primary cause, although the mechanisms involved are not yet known.

The Centers for Disease Control issued a report that 23 million Americans were at risk because of exposure to particles <10 μm in diameter and concentrations >155 μg/m^3 of air (the 24 hr average acceptable level is 150 μg/m^3). The EPA is considering setting new levels at a much lower concentration.

■ CHEMICAL IMPACT OF POLLUTANTS ON THE ENVIRONMENT

Sulfur Dioxide and Acid Rain

Over 100 million metric tons of sulfur dioxide from fossil fuels are emitted annually into the atmosphere around the world. Sulfur dioxide plus water (in atmospheric water vapor) forms sulfurous acid and, eventually, sulfuric acid as follows:

$$SO_2 + OH \rightarrow HSO_3$$

In the presence of ultraviolet radiation and O_2, it forms sulfur trioxide:

$$HSO_3 + O_2 \rightarrow HOO + SO_3$$

Sulfuric acid is then formed:

$$SO_3 + H_2O \rightarrow H_2SO_4$$

and an aqueous solution results from the dissolution of sulfuric acid in water droplets.
Sulfuric acid may also form from NO:

$$SO_2 + NO + H_2O \text{ (in water droplets)} \rightarrow \text{aqueous } H_2SO_4 + NO_2$$

The average retention time for sulfur dioxide in the troposphere is very short (about 2–4 days). The sulfuric acid which it forms is carried to the soil in precipitation (rainwater, snow). A pH as low as 1.7 was recorded in West Virginia in 1979 (battery acid and gastric acid are about pH 1). Core samples of snow in the arctic regions revealed a pH of 6.8, 180 years ago vs. 3.8 in recent years.

The anions in acid rain are SO_4^- (70%) and NO^- (30%). This acid mist may affect the respiratory tract of people (and animals). Asthma sufferers and people with allergies are prone to loss of lung function and respiratory disease.

The absorption of acid into the soil solubilizes metals such as aluminum, cadmium and lead and facilitates their movement into vegetation and water, including drinking water. The accumulation of these metals may contribute to human diseases. Aluminum has been implicated in dementias, lead may affect the development of the central nervous system in infants and children and cadmium can cause kidney disease (see Chapter 6). Acidification of lakes leads to a complete loss of aquatic life. It is estimated that up to 4,000 lakes in Ontario have been so affected.

Paradoxically, although the Mount Pinatubo volcanic eruptions were partly responsible (along with the El Niño) for the extremely cool, wet summer of 1992 in North America, the long-term effects are more likely to contribute to acid rain and global warming. The 1991 eruption injected 15–30 megatons of SO_2 into the stratosphere which, within 1 month, was converted to H_2SO_4. This formed an aerosol which is expected to remain in the atmosphere for up to 3 years. The total aerosol load is estimated to be 10–20× that produced by anthropogenic and other biological sources in the same year. Some models predict that ozone will be rendered more susceptible to degradation by atmospheric chlorine, and reflection of long wavelength infrared may increase global warming (see below). In fact, a marked decline in atmospheric ozone began in 1991, but recovery was noted in 1993 and by 1994, it had returned to essentially normal levels. It is not known whether this is a long-term trend nor whether the effect was attributable to the Mt. Pinatubo eruption.

The Chemistry of Ozone
In the stratosphere, at an altitude of about 20 miles, short-wave ultraviolet radiation converts O_2 to O_3, which, by direct absorption, prevents UV radiation from penetrating the earth's atmosphere. When longer UV wavelengths are absorbed (>242 nm), O_3 is split back into O and O_2 and thus is recycled.

Ozone depletion is of considerable concern because it contributes to climatic change by allowing short-wave UV radiation to penetrate to the earth's surface. Since this band is the ionizing form of UV radiation, ozone depletion is also associated with an increased risk of skin cancer. The layer is thinnest at the equator, so that the incidence of skin cancer in tropical climates is greater than in the temperate zones. Light-skinned people are at greatest risk. The incidence of skin cancer in Texas is 3.8:1,000 compared to 1.2:1,000 in

Iowa. The incidence of skin cancer is also increasing in northern climates as well, and warnings against unprotected sunbathing will be routine as each summer approaches. Sunscreen factors of 20 or more are recommended, as is avoidance of exposure between the hours of 1100 and 1500.

The ozone layer is normally maintained at about 1 ppm, but it can be depleted by the action of certain pollutants. Nitric oxide (NO) is a major offender in this regard. The following reactions can occur:

$$NO + O_3 \rightarrow NO_2 + O_2$$
$$NO_2 + O\cdot \rightarrow NO + O_2$$

Thus, NO will recycle to break down thousands of ozone molecules unless it reacts with another free radical, e.g.:

$$OH + NO \rightarrow HNO_3 \text{ (nitric acid)}$$

Chlorine

Chlorine, its oxides, and chlorine compounds such as chlorofluorcarbons (CCL_3F), widely used as aerosol propellants and refrigerants, also contribute to ozone depletion, e.g.:

$$Cl + O_3 \rightarrow ClO + O_2$$
$$ClO + O^- \rightarrow Cl + O_2$$

The appeal of CFCs was that they are chemically inert under nearly all conditions, and it was not until the impact of supersonic jet transports was studied that the effect of UV light on them was realized. The calculated ozone loss is presently 1%, but this would increase to 10% if the release of 800,000 tons of chlorofluorocarbons annually were to continue.

CFCs are a particular concern because they are heavier than air. They reach the stratosphere by slow percolation vertically, driven by the concentration gradient. This means that even if their use were to stop immediately, their effect will not begin to decline for many years to come and it has not yet peaked. CFCs are themselves greenhouse gases, further contributing to the problem of global warming.

CFCs have been temporarily replaced by hydrofluorocarbons (HFCs) and hydrochlorofluorocarbons (HCFCs) in some jurisdictions. Although these destroy just as much ozone as CFCs, they dissipate much more quickly in the atmosphere. In November of 1992, 86 countries attended the United Nations Ozone Layer Conference in Copenhagen. The conference agreed to an accelerated ban on CFCs by 1995 instead of 1999 as originally proposed. A similar ban or reduction on HFCs and HCFCs was blocked by France.

Ironically, two physicians in Dortmund, Germany, may have rendered all of the HCFC technology obsolete. The substitute, labelled Greenfreeze, is a simple, nonpatentable mixture of butane and propane that can be used in refrigerator units and that is not harmful to the ozone layer. The technology is available to anyone, and Greenfreeze refrigerators are already being marketed in Germany by DKK Schjarfenstein.

The Greenhouse Effect
Water vapor is undoubtedly the greatest single contributor to the greenhouse effect, accounting for up to 75% of heat trapping.

Carbon Dioxide
Another gas of concern regarding global warming is carbon dioxide (CO_2). CO_2 is a product of combustion and biological decay and it is normally consumed by green plants which take it up during the day and convert it to O_2 and carbohydrates by photosynthesis. It is a basic rule that the oxidation of one carbon atom yields one molecule of carbon dioxide:

$$\text{Combustion, decay: } C + O_2 \rightarrow CO_2$$

$$\text{Photosynthesis: } CO_2 \rightarrow O_2 + \text{carbohydrates}$$

$$\text{Reaction in atmosphere: } CO_2 + H_2O \rightarrow O_2 + CH_2O \text{ polymer}$$

This last reaction reverses in winter and returns the CO_2 to the atmosphere. Depletion of rainforests for timber and increased CO_2 production from industrial sources and internal combustion engines have led to a dramatic increase in atmospheric CO_2. Solar energy is either reflected from the surface of the earth or absorbed by it, in which case it heats the earth. Very little heating of the air occurs, which is why air is cooler at high altitude. Most of the reflected energy passes back out into space, but some is at a very long wavelength (50,000 nm) because the earth is much cooler than the sun, and it is reflected back to earth by water vapor and CO_2, the "greenhouse effect". This results in additional global warming.

There has been considerable controversy regarding anthropogenic contribution to global warming, partly because satellite measurements have failed to detect significant evidence of it. Very recently, however, an international panel of over 200 scientists reached a consensus that the effect is real and significant. Barring strict controls on greenhouse gas emissions, global average temperatures will rise by 1.5–3°C by the year 2050. By 2030, sea levels will rise 8–29 cm and continental interiors (the breadbaskets of the world) could go dry in summer. Although considerable uncertainties remain, warming since the past century is real, with increases of 0.3–0.6°C since the late 19th century.

Some lessons can be learned from history. In the late autumn of 1815, a huge volcano called Tambora in Indonesia exploded with a force many times that of Mt. St. Helen. The resulting dust was spread over the surface of the earth and 1816 became known as "the year without a summer". Air temperature dropped an average of 1°C. Crops were severely affected in the U.S., and it snowed in June in Maine. It was so cold and wet in Switzerland that summer that Mary Shelley challenged her husband to a writing contest for something to do. The result was *Frankenstein*. Fortunately, this situation was temporary, but it illustrates how little a change in temperature is required to effect drastic climatic changes.

Methane

Methane is a greenhouse gas, reflecting radiant energy back to the earth. As a fuel, it has a lot of advantages (see below), but uncombusted methane reaches the atmosphere from fermentation in the intestinal tracts of animals (both wild and domestic) and from the fermentation (rotting) of vegetation. Rice paddies are a major source of methane. It has been suggested that the North American penchant for burgers is a contributing factor to the greenhouse effect by encouraging extensive beef production (and rainforest destruction) in South America. Animals and rice paddies each produce about 100 megatons of methane annually, and since rice is one of the world's staple grains, a switch to a vegetarian diet would not likely result in a net methane saving. The biggest source of methane in the world is its termites which produce about 200 megatons annually. Since they digest dead wood, their population is expanding because of deforestation, and this is probably the greatest threat to the environment from methane (see also Chapter 13).

Like ozone levels, levels of polluting gases in the atmosphere, notably methane and carbon monoxide, have been returning to normal in recent years, and this too has been suggested to be the result of the declining influence of the Mt. Pinatubo explosion.

Subtle Greenhouse Effects

Significant impact can occur long before the catastrophic events discussed above. As temperatures climb, diseases of humans, livestock and plants which are normally constrained by subfreezing winter temperatures will advance northward. Those depending on insect vectors will gain a foothold more rapidly as insect populations soar with increases in temperature. The cold barrier to Africanized honey bees will move north, and their spread will be accelerated. Crop failures are bound to occur as crops are assailed by both drought and disease. This will necessitate increasing dependency on pesticides and irrigation, further stressing the environment. There is already evidence that expanding human populations modify the environment and that diseases adapt to the new conditions, in some cases evolving to become new threats to human health.

Global Cooling

New Ice Age?

Studies of ice core samples from the polar ice cap suggest that the last ice age, in which the world was covered with ice to a depth of nearly 1,000 meters as far south as New York, resulted from a mean temperature drop of no more than 1.5°C. Ice ages result from irregularities in the earth's orbit around the sun which take it further away about every 200,000 years. Interglacial periods last about 10,000 years, and there is some evidence that the present one is ending soon. Complicated, is it not?

It is even further complicated by the fact that the oceans act as CO_2 sinks, dissolving vast amounts of the gas. As gas is dissolved, the greenhouse effect is lessened, the temperature drops, and more snow and rain fall, contributing to the polar ice caps. As they expand, they reflect more and more UV rays back into the stratosphere, further lowering the earth's temperature. The reflective property of materials is known as the albedo factor, and the efficiency of ice is nearly 100%.

Sulfur Dioxide

SO_2, a major contributor to acid rain, may also contribute to global cooling. Aqueous-phase oxidation of SO_2 to SO_3 in clouds (rich in water vapor) occurs, and evaporation of the water releases a particulate aerosol of SO which back-scatters solar radiation.

Motor Vehicle Exhaust

As noted above, the internal combustion engine is a major source of CO_2 and other pollutants including the following.

Lead

Polar ice in Greenland has shown a sharp increase in carbon and lead deposits since the 1950s. Lead rose from 0.03 to 0.20 µg/kg snow or ice. Levels corresponding to the era of 800 B.C. were <0.001 µg/kg ice. In urban areas the lead content of the air is 5–50× higher than in the country (1.1 vs. 0.02 µg/m^3). In 1968, the gasoline combustion engine accounted for 181 of the 183 kilotons of lead released into the atmosphere annually in the U.S. Other sources included coal and oil combustion and manufacturing processes including lead smelting. Vehicles using tetraethyl lead gasoline emit, on average, 1 kg of lead/year. Distribution and eventual inhalation of this lead by humans constitute a health hazard. At levels of about 10/µg/dL (ingested), it is a potent neurotoxin to infants and children, causing impaired hearing, slowed neuronal transmission, and a variety of behavioral problems including hyperactivity, learning disabilities and reduced mental capacities (see also Chapter 6). Lead from solder joints in plumbing is also implicated as a source.

Carbon Monoxide

Carbon monoxide (CO) is a product of all forms of combustion, and it has been responsible for many deaths due to the use of heating devices (camp stoves, space heaters, defective furnaces, fireplaces, etc.) in poorly ventilated sleeping areas, including tents. Since the internal combustion engine is the principal source of CO (it has been used as an instrument of suicide many times over), it is convenient to consider CO toxicity here. Its natural concentration in the atmosphere is extremely low (1–2 ppm) and about 100 megatonnes yearly arise from vehicle engines in the U.S. Other sources include natural processes (70 megatonnes) and other human combustion sources (250 megatonnes). Urban concentrations may range from 20–100 ppm depending on traffic and weather conditions. Fire fighters are at risk from CO poisoning if a breathing apparatus is not worn. Combination with free radicals in the environment helps to buffer the atmospheric levels of CO:

$$CO + OH^- \rightarrow CO_2 + H^+$$

CO combines irreversibly with hemoglobin (Hb) to form carboxyhemoglobin, which cannot combine with O_2, so that asphyxiation results. Levels >6.4 thousand ppm cause dizziness and headaches within 2 min and unconsciousness and possibly death in 15 min. Treatment consists of intravenous nitrates which cause the formation of methemoglobin. This has a much greater affinity for CO and binds to free the Hb. If a hyperbaric facility is available, O_2 at high pressure (2 atmospheres) will provide adequate oxygenation through dissolved O_2 until enough Hb is free to assume oxygen transport.

■ NATURAL FACTORS AND CLIMATE CHANGE

Although much has been made of the impact of human activity on climate change, the influence of natural events is often ignored, as is the fact that nearly all computer simulation models are based on a doubling of CO_2 without reference to other factors, including water vapor, discussed above. Methane is another natural greenhouse gas (see above and Chapter 13).

Contrary to statements in many texts, the Earth's orbit is not a fixed elipse, but will vary considerably. Moreover, its tilt may vary by several degrees. Both effects are due to the gravitational pull of Jupiter, and they can have a dramatic impact on climate. Some estimates claim that they are responsible for up to 85% of climate variation. Solar activity such as flares and sunspots also correlates well with climatic changes. The 1988–89 season was very active, with over 400 sunspots recorded. The activity of El Niño, the Pacific upwelling of warm water from the depths, correlates well with sunspot activity, and it has a marked effect on climate.

From 1675 to about 1725, there was a "Little Ice Age", with record cold temperatures and snowfalls. There was ice skating on the Thames River in London, England, and elsewhere in Europe. Since then, the climate has been warming. Should this trend have reversed? Cooling trends tend to occur fairly slowly, whereas warming trends are much more abrupt. Ice cores taken from the Vostock 4 site in Russia, going back 800,000 years, measured CO_2 levels in bubbles in the ice, and these indicated a natural cycle of about 100 years. This cycle appears to be shortening dramatically. The observatory at Moana Loa in Hawaii recorded atmospheric CO_2 levels of 315 ppm in 1958, rising to 355 ppm in 1989. Methane levels are rising even faster, from 400 ppb to 1,800 ppb in the same period. There is little doubt that human activity has contributed to these changes, but the seas are a vast sink for CO_2, and a natural warming trend would also release CO_2 from this sink. Increased water vapor from the oceans would increase cloud cover, which would probably have a buffering effect on the warming trend. Nevertheless, some models predict a 4-ft rise in sea level over the next 50 years unless the trend is reversed, submerging Bangladesh and the Maldives. In northwestern Ontario is the Experimental Lakes Area which has been the subject of climatic, hydrologic and ecological study for over 20 years. Records show that both air and lakewater temperatures have increased by 2°C during this period and that the ice-free season has lengthened by 3 weeks. The thermocline has also deepened. Evaporation has been higher, and precipitation lower, than normal, with the result that lake levels have dropped and pollution has concentrated, all of which have ecological consequences.

Glaciers and small ice caps respond much more rapidly to climate changes than do the polar ice caps, and thus they are useful for computer modeling of short-term effects of warming. In the last 100 years, mountain glaciers have retreated in most parts of the world and have likely contributed to the 10–20 cm rise in sea level during this period. A recent study of 12 such areas where data collection has been in progress for several years indicated that for a 1 kelvin (K) rise in temperature, glacial melting would account for a rise in sea level of 0.58 mm/yr, which is significantly less than earlier estimates.

■ REMEDIES

How all these factors will balance out is anybody's guess, but it is clear that we do not know enough to take chances. As one wag put it, Chicken Little only has to be right once. It is abundantly clear, however, that a reduction in fossil fuel consumption is essential. This means more fuel-efficient cars, changes in transportation habits with more public transit usage, and the development of alternative fuels. One such is natural gas (methane) for heating and for automotive fuel. Although methane obeys the one carbon-one CO_2 rule, and although it is a greenhouse gas itself, it produces more energy per carbon with fewer polluting by-products than petroleum oil-based fuels, so that the negative impact on the environment is much less. Further development of nuclear energy may also be essential. Despite the emotional reactions that it can generate, a well-designed, well-regulated and well-operated nuclear power generator may be the safest source of electrical energy that we have. It is worth remembering that not one proven death has resulted in North America from a nuclear generator (see also Chapter 2), but hundreds have died from failures of hydroelectric dams and thousands from coal mining accidents and black lung disease. In 1979, a power dam in India collapsed, killing thousands. *Time* magazine carried only a few lines, compared to the several pages devoted to Chernobyl. In 1989, in the (then) USSR, a gas pipeline explosion killed hundreds of people in two passing passenger trains. The Atomic Energy Commission of Canada did a study indicating that nuclear energy is even safer than solar energy, since the latter requires materials that need to be mined and refined with energy derived from coal. The calculation determined that solar energy was inherently 500× more dangerous than nuclear energy. Fossil fuels (coal and oil) were 2,000× more dangerous. Natural gas was the safest of all. Hydrogen may also become a source of pollution-free energy for transportation.

A theory that was popular a number of years ago is undergoing a renaissance. Known as the Gaia hypothesis, it considers the earth and its atmosphere as a living organism, with all parts interconnected and in balance, like a mammal in a state of homeostasis. Interference with one component of the system may have far-reaching consequences that may not be foreseen with our incomplete understanding of how the system works. Chapter 13 deals with this topic in more detail. One essential change that must occur is a dramatic reduction in the use of fossil fuels by developed countries, especially in North America. Northern developed countries use 7.5 kilowatts per person per year, vs. 1.0 for southern, developing countries. Given the populations of 1.2 billion for the former and 4.1 billion for the latter, the North consumes 2.2 times as much energy as the South. Add to this the fact that 6 million hectares of arable land are lost annually from erosion, development and salination whereas 4 million are created by clearing, which entails the loss of irreplaceable rainforests, and it becomes evident that major changes are essential. One of these is a check of the population explosion. The doubling time of the earth's population is 30 years and of its energy consumption, 20 years. It is evident that we cannot continue in this manner if we and the planet are to survive.

REVIEW QUESTIONS

1. Which of the following sources, in industrialized countries, accounts for the greatest amount of air pollution?
    a. Industry
    b. Generating electric power
    c. Transportation
    d. Heating
    e. Waste incineration

2. One reason for stopping atmospheric nuclear tests was concern over the deposition of a radioactive isotope in the developing bones of children. The isotope of concern was
    a. Cobalt 60.
    b. Strontium 90.
    c. Iodine 125.
    d. Cesium 133.
    e. Carbon 14.

3. The diameter of droplets or particles defined as aerosols is
    a. >10 µm
    b. From 5 to 10 µm.
    c. Less than 1 µm.
    d. From 1 to 2 µm.
    e. From 2 to 5 µm.

For Questions 4 to 7, use the following code: answer A if statements 1, 2 and 3 are correct; answer B if statements 1 and 3 are correct; answer C if statements 2 and 4 are correct; answer D if only statement 4 is correct; answer E if all statements (1,2,3,4) are correct.

4. Which of the following can be sources of indoor, household pollution?
    1. Radon gas.
    2. Cigarette smoke.
    3. Asbestos fibers.
    4. Formaldehyde gas.

5. Pollutants which can contribute to acid rain include
    1. Sulfur dioxide.

2. Sulfur trioxide.
3. Nitric oxide.
4. Chlorofluorocarbons.

6. Acid rain
    1. May solubilize toxic metals.
    2. Can never be neutralized naturally.
    3. Can kill aquatic life.
    4. Is never the result of a natural phenomenon.

7. Which of the following statements is/are true?
    1. The troposphere is the air mass in contact with the Earth's surface, and it extends vertically for about 10 miles.
    2. Hadley cells are vertical air movements which circulate air northward and southward from the equator.
    3. The stratosphere extends from about 10 to 30 miles above the Earth.
    4. Pollutants are rapidly cleared from the stratosphere.

8. List four groups of individuals who are at greater than average risk of respiratory problems from air pollutants (see pages 5–7 and 5–8).

9. List four conditions known to be associated with air pollution in the workplace (see pages 95–98).

10. List five natural factors likely to contribute to global warming.

11. What do you think is the single most important underlying cause of anthropogenic environmental problems?

12. List three environmental consequences of global warming.

13. Complete the following equations (see pages 98 and 100).

$$SO_2 + H_2O \rightarrow$$

$$NO + OH \rightarrow$$

$$NO_2 + O^{\cdot} \rightarrow$$

$$Cl + O_3 \rightarrow$$

$$ClO + O^{\cdot} \rightarrow$$

$$HSO_3 + O_2 \xrightarrow{uv}$$

$$SO_3 + H_2O \rightarrow$$

$$SO_3 + NO + H_2O \rightarrow$$

$$NO + O_2 \rightarrow$$

ANSWERS

1. c; 2. b; 3. c; 4. E; 5. A; 6. B; 7. A

CASE STUDY #3

In August of 1989, a previously healthy 4-year-old boy developed signs and symptoms which included leg cramps, rash, itching, excessive perspiration, rapid heartbeat, intermittent low-grade fever, personality changes and peripheral neurological disorders. The interior of the house had been painted 1 month earlier using over 60 liters of a latex-based paint. The house was sealed and air-conditioned.

Q. Is there likely to be any connection between the boy's illness and the redecorating?

The signs and symptoms strongly resemble a condition known as acrodynia, a rare form of childhood mercury poisoning. A 24-hr urinary mercury determination revealed a level of 65 µg/L. The boy's mother and two siblings had similar urine mercury levels.

Q. Do these results point to an environmental source of the mercury?

Q. What additional analysis or analytical information would be useful?

CASE STUDY #4

In the same year (1989), two farm workers in their 30s were working in an indoor manure pit 25' × 25' × 5' deep, attempting to clear a blocked pump intake pipe. Several hours later, both men were found face down in several inches of liquified manure.

In an unrelated incident, five men died after consecutively entering a similar pit, each attempting to rescue those who had entered before. The first person to enter was attempting to replace a shear-pin in a piece of equipment.

Q. What is the underlying cause of death in both of these incidents?

Q. What simple safety measure could have prevented these tragedies?

See also, Case Study #2.

CASE STUDY #5

Two machine shop workers were cleaning and degreasing equipment in a confined, poorly ventilated space using a chlorofluorocarbon known as CFC-113. After about 20 minutes, one worker clutched his chest and collapsed. He was cyanotic and not breathing. The second worker called for help and began to administer artificial respiration (he was not trained in CPR) but himself collapsed soon afterwards. Both victims were evacuated. The first was pronounced dead on arrival at hospital; the second recovered in the ambulance after oxygen was given.

Q. What was the portal of entry of the toxicant?

Q. What organ systems were involved in the intoxication?

Q. What was the likely offending agent?

Q. What steps should have been taken to prevent this accident?

■ CASE STUDY #6

Soapstone carving is an economic mainstay for the Inuit of the Canadian North. Recently the Inuit Art Foundation created a comic-book superhero called Sanannguagartiit ("your carving buddy" in the Inuktitut language). With his flying snowmobile and his loyal sled dog Quimmiq, he criss-crosses the Arctic demonstrating masks and respirators and advising carvers to leave their work clothes at the carving hall and not to take them home.

Q. Why would it have been necessary to institute this program?

■ CASE STUDY #7

Workers in a particular industry have been having a very high incidence of respiratory problems including:
1. A high frequency (15.7% of workers) of chronic bronchitis.
2. A sevenfold increase in restrictive lung function and a threefold increase in obstructive lung function compared to the general population.
3. A 20% occurrence of shortness of breath and flu-like symptoms such as fatigue, muscle and joint pain and general malaise.
4. A 30% annual occurrence of absenteeism due to respiratory illnesses.
5. A 35% occurrence of wheezing and tightness in the chest.
6. A 50% frequency of chronic, productive cough and frequent colds.

Q. Is this most likely a primary infection or a pollutant?

Q. If the latter, what is the likely portal of entry?

These men are farm workers in the pork industry.

Q. What contaminants could be present in the air in pig barns?

Q. What corrective measures should or could be taken?

■ CASE STUDY #8

Recently in Wisconsin, an incident was reported to public health authorities in which 11 high school students were treated in two different emergency departments for acute respiratory symptoms including labored breathing (dyspnea), spitting (coughing) blood (hemoptysis), cough and chest pain. Two of them required hospitalization. All of them had participated in an ice hockey tournament the previous evening in an indoor arena. Interviews with other players and spectators revealed that many of them had suffered respiratory problems which became progressively worse as the evening wore on, and many had central nervous system symptoms including headache, dizziness, sleepiness, nausea and vomiting. Of 131 students who were interviewed, 48% reported symptoms. The frequency among players was double that among spectators.

Q. What is the likely portal of entry in this toxic reaction?

Q. What offending agent(s) would you suspect?

Q. What laboratory test would be helpful?

■ FURTHER READING

Abelson, P.H., Global change. Editorial. *Science*, 249, 1085, 1990.
Bates, D.V., Asbestos: the turbulent interface between science and policy. *Can. Med. Assoc. J.*, 144, 554, 1991.
Brasseur, G. and Granier, C., Mount Pinatubo aerosols, chloroflurocarbons, and ozone depletion. *Science*, 257, 1239, 1992.
Coghlin, J., Gann, P.H., et al., 4-aminobiphenyl hemoglobin adducts in fetuses exposed to the tobacco smoke carcinogen in utero. *J. Natl. Cancer Inst.*, 83, 244, 1991.
Dockery, D.W., Pope, C.A., et al., An association between air pollution and mortality in six U.S. cities. *New Eng. J. Med.*, 329, 1753, 1993.
Epidemiologic notes and reports. Nitrogen dioxide and carbon monoxide intoxication in an indoor ice arena — Wisconsin, 1992. *Morbid. Mortal. Weekly Rep.*, 41, 383, 1992.
Fisher, D.E., *Fire and Ice: The Greenhouse Effect, Ozone Depletion and Nuclear Winter*, Harper and Row, New York, 1990.
Friedlander, S.K. and Lippman, M., Revising the particulate ambient air quality standard. *Environ. Sci. Technol.*, 28, 148A, 1994.
Jensen, J., Plastics and pulmonary edema. *Can. Res.*, Nov., 17, 1988.
Kerr, R.A., The greenhouse consensus. *Science*, 249, 481, 1990.
Kerr, A., Pinatubo fails to deepen the ozone hole. *Science*, 258, 395, 1992.
Lelieveld, J. and Heintzenberg, J., Sulfate cooling effect on climate through in-cloud oxidation of anthropogenic SO_2. *Science*, 258, 117, 1992.
Lloyd, S., The calculus of intricacy. *The Sciences*, Sept./Oct., 38, 1990.
McDonald, J.C. and McDonald, A.D., Asbestos and carcinogenicity. *Science*, 249, 844 (Lett.), 1990.

Mossman, B.T., Bignon, J., et al., Asbestos: scientific developments and implications for public policy. *Science*, 247, 294, 1990.
Novelli, P.C., Masarie, K.A., et al., Recent changes in atmospheric carbon monoxide. *Science*, 263, 1587, 1994.
Oelermans, J. and Fortuin, J.P.F., Sensitivity of glaciers and small ice caps to greenhouse warming. *Science*, 258, 115, 1992.
Planet earth — problems and prospects. *Notes, Sesquicentennial Symposium of Queen's University,* Kingston, June 1991.
Population at risk from particulate air pollution — United States, 1992. *Morbid. Mortal. Weekly Rep.*, 43, 290, 1994.
Ramade, F., *Ecotoxicology.* John Wiley & Sons, Chichester, 1987.
Ravishankara, A.R., Solomon, S., Turnipseed, A.A. and Warren, R.F., Atmospheric lifetimes of long-lived halogenated species. *Science,* 259, 194, 1993.
Schindler, D.W., Beaty, K.G., et al., Effects of climatic warming on lakes of the central boreal forest. *Science*, 250, 967, 1990.
Rodhe, H., A comparison of the contribution of various gases to the greenhouse effect. *Science*, 248, 1217, 1990.
Robert, O., The threat of barn dust. *Vet. Mag.*, 3, 29, 1991.
Yanko, D., Are animal disease patterns changing because of global warming? *Vet. Mag.*, 2, 18, 1990.

5 HALOGENATED HYDROCARBONS

■ INTRODUCTION

Halogens are the related elements chlorine (Cl), bromine (Br), fluorine (F) and iodine (I). They may exist as gases (Cl_2, F_2), a liquid (Br_2) or a solid (I_2). Halogenated hydrocarbons, also known as organohalogens, are a group of organic compounds of diverse structure to which one of these halogens has been attached. The core structure may be either simple, consisting of one or two carbons, or it may be a more complex aromatic one. Because they have been implicated almost universally in toxic reactions in mammals and lower species (including carcinogenesis in some), it is appropriate to consider them as a group despite their chemical diversity.

Early Examples of Toxicity from Halogenated Hydrocarbons

One of the oldest and simplest of these compounds is carbon tetrachloride (CCl_4) which was used extensively as a solvent and as a dry cleaning agent until its hepatotoxic nature was discovered. In fact, it was used originally as a treatment for hookworm in humans and domestic animals and as a component in fire extinguishers. These uses may still exist in some parts of the world.

CCl_4 owes its toxicity to the fact that it is converted in the liver to carbon trichloride ($^-CCl_3$) which is a free radical capable of inducing peroxidation of lipid double bonds and of poisoning protein-synthesizing enzymes. Other older halogenated hydrocarbons include trichloroethylene, and the anesthetics halothane and chloroform (now abandoned because of its toxicity). Figure 21 shows the variety of structural formulae included in this class of compounds. The earliest form of poisoning associated with a halogen was probably

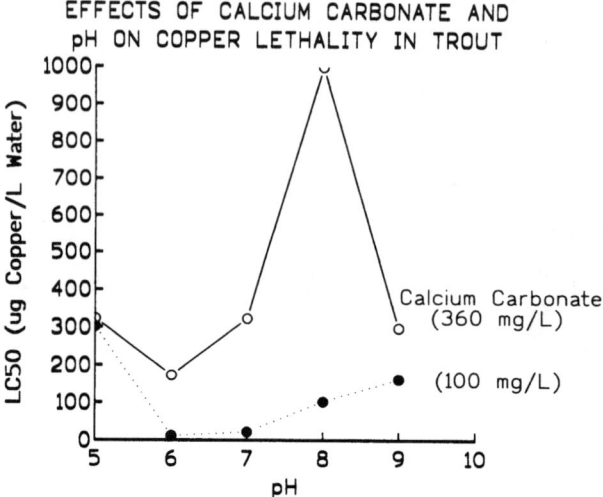

Figure 21 Structural formulae of a number of halogenated hydrocarbons.

"bromism", a condition resulting from the use or abuse of sodium or potassium bromide as a sedative and sleeping potion early in this century. Symptoms included severe headache, stupor, delirium, cardiac problems, very bad breath (from the bromine) and an acneform skin rash of the type now called "chloracne". The expression "bromide" is used now to indicate a soothing but meaningless statement of the sort frequently uttered by certain politicians.

■ PHYSICOCHEMICAL CHARACTERISTICS AND CLASSES OF HALOGENATED HYDROCARBONS

The characteristics of halogenated hydrocarbons that make them useful for a variety of applications are generally the same ones that make them hazardous to the environment and to humans. These include:

1. High lipid solubility.
2. Ability to survive heat >800°C.
3. High resistance to chemical breakdown.
4. Toxicity to microorganisms.

These agents are used for a variety of purposes (see below).

Antibacterial Disinfectants

Hexachlorophene has been used for many years as a surgical scrub and, as a 3% solution, as a hospital disinfectant. It is also used as the active ingredient in deodorant soaps. In the late 1960s, a change was proposed in U.S. FDA regulations to permit the use of hexachlorophene as an antifungal wash for fruit and vegetables. In light of the then-recent thalidomide tragedy, extensive testing was required for approval to be granted. Rats fed high levels of hexachlorophene developed weakness, ataxia, paralysis and evidence of a type of brain pathology known as status spongiosus, indicative of axonal degeneration. In 1971, a study was done in which infant monkeys were washed daily for 90 days in 3% hexachlorophene. Neurological symptoms were observed, and *status spongiosus* was seen in all specimens at post mortem. Significant blood levels have been detected in infants washed with 3% hexachlorophene, and those with severe diaper rash, burns or congenital skin disorders are especially prone to absorb it, as the natural permeability barrier has been disrupted (see Chapter 1). Autopsies of infants dying from a variety of causes and who received high exposures showed evidence of *status spongiosus*. In 1972 hexachlorophene was accidentally added in high concentration to baby powder during its manufacture in France. Forty-one deaths of infants and young children were attributed to this error. A few years later, Dr. Hildegard Halling, a Swedish physician, published a report indicating that nurses who washed frequently in hexachlorophene (10–60× daily) had a higher incidence of birth defects in their offspring (25/460 births) than those who did not (0/233 births). Although this clinical study was criticized for design flaws, others with rats have revealed teratogenic effects. This product is no longer used in nurseries in North America, and pregnant women are advised to avoid it.

Herbicides
This group includes 2,4-dichlorophenoxyacetic acid (2,4-D), 2,4,5-trichlorophenoxyacetic acid (2,4,5-T), and dioxins such as 2,3,7,8-tetrachlorodibenzo-p-dioxin (TCDD = dioxin). Agent Orange, used as a defoliant in Vietnam, was equal parts of 2,4,-D and 2,4,5,-T and contained TCDD as a contaminant. One of the best-documented human exposures to dioxin was the explosion at Seveso, Italy (see Chapter 2). Hundreds of individuals suffered from chloracne, which is the hallmark of toxicity of halogenated hydrocarbons.

The herbicides 2,4,-D and 2,4,5-T are used to control broad-leafed plants along highways, railways and utility rights-of-way. They are hormonal growth promoters and force plants to consume energy at a greater rate than it can be replaced. The humans at greatest risk of toxic exposure are the workers who apply the sprays, and poisoning from dermal and respiratory absorption has occurred as well as from accidental ingestion. Signs and symptoms include peripheral neuritis, muscular weakness and chloracne. Although 2,4,5-T is a weak teratogen in some animals, it is the presence of TCDD, or dioxin, that is the greatest source of public concern.

Dioxin (TCDD) Toxicity

Dioxins are a family of compounds of which TCDD (2,3,7,8-tetrachlorodibenzo-p-dioxin) has received the most public attention. There is significant species variation in TCDD toxicity, with the guinea pig being most sensitive (LD_{50} 1 µg/kg) and the rat quite insensitive (LD_{50} 22 µg/kg). Other toxic manifestations of dioxin toxicity are as follows.

Hepatotoxicity. All species show enlarged livers, and microsomal monooxygenase enzyme induction occurs in most. Rats develop fatty livers with triglyceride deposition. Hepatic fibrosis has been reported in humans. People exposed at Seveso had elevated serum enzyme levels (serum glutamic-oxaloacetic transaminase, SGOT, and serum glutamic-pyruvic transaminase, SGPT), indicating liver damage, for several weeks after the accident.

Porphyria. Porphyrins are pigments widely distributed in nature, and they are present in the body as by-products of heme synthesis. Heme is required for the formation of hemoglobin, myoglobin and cytochromes. Heme is ferrous protoporphyrin IX. Hematin, the iron-containing molecule in catalase and peroxidases, is ferric protoporphyrin IX. The rate-limiting step in the synthesis of heme is ALA synthetase (ALA = gamma-aminolevulinic acid). Dioxin significantly increases the levels of ALA synthetase and hence, ALA levels and the synthesis of porphyrins. This is not enzyme induction, but probably is due to interference with a feedback control system. The excess porphyrins are excreted in the urine, giving it a port wine color, and they are deposited in the skin, producing pigmentation. Because porphyrins are photoreactive, a condition known as *porphyria cutanea tarda* develops, characterized by photosensitivity, blistering, fragility of the skin, pigmentation and hirsutism (hairiness). Congenital defects in porphyrin metabolism cause the same syndrome, and this has been suggested as the explanation for the vampire and werewolf myths of Europe (characterized by hirsutism and the avoidance of sunlight). There is even an explanation of why drinking blood might have a therapeutic effect. Heme is the feedback substance that turns off ALA synthetase. Absorption of sufficient heme might thus inhibit porphyrin synthesis. In the late 1950s, an extensive outbreak of *porphyria cutanea tarda* occurred in Turkey during a famine as the result of consuming seed grain treated with hexachlorobenzene as an antifungal agent. A simplified scheme of the steps involved in porphyrin synthesis is shown in Figure 22.

Chloracne. This skin disorder is typified by rash, cysts and hyperpigmentation and this is the hallmark of poisoning with all halogenated hydrocarbons. It was the predominant toxic manifestation at Seveso.

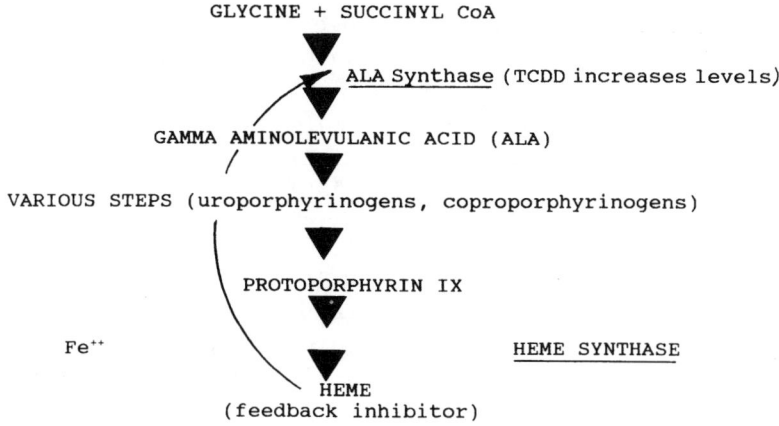

Figure 22 The Ah receptor and enzyme induction.

Cardiovascular effects. A 10-year mortality study of the population exposed to TCDD after the Seveso explosion of 1976 revealed a significantly increased mortality from cardiovascular events.

Carcinogenicity. Dioxin is a potent hepatocarcinogen in mice and to a lesser extent in rats. There is a latency period before the emergence of the liver tumors. Evidence of cancer in several studies of Vietnam veterans, for whom claims of increased incidence of cancer have been made, has been inconclusive. A retrospective cohort study was conducted by scientists at the (U.S.) National Institute for Occupational Safety and Health (NIOSH) on 5,172 workers at 12 U.S. plants in which TCDD was a chemical contaminant of the manufacturing process. Exposure was well documented and serum TCDD levels were obtained from 253 workers. The mortalities from several cancers previously associated with TCDD (stomach, liver, and nasal cancers, Hodgkin's Disease and non-Hodgkin's lymphoma) were not significantly different from the overall population, but the incidence of all cancers taken together was slightly but significantly increased. In a subcohort of 1,520 workers with more than one year of exposure and more than 20 years of latency, mortality from soft tissue sarcoma was significantly higher than for the general population. The authors concluded that the results were not suggestive of the high relative risks of cancer reported for TCDD in previous studies. The slight risk of increased soft tissue sarcoma is weakened by the small numbers involved (only three cases) and confounding factors such as smoking and exposure to other chemicals. In another epidemiologic study of 754 Monsanto employees exposed to high levels of TCDD in a 1949 accident, of whom 122 developed chloracne, there was no increased incidence of cancer in those who developed chloracne, although they were presumably the group with the highest exposure. Conversely, workers who were also potentially exposed to 4-aminobiphenyl, a potent bladder carcinogen, had increased mortality from bladder cancer, lung cancer and soft tissue sarcoma. This suggests that TCDD might act as a cocarcinogen or promoter. Again, the effects of confounders such as smoking and exposure to other chemicals could not be ruled out, but recent experimental evidence supports the suspicion that TCDD could act in this way. Walsh et al. studied the cell toxicity of aflatoxins in cultured human epidermal cells. AFB_1 was markedly toxic at 1 µg/ml. Neither AFB_2 nor AFB_1 dihydrodiol were toxic. TCDD alone was not toxic to the cells but at 5 nM it dramatically stimulated AFB_1 toxicity at levels as low as 0.1 µg/ml. It also increased the formation of AFB_1 epoxides and a 20-fold increase in DNA adduct formation was observed. AFB_1 is the most carcinogenic of the aflatoxins (see Chapter 10).

The most recent report from Seveso is suggestive of a carcinogenic effect in humans. The exposed population was divided into three groups according to their likely level of exposure. Persons who had left the area were traced with a 99% success rate. They were followed from 1977 to 1986, and cancer incidences were compared to a reference population not exposed to high TCDD levels. The population exposed to the highest levels was small, but nearly 5,000 were in the middle exposure area. In these, the relative risk factors compared to the reference population were elevated for several cancers: 2.8 for hepatobiliary cancer, 5.7 for lymphoreticulosarcoma in men, 5.3 for multiple myeloma and 3.7 for myeloid leukemia in women. In the lowest exposure zone, the incidence of non-Hodgkin's lymphomas and soft tissue tumors was elevated, especially among those who had lived in the area for over 5 years. Surprisingly, the incidences of breast cancer and endometrial cancer were below expected levels. This may relate to the effects of TCDD on estrogen receptors (see below).

Weaknesses of this study include the inability to control for confounding factors such as smoking and lack of hard data regarding real exposure levels. Despite the conflicting results of several epidemiological studies, most authorities now agree that there is a high index of suspicion for TCDD carcinogenicity in humans, especially for non-Hodgkin's lymphoma. The definitive word, however, remains to be heard, and the existing evidence comes from high industrial and occupational exposures which may not be relevant to environmental exposures encountered by the general population if a threshold truly exists because of the TCDD receptor story.

The mechanism of carcinogenesis in animals appears to be epigenetic. Recent evidence indicates that TCDD binds to a specific receptor, the Ah (for aryl or aromatic hydrocarbon) receptor, and that a minimum number of Ah receptors must be occupied for TCDD to exert its effect (see below). The implication of this is that there is thus a threshold dose and that the linear multistage carcinogenesis model is inappropriate for TCDD and for any other agent which works by this mechanism such as PCDDs and PCDFs (see Chapter 3 and below). The EPA is now reconsidering the use of the linear multistage model, at least for TCDD and perhaps for a few other agents.

Neurotoxicity. Numerous toxic effects have been observed including impaired vision, hearing and smell, depression, sleep disturbances and others. These were observed in workers exposed to TCDD in the 1949 Monsanto accident.

Reproductive toxicity. Testicular atrophy, necrosis and decreased spermatogenesis have been seen in laboratory animals (see also, TCDD and Estrogen Receptors, page 121).

Metabolic disturbances. Weight loss and depletion of adipose tissue occurs in lab animals.

Enzyme Induction

TCDD is one of the most potent inducers of aryl hydrocarbon hydroxylase (AHH) yet discovered. AHH is synonymous with cytochrome P_1-450, previously known as cytochrome P-448. In some species, TCDD is effective at 1 µg/kg. It induces synthesis of hepatic microsomal cytochrome P_1-450. TCDD uptake into the cell is passive (i.e., concentration gradient-dependent). Intracellularly, it binds to the Ah (aromatic hydrocarbon) cytosolic receptor, which is the product of a regulatory gene. The inducer-receptor complex is translocated to the nucleus where it activates numerous structural genes. The information is transcribed to m-RNA and translated to protein synthesis and the production of P_1-450, which is bound to the cell membrane. This is known as a pleiotropic response, i.e., it results in more than one phenotypic effect. The consequence of this induction is that several drug (xenobiotic)-metabolizing enzymes are induced, and reactive metabolites may be formed which react with proteins and nucleic acids to cause mutations, teratogenesis and carcinogenesis as well as altered drug metabolism. Conversely, the reactive metabolites may be excreted or detoxified, e.g., by conjugation with glucuronide. A schematic representation of this pathway is shown in Figure 23. This system has been most studied in the mouse, where it is inherited as an autosomal dominant pattern. Similar systems have been identified in the rat, rabbit and some fish. It is most heavily concentrated in the liver. In humans, there is considerable variation in the Ah locus.

To date, no endogenous substrate has been identified for the Ah receptor, but it is known that primitive species (bacteria, yeasts) utilize

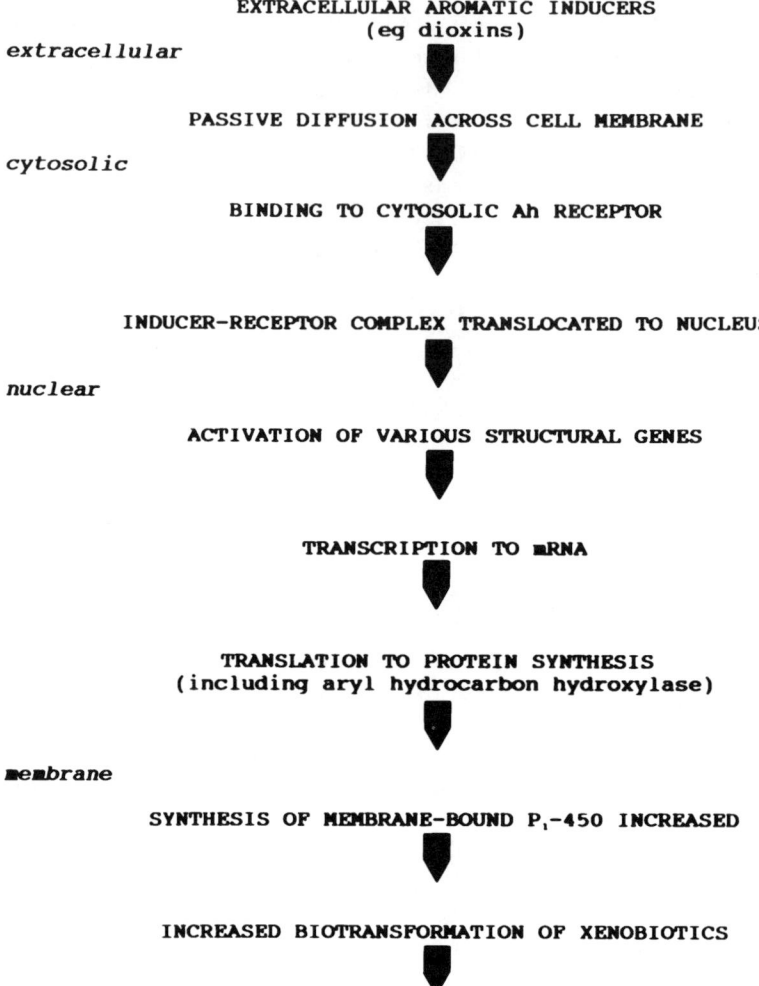

Figure 23 Effect of TCDD on the (simplified) heme synthetic pathway.

polycyclic hydrocarbons as an energy source and possess P-450 metabolizing enzymes for them (camphor in Pseudomonas, benzo-[a]-pyrene in yeast), so these may have evolved as a detoxication system. The mechanism of TCDD toxicity is not known, but if there is a natural substrate for Ah receptors, its displacement by TCDD could be involved in the latter's toxicity. It is not clear whether TCDD itself or its metabolite(s) are responsible.

TCDD also has nonreceptor-mediated effects, including interference with calcium homeostasis and a variety of membrane-related changes. It is interesting that the chloracne associated with TCDD also occurs with bromides, which could not act as Ah ligands.

TCDD and Estrogen Receptors

It has long been known that rats fed low levels of TCDD (1 µg/kg/day) for long periods have a higher incidence of hepatocellular carcinomas in female, but not in male, rats and a lower incidence of spontaneous mammary and uterine tumors. TCDD has also been shown to inhibit several estrogen-dependent responses in rats such as increased uterine weight and estrogen receptor levels. The structure-activity relationships for TCDD and related compounds as antiestrogens are similar to their Ah receptor binding affinities. It has recently been shown that TCDD is more potent than the antiestrogenic agent tamoxifen, used to treat estrogen-dependent breast cancer, in suppressing the estrogen-induced expression of pS2, a prognostic marker for breast cancer. The mechanism of action is believed to relate to the fact that the Ah receptor serves as a ligand-activated transcription factor for many genes involved in cell growth and differentiation, including the gene for the estrogen receptor. This characteristic of TCDD and other halogenated aromatic hydrocarbons could explain the apparently paradoxical reduction in the incidence of breast cancer seen in the Seveso women.

These recent observations have stirred considerable debate concerning the potential threat of estrogen-like substances in the environment. The EPA now feels that the effects of TCDD on reproduction and development may be of greater concern than its cancer-causing potential. As is often the case, the news media have sensationalized the situation, epidemiological studies have been criticized, the scientific community is divided on the issue and the public is confused.

Paraquat Toxicity

This herbicide (1,1'-dimethyl-4,4'-bipyridinium ion) is highly water-soluble, therefore poorly absorbed across the skin or gastrointestinal mucosa. It is extremely toxic to humans when inhaled, however, and 5 g may be fatal. Pulmonary congestion, edema and hemorrhage may result in almost complete functional destruction of the lung. In severe cases, lung transplantation has been tried as a last resort, with disappointing results. Although paraquat is poorly absorbed from the oral route, its highly toxic nature can result in lung toxicity days or weeks later. It is thus one of the "hit-and-run" class of toxicants. Liver and kidney damage and neurological damage also occur (see Chapter 9, Pesticides).

Insecticides

Chemical insecticides are used to increase food crop production, to protect livestock and household pets against insect pests (warble fly, bot fly, screw-worm, fleas), to control disease-carrying insects (anopheles mosquitoes that carry malaria), and to control destructive insects like termites. It was a chlorinated hydrocarbon insecticide, dichlorodiphenyltrichloroethane (DDT), that first raised concerns over the impact of pesticides on the environment. In 1961, the author Rachel Carson brought out her book *Silent Spring* in which she documented the devastating effect that this chemical had on bird life because it weakened eggshells so that the eggs collapsed in the nest or the chicks were abnormal at hatching. The persistence of the chemical (it and its metabolite DDE have a biological T1/2 of 50 years) and its high lipid solubility result in its biomagnification up the food chain.

Predatory and fish-eating birds are especially vulnerable to DDT. The product was banned in the U.S. and Canada in 1972, but it is still used elsewhere in the world, including Mexico, and trace levels may be present in imported products. Levels of from 1 part/billion (ppb) to 1 part/million (ppm) may be present in fish, oysters and other seafoods and can contribute to human tissue levels of up to 10 ppm.

Halogenated hydrocarbon insecticides (chlorinated hydrocarbons) are principally neurotoxic, interfering with axonal transmission by altering sodium and potassium transport across the axonal membrane to prevent normal repolarization. Evidence of carcinogenicity also has been obtained in animal experiments (see Chapter 9 on pesticides for more details).

Industrial and Commercial Chemicals

Biphenyls

Polybrominated biphenyls (PBBs) are used as fire retardants in thermoplastics for TV and office machine casings. The more familiar polychlorinated biphenyls (PCBs) are highly stable and resistant to degradation in acids, bases, by oxidation and by heat (to 800°C). These characteristics make them ideal insulators in transformers in the electric power industry and as hydraulic fluid and in brake linings. They are also used as plasticizers in polymer films. The same characteristics, however, make these agents very persistent in the environment. Exposure to these compounds is largely an occupational hazard, but exposure in the environment can occur as a result of contamination of ground water from spills, improper storage of waste PCBs from old transformers and capacitors, or from fires in storage sites. Although the manufacture of PCBs was banned in the U.S. in 1977, they remain a problem because of their persistence and resistance to destruction. Forty percent of North Americans have body fat levels of 1 ppm or higher (these agents have high lipid solubility). Prior to 1970, 500,000 tons of PCBs were produced in North America.

Toxicity. Animal — the LD_{50} in rats may be 1–10 g/kg. Chronic toxicity involves skin lesions, hepatotoxicity, immunosuppression and reproductive dysfunctions. Carcinogenicity has also been reported. Recently, genetic damage in cetaceans (whales, dolphins) and seals in the Baltic Sea has been ascribed to high levels of PCBs. Human — characteristic chloracne, impaired immune response, liver damage, gastrointestinal disturbances (nausea, vomiting, loss of appetite), CNS disturbances (weakness, ataxia) as well as reproductive problems and cancer have been reported.

Pharmacokinetics and Metabolism. Because of the high lipid solubility, these compounds are well absorbed from the gastrointestinal tract (>90%) and stored in body fat. They are secreted in milk (toxicity has been shown in nursing mice and rats), and they constitute a hazard for the nursing infant (see study of Michigan mothers in Chapter 3). They have been shown to induce cytochrome P-448 (now P_1-450) through the Ah receptor pathway and to form reactive intermediates and to deplete glutathione. Conjugation with glucuronide and renal excretion are the final detoxification mechanisms.

Biodegradation. Recent studies indicate that solar photolysis (near UV light) can reduce the T1/2 in surface water to 1–2 years by dechlorination of PCBs, but this has no effect on bottom sediments which may contain high levels. Fortunately, chemical dechlorination can occur here, and recently, subsequent oxidation by anaerobic bacteria has been discovered in sediments of the Hudson River. These discoveries offer some hope that biodegradation of PCBs may occur at a faster rate than anticipated.

Accidental Human Exposures. In 1968, nearly 1,700 people in the Fukuoka region of Japan developed chloracne as a result of using rice cooking oil contaminated with PCBs. The PCBs in turn were contaminated with tetrachlorodibenzofuran which is structurally and toxicologically similar to TCDD. These "Yusho" patients (Yusho means "oil disease") constitute the largest human population known to be exposed to toxic levels of PCBs, and their health continues to be monitored closely to identify delayed effects. Five years later, 22 deaths were reported in 1,200 of these patients, 9 from cancer, 2 involving the liver. Calculated body burdens were 5.9 µg/kg for chloracne and 4.4 µg/kg for nausea and anorexia. These are 200× higher than average current levels found in North American populations. Similar mortality statistics occurred after an outbreak in Taiwan. In neither case were the cancer deaths considered excessive, and they were not age-adjusted (see also Chapter 3). Several cases of contamination of animal feeds have occurred. In North Carolina, a leaky heat exchanger contaminated 16,000 tons of chicken feed, only 10% of which was recovered. No human health problems were directly attributed to the accident, but there was some evidence that children might have been affected *in utero* (see also Chapter 3).

A major human exposure to PBBs occurred in Michigan in 1973 (the year of Seveso). A PBB product intended for use as a fire retardant was accidentally bagged as a magnesium supplement for dairy cattle. Whole herds of dairy cattle were afflicted with loss of appetite, open sores, weight loss, loss of milk production, sterility and stillbirths. It took several months of detective work to trace the source of the problem; meanwhile, human food supplies were contaminated by milk containing the PBBs. Meat and eggs were also contaminated. By 1976, nearly 30,000 of Michigan's best dairy cattle had died or been destroyed, along with thousands of sheep and hogs and millions of chickens. Thousands of Michiganders had consumed unknown quantities of contaminated food, and hundreds began to complain of headache, fatigue, joint pain and numbness in fingers and toes. Farm families were the most severely afflicted. Little was then known about the human toxicity of PBBs, and long-term effects have not been identified, but could yet emerge (see also Chapter 3).

The Problem of Disposal. Recent evidence from a disposal site in Great Britain suggested that current incineration temperatures may not be adequate, as levels of PCBs were detected in the soil around the site. A complicating factor was the presence of a municipal incinerator within 100 meters, but some authorities feel that incineration temperatures should reach 2,700°C. In Canada in 1980, it was estimated that 25 million kg of PCBs were still in use, and no secure storage or disposal sites were available. The PCB fire (deliberately set) at St. Basile-le-Grand in 1988 highlights the dangers of unsecured storage. Public concern over PCB hazards has led to the NIMBY response (Not In My Back Yard) and resulted in resistance to the establishment of proper disposal sites and to refusal to allow ships containing toxic wastes to unload. Such ships wandering the seas in search of a berth create a potential for a marine environmental disaster, with contamination of food fish, that could be far worse than the hazards associated with well-run disposal sites.

Solvents

Carbon tetrachloride, chloroform and methylene chloride (dichloromethane) are still popular industrial solvents because they are not flammable. Exposure is mainly an industrial problem, but these agents may still appear as cleaning fluids in the home. A source of some concern is the presence of trace amounts of chloroform in drinking water as a result of the chlorination process.

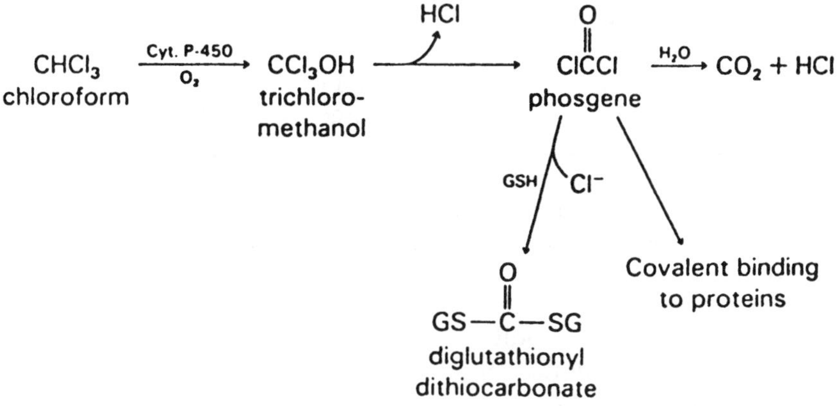

$CHCl_3$ chloroform $\xrightarrow[O_2]{Cyt.\ P\text{-}450}$ CCl_3OH trichloromethanol $\xrightarrow{-HCl}$ $ClCCl$ (O=) phosgene $\xrightarrow{H_2O}$ $CO_2 + HCl$

phosgene → (GSH) → $GS-C(=O)-SG$ diglutathionyl dithiocarbonate

phosgene → (Cl^-) → Covalent binding to proteins

Figure 24 Biotransformation of chloroform.

Toxicity
These chemicals are hepatotoxic, causing central lobular necrosis with fatty degeneration of adjacent areas. They also can cause renal damage, and chronic exposure has been linked to neoplasms of the lung and liver. Cardiac arrhythmias have been reported, as has nausea and vomiting. The cardiac arrhythmias result from sensitization of the heart to catecholamines such as adrenaline and noradrenaline. Other halogenated anesthetics like cyclopropane and halothane will also do this, and hepatotoxicity has been reported as a toxic effect of halothane in some patients as well as in anesthetists routinely exposed to the drug. Liver toxicity can be increased in mice exposed to inducers of microsomal enzymes, suggesting that a toxic metabolite may be involved.

Mechanism of Toxicity. These substances are metabolized by cytochrome P-450 enzymes. Chloroform is converted to chloromethanol, phosgene, and CO_2 + HCl. Phosgene is normally conjugated with glutathione for excretion, but if glutathione is depleted, covalent binding to proteins may lead to liver and kidney necrosis. The metabolic pathway for chloroform is shown in Figure 24. Methylene chloride is metabolized to CO_2 + CO. Carbon monoxide poisoning through the formation of carboxyhemoglobin can occur. Chloroform is an example of a trihalomethane.

Trihalomethanes (THMs)
These are halogen-substituted, single-carbon compounds having the general formula CHX_3, where X may be chlorine, fluorine, bromine, iodine or a combination of these. They are formed during the process of water chlorination from naturally occurring organic compounds. The most common agents found in drinking water are chloroform

($CHCl_3$), bromodichloromethane ($CHBrCl_2$), chlorodibromomethane ($CHClBr_2$) and bromoform ($CHBr_3$). These are liquid at room temperature, rather volatile, and only slightly soluble in water. Their octanol-water partition coefficients range from 1.97 for chloroform to 2.38 for bromoform. All are sensitive to decomposition in air and sunlight. THMs are also released into the environment from industrial sources. Chloroform is the most common THM found in drinking water. Its major toxicity and metabolic transformation are noted above.

As a group, THMs are rapidly and well absorbed from the gastrointestinal tract, metabolized through dihalocarbonyl compounds via the cytochrome P-450 dependent mixed function oxidases to CO_2 and CO and eliminated through the lungs. Because of their high lipid solubility, they accumulate in adipose tissue > brain > kidney > blood.

Several epidemiological studies have shown a correlation between chlorination of surface or ground water and the incidence of many cancers, but correlations with actual measured levels of THMs have been harder to demonstrate. Exceptions are pancreatic cancer in white males, rectal cancer in males only and stomach cancer in both sexes. When population migration patterns were considered, however, the correlation with stomach and rectal cancer could not be demonstrated, and other studies have suggested that other water quality parameters may be involved.

In animal studies, chloroform has been shown to be carcinogenic in rats and mice. Chlordibromomethane has been reported to be hepatotoxic in mice, but no evidence of carcinogenicity was obtained. These agents are probably mutagenic and teratogenic, as indicated in some studies.

In light of these facts, there are efforts directed to limiting the levels of THMs in drinking water. Standards vary widely throughout the world. In Canada, the current maximum standard is 350 µg/L, not to be exceeded. In the U.S., the EPA has set a limit of 100 µg/L. This is an average based on quarterly samples and is therefore more enforceable. The WHO sets a guideline of 30 µg/L, but with a warning that disinfecting efficiency should not be compromised in the pursuit of lower levels. The European Economic Community passed a directive that haloform levels in drinking water should be "as low as possible", which is unenforceable.

It is thus evident that the problem of THMs in drinking water is another example of how cost-benefit analysis must be performed to weigh the potential risks of cancer from the chemicals against the known risks of epidemic infections if water supplies are not treated.

REVIEW QUESTIONS

For questions 1 to 6, use the following code: answer A if statements 1, 2 and 3 are correct; answer B if statements 1 and 3 are correct; answer C if statements 2 and 4 are correct; answer D if statement 4 only is correct; answer E if all statements (1,2,3,4) are correct.

1. Halogenated hydrocarbons are characterized by
   1. High lipid solubility.
   2. Susceptibility to chemical breakdown.
   3. Toxicity for microorganisms.
   4. Decomposition at temperatures greater than 200°C.

2. Dioxin (TCDD) toxicity is characterized by
   1. Chloracne.
   2. Hepatotoxicity.
   3. Porphyria.
   4. None of the above (1,2,3).

3. Dioxin (TCDD)
   1. Induces the enzyme gamma aminolevulinic acid (ALA) synthetase.
   2. Interferes with feedback inhibition of ALA synthetase.
   3. Inhibits porphyrin synthesis.
   4. Induces aryl hydrocarbon hydroxylase.

4. Regarding chloroform:
   1. Phosgene is a major metabolite.
   2. Phosgene is detoxified by conjugation with glucuronide.
   3. Liver necrosis can occur if phosgene escapes the detoxification process.
   4. Phosgene is the only toxic metabolite of chloroform.

5. Regarding the detoxification of PCBs while in the environment (biodegradation):
   1. Sunlight may break down PCBs in surface water.
   2. No breakdown of PCBs occurs in bottom sediments.
   3. Bacteria may detoxify them by oxidation.
   4. Chemical dechlorination does not reduce the toxicity of PCBs.

6. Regarding the aromatic hydrocarbon (Ah) receptor:
   1. TCDD attaches to it.
   2. Occupation of a certain minimum number of receptors is necessary for TCDD to be carcinogenic.
   3. The linear multistage assessment model for carcinogens may not be appropriate for TCDD.
   4. No other chemical is known to attach to the Ah receptor.

For questions 7 to 11, match the chemical listed below to the appropriate use.
   a. Hexachlorophene
   b. Polybrominated biphenyls (PBBs)
   c. Polychlorinated biphenyls (PCBs)
   d. 2,4-dichlorophenoxyacetic acid (2,4-D)
   e. Dichlorodiphenyltrichloroethane (DDT)

7. ___. Insecticide.

8. ___. Disinfectant.

9. ___. Transformer insulator.

10. ___. Fire retardant.

11. ___. Herbicide.

For questions 12 to 15, answer True or False.

12. Dioxin (TCDD) can cause behavioral abnormalities.____

13. Porphyrins are by-products of heme synthesis.____

14. Victims of TCDD poisoning at Seveso showed no evidence of hepatotoxicity.____

15. Pentachlorophenol is used as a wood preservative. From its name, one would predict that it would cause chloracne if accidentally consumed.____

ANSWERS

1. B; 2. A; 3. E; 4. A; 5. B; 6. A; 7. e; 8. a; 9. c; 10. b; 11. d; 12. T; 13. T; 14. F; 15. T.

CASE STUDY #9

Part 1
Three members of a family became dizzy and nauseated within 1 hour of eating snacks (taquitos) consisting of tortillas wrapped around a meat filling. Two of them subsequently had grand mal epileptic seizures. The snacks were commercially prepared and sold in sealed bags of 48. They had been purchased a few days earlier. In an unrelated case a few weeks later, a 17-year-old male had four closely spaced seizures 30 minutes after consuming taquitos from the same manufacturer and purchased from the same store. The boy was on long-term antiepileptic therapy because he had been diagnosed as an epileptic the previous year. Following the initial episode, the

manufacturer had voluntarily removed from shop shelves and destroyed all existing packages of the product.

Q. What organ system seems to be the primary site of toxicity?

Q. What is the most likely cause of the reaction (a) a bacteria or (b) a chemical?

Q. What is the likely site of contamination (a) the factory or (b) the retail store?

Part 2
Analysis of some remaining taquitos from the first case revealed traces of endrin. No source or trace of endrin was found at the factory.

Q. To what class of compound does endrin belong?

Review the toxicity of this class of chemicals.

A state-wide press release turned up several other cases of seizures including five persons who had experienced seizures within 12 hours of consuming taquitos purchased from the same store.

Q. What preventive or remedial measures might you recommend?

CASE STUDY #10

A maintenance employee in a factory died after acute exposure to solvent fumes. He had been using a mixture of chlorinated solvents to remove grease from machinery. The principal component was trichloroethane (methyl chloroform).

Q. What was the immediate cause of death?

Q. What steps might you suggest to prevent this type of accident?

Q. This substance is similar to carbon tetrachloride. What would have been the nature of the toxic response if the exposure had been chronic rather than acute?

FURTHER READING

Albert, A., *Xenobiosis*. Chapman and Hall, New York, 1987.
Axelson, O., Editorial: Seveso: Disentangling the dioxin enigma? *Epidemiology*, 4, 389, 1993.
Bertazzi, P.A., Pesatori, A.C., et al., Cancer incidence in a population accidentally exposed to 2,3,7,8,-tetrachlorodibenzo-para-dioxin. *Epidemiology*, 4, 398, 1993.
Bertazzi, P.A., Zochetti, C., et al., Ten-year mortality study of the population involved in the Seveso incident in 1976. *Am. J. Epidemiol.*, 129, 1187, 1989.

Collins, J.J., Acquavella, J.F. and Friedlander, B.R., Reconciling old and new findings on dioxin. *Epidemiology*, 3, 65, 1992.
Consultation package on trihalomethanes. Environmental Health Directorate, Health and Welfare Canada, 1989.
Collins, J.J., Strauss, M.E., Levinskas, G.J. and Conner, P.R., The mortality experience of workers exposed to 2,3,7,8,-tetrachloro-p-dioxin in a trichlorophenol process accident. *Epidemiology*, 4, 7, 1993.
Fingerhut, M.A., Halperin, W.E., et al., Cancer mortality in workers exposed to 2,3,7,8-tetrachlorodibenzo-p-dioxin. *New Eng. J. Med.*, 324, 212, 1991.
Fingerhut, M.A., Steenland, K., et al., Old and new reflections on dioxin. *Epidemiology*, 3, 69, 1992.
Gierthy, J.F., Bennett, J.A., et al., Correlation of in vitro and in vivo growth suppression of MCF-7 in human breast cancer by 2,3,7,8,-tetrachlorodibenzo-p-dioxin. *Cancer Res.*, 53, 3149, 1993.
Gorman, J., *Hazards to Your Health: The Problem of Environmental Disease*, New York Academy of Science, New York, 1979.
Gough, M., Agent Orange studies. *Science* (Letter to Ed.), 245, 1031, 1989.
Hodgson, E. and Levi, P.E. (Eds.), *Textbook of Modern Toxicology*, Elsevier, New York, 1987.
Johnson, E.F. A partnership between the dioxin receptor and a basic helix-loop-helix-protein. *Science*, 252, 924, 1991.
Jones, G.R.N., Polychlorinated biphenyls: where do we stand now? *Lancet*, 2, 791, 1981.
Landers, J.P. and Bunce, N.J., The Ah receptor and the mechanism of dioxin toxicity. *Biochem. J.*, 276, 273, 1991.
Okey, A.B., Riddick, D.S. and Harper, P.A., Molecular biology of the aromatic hydrocarbon (dioxin) receptor. *TiPS*, 15, 226, 1994
Poland, A. and Knutson, J.C., 2,3,7,8,-tetrachlorodibenzo-p-dioxin and related halogenated aromatic hydrocarbons: examination of the mechanism of toxicity. *Ann. Rev. Pharmacol. Toxicol.*, 22, 517, 1982.
Richardson, M.L. (Ed.), *Risk Assessment in the Environment*, Royal Society of Chemistry, London, 1988.
Roberts, L., EPA moves to reassess the risk of dioxin. *Science*, 252, 911, 1991.
Roberts, L., Dioxin risks revisited. *Science*, 251, 624, 1991.
Ryan, J.J., Gasiewicz, T.A. and Brown, J.F., Human body burden of polychlorinated dibenzofurans associated with toxicity based on the Yusho and Yucheng incidents. *Fundam. Appl. Toxicol.*, 14, 722, 1990.
Safe, S., Astroff, B., et al., 2,3,7,8-tetrachlorodibenzo-p-dioxin (TCDD) and related compounds as antioestrogens: characterization and mechanism of action. *Pharmacol. Toxicol.*, 69, 1, 1991.
Schneider, M.-J., *Persistent Poisons,* New York Academy of Science, New York, 1979.
Stone, R., News and comment: environmental estrogens stir debate. *Science*, 265, 308, 1994.
Walsh, A.A., Hsieh, P.H. and Rice, R.H., Aflatoxin toxicity in cultured human epidermal cells: stimulation by 2,3,7,8,-tetrachlorodibenzo-p-dioxin. *Carcinogenesis*, 13, 2029, 1992.
Zacharewski, T.R., Bondy, K.L., et al., Antiestrogenic effect of 2,3,7,8,-tetrachlorodibenzo-p-dioxin on 17B-estradiol-induced pS2 expression. *Cancer Res.*, 54, 2707, 1994.

6 TOXICITY OF METALS

"Mad as a Hatter"

INTRODUCTION

The process of felting, employed in making hats many years ago, required the use of mercurial compounds, and many hatters suffered from the CNS disturbances, including behavioral disorders, associated with mercury toxicity. Metal intoxication as an occupational disease may be 4,000 years old. Lead was produced as a by-product of silver mining as long ago as 2000 B.C. Hippocrates described abdominal colic in a man who worked as a metal smelter in 370 B.C., and arsenic and mercury were known to the ancients even if their toxicity was not. In 1810, a remarkable case of mass poisoning with mercury occurred. The 74-gun man-o'-war HMS Triumph salvaged 130 tons of mercury from a Spanish vessel wrecked while returning from South America, where the mercury had been mined. The mercury was contained in leather pouches, which became damp and rotten, allowing it to escape and vaporize. Within 3 weeks, 200 men were affected with signs of mercury poisoning including profuse salivation, weakness, tremor, partial paralysis, ulcerations of the mouth and diarrhea. Almost all animals on board died, including mice, cats, a dog and a canary. Five men died. When the vessel put in at Gibraltar for cleaning, all those working in the hold salivated profusely.

The common 19th century practice of adulterating foods and beverages (wine, beer, etc.) to increase profit led Accum to publish a treatise on the subject in 1820. Lead, copper and mercury were frequently detected. Methods were not yet in place to detect arsenic, which was found to be a widespread adulterant later in the century.

In 1875, the British parliament passed the first Food and Drugs Act as a result of these investigations.

In the past, it was common to refer to heavy metal toxicity, as it was those metals that first emerged as industrial hazards. Heavy metals are arbitrarily defined as those having double-digit specific gravities, and they include platinum (21.45), plutonium (19.84), tungsten (19.3), gold (18.88), mercury (13.55), lead (11.35), and molybdenum (10.22). These are in contrast to iron (7.87), manganese (7.21), chromium (7.18), zinc (7.13), selenium (4.78) and aluminum (2.70). Intermediate are copper (8.96) and cadmium (8.65).

In general, it can be seen that metals with specific gravity less than 8 are mostly essential trace nutritional elements (copper also is one and therefore the exception, as is aluminum, which is not a nutritional element), whereas those having specific gravity greater than 8 are the more toxic ones. It must be stressed once again that dose is all-important. Aluminum, with a specific gravity of 2.70 has toxic properties. Arsenic exists in two solid forms: yellow arsenic (1.97) and grey or metallic arsenic (5.73). Both are highly toxic.

■ LEAD (Pb)

The Latin word for lead is *plumbum*, hence the chemical designation Pb. This word also gave origin to such English ones as plumbob (a mason's line with a metal ball attached for establishing vertical trueness), plummet (to fall as if leaden), and aplomb (a state of calm as undeviating as a plumb line). Lead was obviously well-known to the ancients. In fact, they spent a lot of time trying to turn it into gold (alchemy). Lead toxicity was also familiar to them. Diascorides described its CNS toxicity as delirium.

Despite early knowledge of lead's toxic effects, the low melting point of the metal, coupled with its density, made it a popular and useful one, so that well into the 1940s and early 1950s, it was possible to buy lead toys, and kits were available to cast lead soldiers and lead fishing weights. An 1885 description of chronic lead poisoning is as good as any to be found in a modern text:

> "The chief signs of chronic poisoning are those of general ill health; the digestion is disturbed, the appetite lessened, the bowels obstinately confined, the skin assumes a peculiar yellowish hue, and sometimes the sufferer is jaundiced. The gums show a black line from two to three lines in breadth, which microscopical examination and chemical tests alike show to be composed of sulphide of lead; occasionally the teeth turn black. The pulse is slow and all secretions are diminished. Pregnant women have a tendency to abort. There are also special symptoms, one of the most prominent of which is lead colic."

This colic is paroxysmal and excruciating.

Modern-day sources of lead are numerous. In the 18th century, the industrial West discovered what the Chinese had known for centuries, namely that lead glazes produce crockery with a richer, smoother look. From this source and from lead solder in cans and kettles and water pipes leached by soft (but not hard) water, we consume about 150 µg/day. In some areas, the figure may reach 1–2 mg. Children are more vulnerable, as all dirt and dust contain lead, especially in cities where lead from auto exhaust (tetraethyl lead) settles out on the ground. This will persist long after the conversion to lead-free auto fuel. Children may also consume old lead-based paint, common in older buildings and which may also be on cheap wooden toys. In children, CNS toxicity is the dominant feature. This starts with vertigo and irritability, progressing to delirium, vomiting and convulsions. The mortality rate is about 25% if treated and about 65% if untreated. In infants, exposure produces progressive mental deterioration after 18 months, with loss of motor skills, retarded speech development and hyperkinesis in some cases. In the United States, the Lead Paint Poison Prevention Program was introduced in 1970. Since that time, the mean blood lead level of U.S. children has fallen from over 1 µmol/L (20.7 µg/dL) to less than .25 µmol/L (5.2 µg/dL). Only two deaths in children from acute lead encephalopathy have been reported in the past 20 years.

In all exposed individuals, subchronic toxicity can involve interference with mitochondrial heme synthesis at several levels, with resulting hypochromic (pale) microcytic (small) anemia. The pathway involved in this is illustrated in Figure 25.

Toxicokinetics of Lead
Elemental lead is not absorbed by the skin or through the alveoli of the lungs. Inhaled particulate lead is returned to the pharynx by the bronchial cilia and swallowed. Tetraethyl lead, however, may be absorbed across the skin and alveoli and readily penetrates CNS. Most of it is destroyed in exhaust emissions, but sniffing leaded gasoline can result in severe CNS damage.

Gastrointestinal absorption of lead probably occurs via calcium channels as lead is a divalent cation (Pb^{2+}). It first appears in red blood cells, then hepatocytes, then the epithelial cells of the renal tubules. It is gradually redistributed to hair, teeth and bones where 95% of it is stored harmlessly. The T1/2 in blood is about 30 days; in bone, 25 years. Little reaches the adult brain, but much more enters the infant brain. Renal excretion is the main route of elimination.

Cellular Toxicity of Lead
Lead affects oxidative phosphorylation and ATP synthesis in the mitochondrion. It also increases red cell fragility and inhibits sodium/potassium ATPase. Kidney tubular cells become necrotic, and

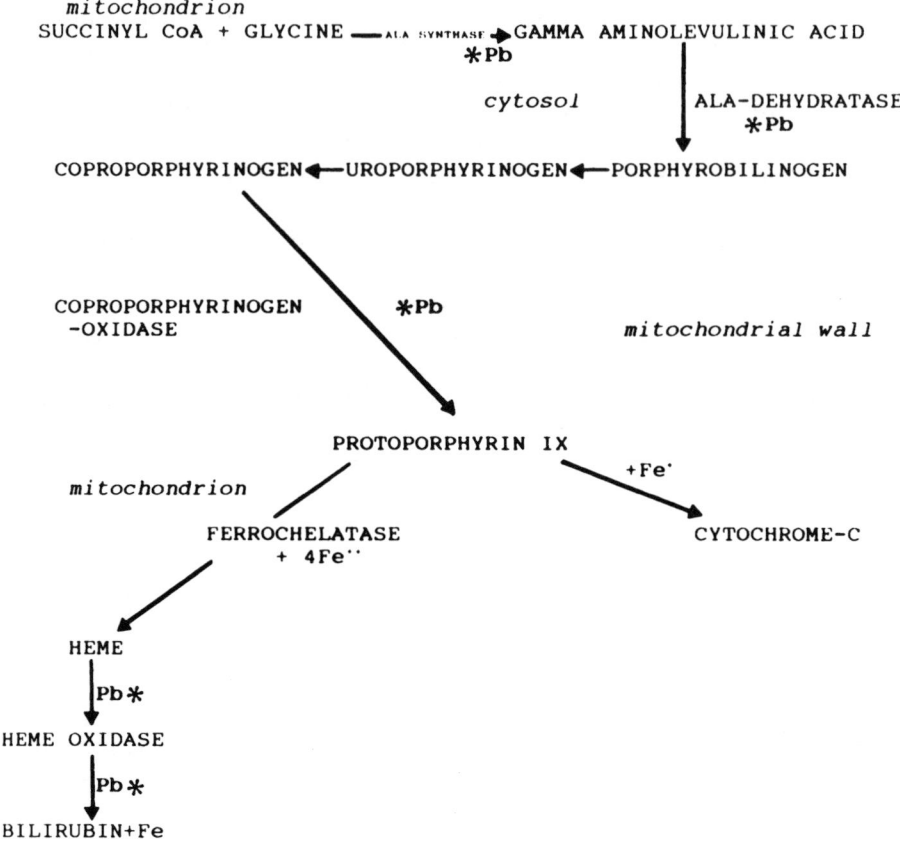

Figure 25 A simplified scheme showing points of interference of lead in heme synthesis. See also Figure 22 for ALA synthase and heme inhibition.

chronic exposure may lead to interstitial nephritis. Nuclear inclusion bodies, consisting of lead bound to a protein, may be formed in renal cells. This may be considered as a protective mechanism. Carcinogenesis has been demonstrated in experimental animals and chromosomal abnormalities have been observed, but evidence of tumor production in humans is scarce. Most of the toxic effects of lead and other heavy metals can be explained by their affinity for thiol groups. This is also the basis of chelation therapy.

Fetal Toxicity
A characteristic of all metals is their ability to penetrate the placental barrier, so that fetal toxicity can occur as a result of maternal exposure. Lead is considered to be a human carcinogen, and pregnant women are generally removed from jobs where exposure may occur.

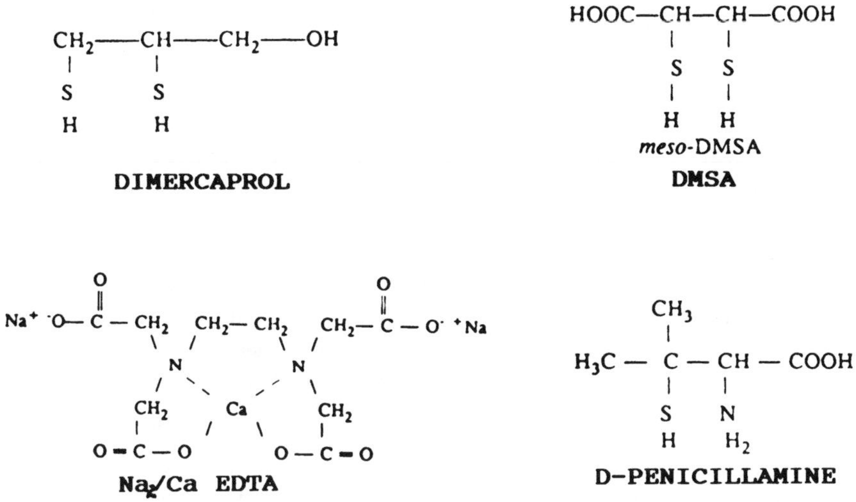

Figure 26 Metal chelators.

Treatment

Lead chelators are the treatment of choice. These bind lead (and other divalent cations) so that it can be excreted. Calcium/sodium ethylene diamine tetra-acetate (CaNa$_2$EDTA) and dimercaprol (British antilewisite, BAL) are given intramuscularly, followed by oral penicillamine for several weeks. BAL was developed during World War II as a treatment for lewisite, a vesicant arsenical poison gas. A newer chelator is meso-2,3,-dimercaptosuccinic acid (DMSA). The chemical structures of these chelators are shown in Figure 26. In the case of EDTA, Pb is exchanged for Ca, whereas with the others, the Pb is bound to sulfhydryl groups. The complexes are excreted, mostly in urine. A disadvantage of chelation therapy is that it does not remove lead from the brain very efficiently.

Despite 50 years of use, objective evidence for benefit of chelation therapy for lead poisoning is scanty. It is widely agreed that it has drastically reduced the mortality from lead encephalopathy if diagnosis and treatment are started early. It also relieves lead cholic, malaise, basophilic stippling, and it rapidly restores red-cell ALA dehydratase. It does not influence the residual manifestations of chronic lead poisoning such as peripheral neuropathy.

■ MERCURY (Hg)

Mercury exists in three forms: elemental mercury, inorganic compounds and organic compounds. Elemental mercury causes toxicity when the mercury vapor is inhaled, as exemplified by the episode described at the beginning of this chapter. The major source of elemental mercury in the environment is the natural degassing of the earth's crust. Estimates of the level of mercury reaching the atmosphere range from 25,000–150,000 tons/yr, and the atmosphere represents a major mechanism for global transport of metallic mercury. Conversely, anthropogenic sources account for only 10,000 tons/yr, but because industrial effluent tends to be concentrated, these are the sources usually associated with toxicity. Metallic mercury and its vapor can be an industrial hazard. Mercury is used in the manufacture of chlorine and sodium hydroxide by the mercury cell process, in paint preservatives and in the electronics industry. It is a by-product of smelting processes (most mineral ores contain mercury), and it is released during fossil fuel combustion.

Elemental Mercury Toxicity

In vapor form, elemental mercury is well absorbed across both the alveoli of the lungs and the blood-brain barrier. Acute poisoning usually occurs within several hours. Weakness, chills, metallic taste, salivation, nausea, vomiting, diarrhea, labored breathing, cough and tightness in the chest may ensue. If the exposure is more prolonged, interstitial pneumonitis may develop. Recovery is usually complete except that residual loss of pulmonary function may persist. Chronic exposure to mercury vapor results in CNS disturbances, including tremor and a variety of behavioral changes that can include depression, irritability, shyness, instability, confusion and forgetfulness. Mercury vapor from mercury nitrate formerly used in the felting process accounted for the "mad hatter" syndrome. The behavioral abnormalities of the "Mad Hatter" in Lewis Carroll's *The Adventures of Alice in Wonderland* were really quite mild, compared to the other characters, which is in keeping with the topsy-turvy world that Carroll created. Thyroid disturbances may also be present.

Inorganic Mercurial Salts

Inorganic salts such as mercuric chloride can cause severe, acute toxicity. The proteins of mucous membranes are precipitated, giving them an ash-grey color in the mouth and pharynx. Intense abdominal pain and vomiting are common. Loss of blood and fluid from the gastrointestinal tract results from sloughing of the mucosa in the stool and may lead to hypovolemia and shock. Renal tubular necrosis occurs after acute exposure, and glomerular damage is more common after chronic exposure. A phenomenon called "pink disease" or acrodynia commonly follows chronic exposure to mercury ions. It is a flushing of the skin which is believed to have an allergic basis.

Organic Mercurials

Methylmercury is the most common cause of organic mercurial poisoning and the most important one environmentally. It is extremely well absorbed from the gastrointestinal tract (90%) and deposited in the brain. Because of its high affinity for SH groups, methylmercury binds to cysteine, and this may then substitute for methionine and be incorporated into proteins. This can result in the formation of abnormal microtubules required for cell division and neuronal migration. The main signs and symptoms are neurological and consist of visual disturbances, weakness, incoordination, loss of sensation, loss of hearing, joint pain, mental deterioration, tremor and in severe cases, paralysis and death. Infants exposed *in utero* may be deformed and retarded.

Mercury is a waste product of many industrial processes. It is methylated by sediment bacteria and cyanocobalamin. Several outbreaks of methylmercury poisoning have occurred. The most widely known began in Minimata, Japan, in 1953 near a plant that manufactured acetaldehyde and which discharged methylmercuric chloride (MeHg$^+$Cl$^-$) into Minimata Bay. People who ate molluscs and large fish from the bay developed the symptoms that came to be known as Minamata Disease. Nine hundred cases developed, and there were 90 fatalities. Because of the high fetal toxicity of mercury, many deformed infants were born. Another source of mercury toxicity is the consumption of seed grains treated with methylmercuric chloride as a fungicide. Several mass poisonings have occurred around the world. In Iraq in 1972, one such episode resulted in over 6,500 cases of poisoning and 500 deaths.

Mechanism of Mercury Toxicity

Mercury toxicity can be explained entirely by its ability to bind with the hydrogen of sulfhydryl (SH) groups to form mercaptides, i.e., X-Hg-SR and HgSR$_2$ where X = an electronegative radical and R = a protein. Organic mercurials such as methylmercury form mercaptides, R-Hg-SR'. The term "mercapto" means "to capture mercury" and refers to sulfur-containing groups. Since SH groups are important components of many enzymes, mercury acts as an enzyme poison and interferes with cell function at many levels. Mercury can also combine with other physiologically important ligands, such as phosphoryl, carboxyl, amide and amine groups. Metallic Hg vapor may be oxidized by catalase enzyme in red blood cells to the less-toxic divalent form. Alcohol competitively inhibits this process. Historically, mercury was an important pharmaceutical agent for centuries, and its pharmacological properties also depend on its affinity for SH groups. It was used as an antibacterial agent (for syphilis), as a laxative, in skin creams and in diuretics.

Mercurial diuretics were still in use in the 1960s, until they were replaced by safer agents. Aminomercuric chloride may still appear in freckle-removing creams, and daily application for years may result in increases in 24-hr urine mercury excretions from 10 µg to 1 mg and the development of symptoms such as excessive salivation and insomnia.

Treatment of Mercury Poisoning
Chelation therapy is recommended for elemental, inorganic mercury poisoning. Dimercaprol and penicillamine are SH-containing chelators. Dimercaprol is given intramuscularly and penicillamine, orally. Hemodialysis may also be used, and vomiting may be induced if there has been recent ingestion of mercury. These treatments are of little use in methylmercury poisoning, however. Dimercaprol actually increases brain levels of methylmercury, and penicillamine and hemodialysis do not relieve symptoms. Some success has been achieved with binding resins taken orally. Since there is a significant enterohepatic recirculation of methylmercury (i.e., it is excreted in the bile and reabsorbed from the intestinal tract), binding it to a polythiol resin prevents its reabsorption because it is excreted in the feces.

The Grassy Narrows Story
In 1969, Norvald Fimreite, a Ph.D. candidate in the Department of Zoology at the University of Western Ontario, first made public his findings on the mercury contamination of fish in Canadian and border lakes. The highest levels were recorded from a small lake, Pinchi, in British Columbia (10 ppm) and from Lake St. Clair (7.03 ppm) in the Great Lakes waterway. The (Canadian) federal standard for export and consumption was 0.5 ppm. His report was a bombshell, coming on the heels of reports of Minamata Disease from Japan. Fimreite estimated that Canadian industry was releasing 200,000 lb of mercury annually into the environment. Most of it came from chlor-alkali plants and from pulp and paper mills which used mercurials as antisliming (antialgal) agents and the chlorine and alkali as bleaching agents. The question of mercury discharge from the Dow (Canada) Chemical plant had been raised 6 years earlier in the Ontario legislature, but nothing had been done. In 1970, the Ontario Water Resources Commission took steps to reduce Dow's output, but in Dryden, near the Manitoba border, the Dryden Pulp and Paper Company (owned by the British Reed Group) had been emitting mercury vapor since 1962, and some workers developed bleeding gums and muscle twitches. By 1970, it had pumped an estimated 20,000 lb of mercury into the surrounding environment, including discharges described as a brown froth into the Wabigoon River. Raw sewage

also was discharged into the Wabigoon, providing a rich source of anerobic bacteria to methylate elemental mercury. The Wabigoon is part of the English River system, and about 50 km downstream lay the Grassy Narrows and White Dog Indian Reserves. The residents gleaned a slim but adequate living as fishing guides and lived largely off the land, eating fish, deer and moose supplemented with garden vegetables. In March of 1970, contamination of fish in Lake Erie was detected, and the Lake St. Clair and Lake Erie fisheries were closed. Chlor-alkali plants and pulp mills were ordered to stop using mercury by the end of May, after a concerted attack in the Ontario legislature by opposition parties. Mercury, however, is not biodegradable, and it is only when it is buried by uncontaminated sediment that it ceases to be a threat. In June of 1970, the Lamms, owners of Ball Lake Fishing Lodge, hired Fimreite to conduct a survey of mercury levels in the fish of the English-Wabigoon system. The findings were appalling. Levels ranged from 13–30 ppm, as high as those from Minamata Bay. The government lifted Fimreite's license to collect specimens for scientific purposes and ignored his appeals to test the residents of the reserves until his data were made public, when it conceded that it had similar findings. A ban was placed on eating fish from the contaminated area, but otherwise the government continued to downplay the problem. Tourist fishing dried up, and the Indians went on welfare. Blood levels of mercury were not seriously studied until 1973 and ranged from 45–289 ppb (normal is about 20 ppb for a city dweller). Some residents were showing signs of mercury poisoning, and the incidence of stillbirths was going up.

The social costs of this tragedy were perhaps even greater than the direct effects of mercury. In the years surrounding the discovery of mercury in the Grassy Narrows area, the death rate rose to 1 in 50, three times the national average. Most were alcohol-related. Many of the deaths were newborn or very young infants. Violence became rampant. Dr. Peter Newbury, also a graduate of U.W.O., conducted a study for the Society of Friends (Quaker) and the National Indian Brotherhood and felt that the CNS effects of mercury were a contributing factor in the violence. Gasoline sniffing became common among young people (it remains a problem on many reserves). The Grand Council of Treaty Three District, which includes Grassy Narrows and Kenora, completed a study in 1973. They found that in the preceding 42-month period, there had been 189 violent deaths of native people. They reported 38 from gunshot, stabbing or hanging; 30 in fires; 42 drownings; 25 from exposure and 16 from car accidents. In the same year, members of the Ojibwa Warrior Society occupied Anicinabe Park on the outskirts of Kenora. Barricades were erected and manned by armed warriors. The park was claimed as Indian land. The standoff lasted for several weeks but achieved little.

CADMIUM (Cd)

Cadmium is present naturally in the environment in very low levels, being solubilized during the weathering of rock (levels are about 0.03 µg/g of soil, 0.07 µg/ml of fresh water, and 1 ng/m^3 of air). Dissolved cadmium may form a number of soluble and insoluble organic and inorganic compounds. Cadmium is chemically similar to zinc, and it is present in zinc ore in a ratio of about 1/250. Most cadmium is produced as a by-product of electrolytic zinc plants. It is used in metal plating, in the manufacture of nickel-cadmium batteries, in the manufacture of pigments, in plastic stabilizers, and small amounts are used in photographic chemicals, in catalysts and in fungicides used on golf courses. Environmentally significant emissions come mainly from smelting operations for copper, lead and zinc, from auto exhaust and manufacture of pigments and alloys (most nickel-cadmium batteries are imported into Canada). Cadmium is readily taken up by plants and stored in the leaves and seeds. It is present in sewage sludge fertilizers (recommended maximums 20 ppm). Water pollution with cadmium may result in high levels in fish and especially in molluscs. The main sources in the human diet are organ meats (cadmium accumulates in liver and kidney), cereal grains, shellfish and crustaceans.

Cadmium Pharmacokinetics

Cadmium intake in Canada averages 50–100 µg/day from inhaled and ingested sources. Inhaled, unpolluted air may contribute up to 0.15 µg/day, whereas breathing air near a smelter can raise the level to 10 µg/day. Cigarettes contain cadmium and smoking increases exposure still further. About 50% of inhaled cadmium is absorbed. Only about 6% of ingested cadmium is absorbed, but it contributes most of the daily load. The FAO/WHO recommends a maximum weekly intake of 500 µg. Absorbed cadmium is bound to plasma albumin and cleared rapidly from the plasma. It is found in red cells only after high exposures. It is rapidly distributed to the liver, pancreas, prostate and kidney, with slow redistribution to the kidney until, over time, it contains most of the cadmium. Renal levels increase up to age 50 and depend on the cumulative exposure. The T1/2 in humans is about 20 years. Cadmium is trapped in the kidney and liver by a cysteine (i.e., SH)-rich protein called metallothionein with a high affinity for cadmium and zinc. Cadmium normally binds to metallothionein, the synthesis of which is induced by the presence of the cadmium. High doses, however, exceed the binding capacity of the protein, and the cadmium is free to bind to other essential cell components, such as the basement membrane of the renal glomerulus.

Cadmium Toxicity

The kidney is the major organ of toxicity. About 200 μg/g wet weight of kidney appears to be the critical concentration in the renal cortex for damage to occur in the form of proximal tubule dysfunction. Once renal disease develops, cadmium is lost from the kidney. Nutritional deficiencies of zinc, iron and calcium may predispose cadmium toxicity by increasing absorption from the gastrointestinal tract. Calcium deficiency increases the synthesis of calcium-binding proteins and cadmium absorption. Workers in metal refineries may be exposed to high levels of cadmium fumes and develop respiratory difficulties. Chronic exposure may lead to obstructive pulmonary disease and emphysema. A major exposure occurred in Japan in the late 1940s. Effluent from a lead processing plant washed into adjacent rice paddies over decades, and the rice accumulated high levels of cadmium. Because the people were calcium-deficient due to a poor diet, they developed acute cadmium toxicity with severe muscle pain, malabsorption, anemia and renal failure. The outbreak was named "Itai-Itai" (ouch-ouch) disease. The fetus appears to be protected from cadmium toxicity by placental synthesis of metallothionein, but heavy exposures can overwhelm this defense.

Animal studies have shown cadmium to be carcinogenic, and there is a suggestion that it may increase the incidence of prostate cancer in elderly men. Other metals, notably arsenic, chromium and lead, also have been implicated as carcinogens.

Treatment

Chelation therapy is not effective. Treatment consists of removing the patient from the source of exposure and supportive measures.

ARSENIC (As)

Arsenic is an age-old pharmaceutical preparation. It was believed to be a tonic because it causes facial flushing (rosy cheeks) and fullness (edema). It is still used in the treatment of trypanosomiasis. Ehrlich studied organic arsenicals and developed the first effective treatment for syphilis (Ehrlich's 606). The chemistry of arsenic is exceedingly complex, since it can exist as a metallic form and as trivalent and pentavalent compounds. Trivalent forms include arsenic trioxide, arsenic trichloride and sodium arsenite. Pentavalent forms include arsenic pentoxide, arsenic acid, lead arsenate and calcium arsenate. Organic arsenicals also may be trivalent or pentavalent. Arsenic in the environment arises from weathering of rock and from emissions from smelting of gold, silver, copper, zinc and lead ores, combustion of fossil fuels, and the use of arsenicals in agriculture as herbicides and pesticides. Airborne particles may travel considerable distances and penetrate deeply into the lungs. Arsenic is taken up by plants, and the degree of uptake varies with the soil type. Fine soils high in clay and organic material inhibit uptake. Arsenic also enters the water system through runoff and fallout. Wells drilled through arsenic-containing rock will yield water high in arsenic. Chronic poisoning may result, and this is a problem in some parts of Nova Scotia. Tobacco also contains arsenic. The average daily intake of arsenic in North America is about 25 µg.

Pharmacokinetics of Arsenicals
Arsenic may be absorbed from the gastrointestinal tract, the lungs and across the skin and mucous membranes. It penetrates intracellularly by an uptake mechanism used in phosphate transport. Like mercury, arsenic binds to sulfhydryl and disulfide groups to poison numerous cell enzymes and respiration. Chromosomal breakage has been observed experimentally. In general, the order of toxicity is organic arsenicals > inorganic arsenicals > metallic arsenic. The trivalent arsenite has a high affinity for SH groups and interferes with the enzyme pyruvate dehydrogenase. Plasma pyruvate levels will increase. Some biotransformation between trivalent and pentavalent forms may occur. The T1/2 of arsenic is about 10 hr, and excretion is mainly by the kidneys. Arsenic is an effective uncoupler of oxidative phosphorylation.

Toxicity
As noted, the most poisonous forms are arsenic trioxide (As_2O_3) and sodium arsenite ($NaAsO_2$). Arsenic tends to accumulate in the liver, kidney, heart and lung. It is also deposited in bone, teeth, hair and nails, and these become important tissues for diagnostic and forensic analysis. The average human intake is about 300 µg/day, but it may be much higher if fish is a large part of the diet, as they accumulate the poison through biomagnification.

Acute arsenic poisoning causes severe abdominal pain, and it is rare. Chronic poisoning results in muscle weakness and pain, skin pigmentation, gross edema, gastrointestinal disturbances, kidney and liver damage and peripheral neuritis with eventual paralysis. The fingernails develop white lines, called Mee's lines, that can be used to determine when exposure occurred.

Arsine is the gaseous form of arsenic resulting from electrolytic processes, and it is extremely toxic, producing rapid and often fatal hemolysis. It has a garlic-like odor.

Treatment
Chelation therapy is used for arsenic poisoning. Both dimercaprol and penicillamine have been used.

Environmental Effects of Arsenic

Arsenic is toxic to a wide range of plants and animals, including marine species. Of the plants, beans, peas and rice are especially sensitive. Algae are sensitive, as well as some protozoa such as *Daphnia magna*. Finned fish are quite susceptible.

■ CHROMIUM (Cr)

Chromium is used in the production of stainless steel, chrome plating, pigments, and in the chemical industry. Chromium has two oxidation states: Cr^{+3} and Cr^{+6}. The latter is much more toxic, causing severe respiratory irritation when inhaled, and possibly lung cancer after long chronic exposure. Kidney damage also occurs. The trivalent form binds readily to electron-donating ligands, such as macromolecules like RNA, but it does not readily cross the cell membrane. Conversely, the hexavalent form readily crosses cell membranes and is reduced to the trivalent form intracellularly. Toxicity is thus normally related to the presence of the hexavalent form in the environment. As the oxyanion, it is taken up by the cell, probably by a sulfate transport system as shown below. Air levels as low as 10 $\mu g/m^3$ of Cr^{+6} can produce respiratory irritation. Chromium is distributed in the biosphere much like arsenic and can have similar effects.

■ OTHER METALS

Virtually any metal, if taken in excessive amounts or by an unusual route, can manifest toxicity. Thus, the inhalation of metal dusts, including aluminum, can cause pulmonary fibrosis. Aluminum also has been incriminated in a form of mental deterioration similar to Alzheimer's disease.

Uranium is nephrotoxic. It binds to albumin and to bicarbonate anion, which is filtered by the kidney where it dissociates. The free uranyl cation binds to proteins in the proximal tubule and damages them. Toxic metals may also substitute for physiological ones, as when lead and strontium 90 are deposited in bones and teeth like calcium.

Even essential metals like iron can be very toxic, especially to young children who may ingest iron-containing vitamin preparations, mistaking them for candy. Vomiting occurs, and vomitus and stool may contain blood. Acidosis and shock develop. Kidney and liver damage can occur. In adults, iron overload sometimes occurs, and hemosiderin is deposited in tissues including muscle. Desferroxime is a specific iron chelator with a low affinity for calcium that is used to remove systemic iron in both types of poisoning.

Antimony is an industrial contaminant with distribution and toxicity essentially like that of arsenic.

CARCINOGENICITY OF METALS

Arsenic has long been recognized as being associated with an increased risk of skin and respiratory cancer. In 1930, workers in a factory making an arsenical sheep dip were identified as having an excessive incidence of skin cancer. Arsenic levels in air >54.6 µg/m^3 were associated with an increase in lung cancer incidence. High water content of arsenic also has been associated with increased cancer risk.

In 1976, a NIOSH study of 300 workers in a cadmium smelter revealed a significantly higher incidence of cancer. The incidence of lung cancer was 2× higher than normal, and prostate cancer also was high. Some of these workers, however, were also exposed to arsenic. Inhalation of chromium dust by workers has been associated with an increased incidence of lung cancer. Nickel is also a respiratory carcinogen, but it lacks other chronic toxic effects. There is some evidence suggesting that lead may be carcinogenic, or perhaps a cocarcinogen owing to its persistence in tissues. Case reports and epidemiological studies are difficult to sort out because exposures frequently involve more than one metal. This is true in the steel industry, where increased cancer frequencies have been observed.

Many cationic metals will form complexes with thiol groups of cell components, and the complex will mimic natural substrates to interfere with cell processes, mostly transport systems. Thus, methylmercury-cysteine complex mimics methionine, and the complex is taken up into the brain by a transport system for neutral amino acids. Inorganic and organic mercury complexes with glutathione and is transported from liver cells into bile. Arsenic and copper do the same thing. Lead can substitute for Ca^{2+} in a number of transport and receptor-mediated processes. Voltage-activated calcium channels will admit a number of metallic cations, a fact which is exploited in research. Cadmium (Cd) and lanthanum (La) act in this way.

UNUSUAL SOURCES OF HEAVY METAL EXPOSURE

In 1988, the Texas Department of Health investigated illegal sales of drugs manufactured in Hong Kong. The tablets, sold as "chuifong tokuwan", contained a veritable pharmacy of drugs, including diazapam, indomethacin, hydrochlorothiazide, mefanemic acid, dexamethasone, lead and cadmium! This potpourri of tranquilizer, diuretic, antiinflammatory agents, corticosteroids and heavy metals was repackaged and marketed as "The Miracle Herb — Mother Nature's Finest". Twenty-four percent of 93 persons who took this preparation had elevated urine cadmium levels (1.8 µg/mL compared to 0.5 µg/mL for random controls). The upper limit of normal is considered to be 2.5 µg/mL. No elevated (>25 µg/dL) blood lead levels were detected, but 42% of these individuals had elevated urine levels of retinol-binding protein, indicative of renal tubular damage.

Some of the "health" supplements such as bone meal (for calcium) contain high amounts of lead, and some zinc supplements are contaminated with cadmium.

In a case in Ohio, several members of a household were hospitalized with a diagnosis of acrodynia, a form of metallic mercury poisoning in which neurological and psychological disorders occur as well as hypertension, rash, sweating, cold intolerance, tremor, irritability, insomnia, anorexia and diminished performance at school. Twenty-four-hr urine collections revealed mercury levels of 850–1500 µg/mL (normal <20 mg/mL). Careful inquiry of neighbors indicated that a previous tenant had spilled a large jar of elemental mercury in the apartment. Treatment was instituted with the oral chelating agent 2,3-dimercaptosuccinic acid (DMSA). Some neurological disorders persisted.

■ REVIEW QUESTIONS

Answer the following questions True or False.

1. The term "heavy metal" usually refers to those with double-digit specific gravities._____

2. Aluminum, which has a specific gravity of only 2.7, has no toxic properties._____

3. All forms of arsenic are equally toxic._____

4. Lead poisoning may be manifested as both gastrointestinal and central nervous system toxicity._____

5. Acute, paroxysmal, colicky pain is common in severe lead poisoning in adults._____

6. Mental retardation does not occur in lead poisoning in children._____

For questions 7 to 11, use the following code: answer A if statements 1, 2 and 3 are correct; answer B if statements 1 and 3 are correct; answer C if statements 2 and 4 are correct; answer D if statement 4 only is correct; answer E if all statements (1,2,3,4) are correct.

7. 1. Subchronic lead poisoning involves interference with mitochondrial heme synthesis at several levels.
   2. Anemia may accompany lead toxicity.
   3. Elemental lead is not absorbed through the skin or lungs.
   4. Elemental lead is not absorbed from the intestinal tract.

8. 1. All metals can cross the placenta and cause fetal toxicity.
   2. Examining a blood sample is the only way of detecting lead in the body.
   3. Chelation treatment with dimercaprol may be used in lead poisoning.
   4. Chelation therapy is of no use in elemental mercury poisoning.

9. 1. Elemental mercury poisoning never occurs.
   2. Methylmercury is the most important environmental source of mercury poisoning.
   3. Chelation therapy is useful in the treatment of methylmercury poisoning.
   4. Methylmercury poisoning may involve a variety of central nervous symptoms as well as signs of arthritis.

10. 1. Cadmium accumulates in the liver and kidney.
    2. Cadmium induces the synthesis of the cadmium-binding protein metallothionein.
    3. The kidney is the major organ of toxicity for cadmium.
    4. Chronic inhalation of cadmium fumes may lead to pulmonary disease.

11. 1. Like mercury, arsenic binds to sulfhydryl and disulfide groups.
    2. Trivalent forms of arsenic are the most toxic.
    3. Arsenic is readily absorbed from virtually all portals of entry.
    4. Arsenic does not accumulate up the food chain.

For questions 12 to 16, select, from the following list, the correct mechanism for the stated metal. (An answer may be used once only.)

   a. Chelates with organic ligands containing SH groups.
   b. Complexes with cysteine and competes with methionine in protein synthesis.
   c. Mimics calcium and is deposited in bone.
   d. Carried to the kidney by bicarbonate where it dissociates to cause renal damage.
   e. The hexavalent form enters cells and is converted to the trivalent form which binds to, and poisons, macromolecules.

12. ____. Lead, arsenic

13. ____. Lead only

TOXICITY OF METALS 147

14. ____. Uranium (uranyl cation)

15. ____. Methylmercury

16. ____. Chromium

ANSWERS

1. T; 2. F; 3. F; 4. T; 5. T; 6. F; 7. A; 8. B; 9. C; 10. E; 11. A; 12. a; 13. c; 14. d; 15. b; 16. e.

CASE STUDY # 11

In early spring, two of five workers employed in demolishing an old iron bridge visited the company's consulting physician complaining of muscle pain (myalgia), joint pain (arthralgia), headache and nausea. These workers had been cutting up sections of the bridge using oxyacetylene torches. Large sections were lowered to a barge moored in the river below the bridge to be cut into smaller sections for hauling away. When the remaining three workers were questioned, it was discovered that they too had been suffering from similar symptoms as well as memory loss and irritability. A supervisor and a secretary who worked in the construction shack on shore were not affected, nor were four men involved in loading trucks or operating a small boom crane.

Q. What organ systems are involved in the affected workers?

Q. Could this be a toxicant causing the illness and, if so, what are its possible sources?

Q. What is the likely portal of entry?

Q. What diagnostic tests might be useful?

Q. What specific treatment might be appropriate?

CASE STUDY #12

A 67-year-old man consulted his physician because of severe abdominal pain, weight loss and fatigue. The doctor initially suspected gastric carcinoma, but the patient was severely anemic, his red cells had basophilic stippling, and he had a blood lead level of 70 µg/dL. Six other household members also were affected, including an 8-yr-old child. All had elevated blood lead levels. The home was located in a suburban residential area, not near any industrial site.

Q. What is the probable portal of entry of the lead in these people?

Q. A search of the home revealed a ceramic jug as the offending agent. Why was it suspect?

CASE STUDY #13

During a routine pre-employment medical examination, a 46-yr-old male was found to have a blood lead level of 50 µg/dL. He was subsequently investigated by a university hospital toxicology clinic which confirmed the same blood lead level 1 month later. Symptoms included numbness of the fingers and palms, tinnitus (ringing in the ears) and an apparent decrease in ability to do mental arithmetic and mild memory deficits. He had been taking ranitidine for indigestion. This is a histamine H2 receptor blocker that suppresses gastric acid secretion.

A detailed personal and employment history was obtained. He had spent 20 years as an electronics technician in the army and in civilian life, but had had little exposure to lead from soldering or welding. He had no hobbies that could serve as a source of lead, no history of bullet or birdshot wounds, and he denied drinking bootleg alcohol or using lead additives in his car.

Q. Is the source of lead likely to be work-related or from the home environment? How could this be determined?

Q. What is the probable significance of the gastric distress?

Q. What therapeutic approach would be appropriate?

FURTHER READING

Albert, A., *Xenobiosis*, Chapman and Hall, New York, 1987.

Angle, C.R., Childhood lead poisoning and its treatment. *Ann. Rev. Pharmacol. Toxicol.*, 32, 409, 1993.

Blyth, A.W., *Poisons; Their Effects and Detection*, Wm Wood & Co., New York, 1885.

Cadmium and lead exposure associated with pharmaceuticals imported from Asia. *Morbid. Mortal. Weekly Rep.*, 38, 612, 1989.

Cancer and the Worker. Proceedings of a Conference. New York Academy of Science, New York, 1977.

Klassen, C.D., Amdur, M.O., Doull, J. (Eds.), *Casarett and Doull's Toxicology*, Macmillan, New York, 1986.

Clarkson, T.W., Molecular and ionic mimicry of toxic metals. *Ann. Rev. Pharmacol. Toxicol.*, 32, 545, 1993.

Elemental mercury poisoning in a household. *Morbid. Mortal. Weekly Rep.*, 39, 424, 1990.

Fimreite, N., *Mercury Contamination in Canada and its Effects on Wildlife*. Ph.D. Thesis, UWO Library, 1970.

Hutchison, G. and Wallace, D., *Grassy Narrows*, Van Nostrand Reinhold, Toronto, 1977.

Jaworski, J. (Ed.), Effects of chromium, alkali halides, arsenic, asbestos, mercury and cadmium in the Canadian environment. *NRC Executive Reports,* Publ. # NRCC 17585, 1980.

Klaassen, C.D., Heavy metals and heavy metal antagonists, in *Goodman and Gilman's The Pharmacological Basis of Therapeutics,* 8th Edit. Gilman, A.G., Rall, T.W., Nies, A.S. and Taylor, P. (Eds.), Collier Macmillan Canada, Toronto, 1990, chap. 66.

7 ORGANIC SOLVENTS AND RELATED CHEMICALS

'Tis a sordid profit that's accompanied by the destruction of health.
Treatise on the Diseases of Tradesmen, B. Ramazzini, 1705

INTRODUCTION

Organic solvents are common in the workplace where they may constitute an occupational hazard, but they also occur in the home in the form of cleaning solutions, paint strippers and brush cleaners and can thus be a source of household poisoning as well as being an environmental hazard. Since they are all fat solvents, they may cause local defatting of tissue and local irritation on contact. Many also are systemic toxicants affecting the CNS like volatile anesthetics or, in some cases, the hematopoietic (blood-forming) system. Commercial solvents are frequently complex mixtures having nitrogen- and sulfur-containing elements (e.g., gasoline and other petroleum-based products).

Industrially, the uses of solvents are many and varied. They are used in extraction processes in the food and pharmaceutical industries (e.g., ethyl alcohol and acetone), for the removal of impurities, for degreasing and vapor cleaning (e.g., trichloroethylene, 1,1,1,-trichloroethane), as vehicles for paints, carriers for pesticides, for printing inks (toluene, ethyl acetate), in adhesives (hexane, toluene, methylethyl ketone). In short, volatile solvents are used whenever a fast drying property is desired in order to leave a coating on a surface. They are also used in the chemical industry for a variety of manufacturing processes.

CLASSES OF SOLVENTS

Aliphatic Hydrocarbons

Aliphatic hydrocarbons are straight-chain or branched carbon-hydrogen compounds. They are often present in complex mixtures in common commercial products such as gasoline, mineral turpentine and kerosene. These mixtures can also contain smaller amounts of unsaturated and cyclic carbon-hydrogen substances. While the vapors of these agents are generally less toxic than those of most organic solvents, inhaling the fumes can not only cause disorientation and stupor through effects on the CNS, but they can sensitize the heart to adrenaline. The result may be ventricular fibrillation which can be fatal if not corrected quickly. Fatalities have occurred in workmen cleaning storage tanks and rail tank cars. Chemical pneumonia often occurs from aspiration of the low-viscosity liquid (an occupational hazard for those who syphon gasoline from others' cars!). One member of this solvent group of special note is n-hexane, the main ingredient of petroleum ether. With a boiling point of 60–80°C, it evaporates readily and produces an insidious form of poisoning. An outbreak of industrial poisoning occurred in Japan in 1973, when workers exposed to fumes from a glue used in making sandals lost sensation and function in fingers and toes from impaired nerve conduction. When the exposure was terminated, they all recovered slowly over several months. Hexane neuropathy has both a central and peripheral component resulting from demyelination of nerve fibers. The actual toxicant is a metabolite of n-hexane, hexane-2,5-dione. An intermediate, 2-hexanone, is also more toxic than the parent substance. If the neurotoxicity of n-hexane is rated as 1, then 2-hexanone would be 10, and hexane-2,5-dione would be 40.

$$CH_3(CH_2)_4CH_3 \xrightarrow{oxidation} CH_3-\overset{\overset{O}{\|}}{C}-(CH_2)_3CH_3 \xrightarrow{oxidation} CH_3-\overset{\overset{O}{\|}}{C}-(CH_2)_3-\overset{\overset{O}{\|}}{C}-CH_3$$

n-HEXANE   2-HEXANONE   HEXANE-2,5-DIONE

Metabolites are produced in the liver by cytochrome P-450 oxidases. They appear to condense with lysine in the myelin, causing disorganization of the membrane.

Halogenated Aliphatic Hydrocarbons

This group of substances containing chlorine substituents includes methylene dichloride, chloroform, carbon tetrachloride and chlorinated ethylenes (e.g., trichloroethylene) as well as a number of chemicals that contain other halogens such as bromine and fluorine. They are used as anesthetics and refrigerants. Hepatotoxicity and nephrotoxicity characterize this group. The liver converts carbon tetrachloride (CCl_4) to the free radical $^-CCl_3$ which attacks the endoplasmic reticulum, causing protein synthesis to cease. Recovery will usually occur from a single toxic exposure, but repeated exposures lead to cirrhosis. Chloroform is converted to phosgene ($COCl_2$) which has similar toxicity to CCl_4. Phosgene was used as a poison gas in World War I. The chemical reactions involved in this process are illustrated in Chapter 5 (Figure 24).

Trichloroethylene was used as a light anesthetic in childbirth. It and similar agents are used as industrial solvents. Fatalities have occurred when workers were overcome in enclosed tanks. Death usually results from aspiration of vomit. It is also hepatotoxic.

Dichloromethane (methylene chloride) was used for many years as a paint stripper. It has a boiling point of 40°C and thus readily forms vapor. It is converted to carbon monoxide (CO) which forms carboxyhemoglobin in the red blood cells (see Chapter 5).

Bis(chloromethyl) ether is discussed below under "Cancer in the Workplace".

Other halogenated hydrocarbons are discussed in Chapter 5, and the chlorofluorocarbons (CFCs) and their impact on the environment are discussed in Chapter 4. Some anesthetics are halogenated hydrocarbons. One such is halothane, chemically a haloalkane, which is a volatile liquid. Although most halothane is eliminated in the expired air, some undergoes hepatic biotransformation to a reactive metabolite, an alkylating radical, which causes lipid peroxidation and which is hepatotoxic. It does not deplete glutathione, as do some hepatotoxic agents. The reaction is mediated by cytochrome P-450, and toxicity is increased by hypoxia. The incidence of this toxic reaction is low, perhaps one per 38,000 patients, and it appears to be genetically determined. Other halogen-containing general anesthetics are methoxyflurane, enflurane and isoflurane. Methoxyflurane is rarely used in North America now because it releases free fluoride during biotransformation. Free fluoride is toxic to the kidney.

Aliphatic Alcohols

Ethanol (ethyl alcohol) is present in alcoholic beverages and also in some lotions, perfumes, mouthwashes, cough syrups, etc. Although there is a tendency to trivialize ethanol's toxicity, 3–6 ml/kg of pure ethanol may be fatal. Since 70-proof liquor contains 40% ethanol, this fatal adult dose equates with as little as 525 ml of liquor. In severe poisoning, the classical signs of inebriation (difficulty with balance, locomotion and talking) progress to coma, metabolic acidosis, hypothermia, hypotension and severe respiratory depression. Hypoglycemia may be present, especially in young children. Treatment is largely supportive, with correction of acidosis and hypoglycemia (5% glucose i.v.)

and hemodialysis to remove ethanol and its metabolites. Other simple alcohols and glycols produce early effects resembling ethanol intoxication, which is why they are often abused by alcoholics who are at the bottom of the socioeconomic scale.

Methanol is by far the most toxic of the alcohols. In humans and other primates, it is oxidized to formaldehyde, a very reactive substance which the eye cannot convert to the harmless formate anion. As little as 4 ml has caused blindness through retinal damage. The transformation is illustrated below.

$$CH_3OH \xrightarrow{\text{alcohol dehydrogenase}} HCHO$$

Methanol is present in paint thinners and removers, windshield washer fluids and in fuels for small engines. It is also present in pineapple, where the levels may exceed the recommended limit of 2 ppm. Outbreaks of poisoning have occurred as a result of deliberate adulteration of wine. Home-distilled liquor also has caused poisonings, and "skid row" alcoholics may consume substances containing methanol as a substitute for ethanol. Initial symptoms are those of ethanol intoxication. A fairly long latency period may occur (6–30 hr), followed by acidosis, delirium, coma, visual disturbances, which may or may not progress to permanent blindness, cardiac disturbances and death. Treatment consists of the administration of activated charcoal by mouth, ethanol (orally or i.v.) as a competitive substrate for alcohol dehydrogenase, and hemodialysis to remove methanol, formaldehyde and formic acid (which causes acidosis). A fatal dose is 60–250 ml. An experimental approach to treatment involves the administration of inhibitors of alcohol dehydrogenase, such as 4-methyl pyrazole.

Isopropyl alcohol is found in rubbing alcohols, aftershave lotions and window-cleaning fluids, and it is a widely used industrial solvent. It is oxidized to acetone and causes acetonemia but not acidosis. Its toxicity lies between that of ethanol and methanol, being about twice as toxic as ethanol, with coma and respiratory arrest resulting from CNS depression.

Higher alcohols are generally less toxic. N-butanol vapors can produce eye irritation (conjunctivitis, keratitis), and inhalation may cause pulmonary edema.

Acetone (C_3H_6O), although a ketone rather than an alcohol, produces similar toxic symptoms to those of ethanol, and the treatment is the same. Acetone, like the aliphatic hydrocarbons, dissolves lipids from the skin and can be extremely damaging to the cornea by virtue of defatting the epithelial cells.

N.B. The hepatotoxicity of the halogenated hydrocarbons can be potentiated by numerous alcohols including ethanol, methanol and isopropyl alcohol, and acetone. The mechanism is unclear.

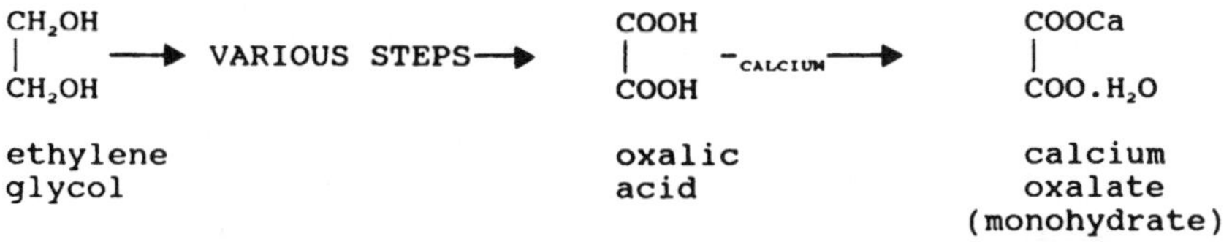

$$\begin{array}{c}\text{CH}_2\text{OH}\\|\\ \text{CH}_2\text{OH}\end{array} \xrightarrow{\text{VARIOUS STEPS}} \begin{array}{c}\text{COOH}\\|\\ \text{COOH}\end{array} \xrightarrow{-\text{CALCIUM}} \begin{array}{c}\text{COOCa}\\|\\ \text{COO}\cdot\text{H}_2\text{O}\end{array}$$

ethylene glycol / oxalic acid / calcium oxalate (monohydrate)

Figure 27 The conversion of ethylene glycol to calcium oxalate.

Glycols and Glycol Ethers

Ethylene glycol is present in antifreeze. A dose of 100 ml orally may be fatal for an adult. Early signs and symptoms resemble ethanol poisoning. After 24 hr, there may be pulmonary edema and myocardial depression and after 48 hr, renal tubular necrosis and renal failure. Hypoglycemia and hypocalcemia may occur. Treatment is as for methanol.

The toxicity of ethylene glycol is due to its metabolites. It is metabolized by alcohol dehydrogenase, which is why ethanol is given orally or i.v., as it is for methanol poisoning. When hemodialysis is used in conjunction with i.v. ethanol, ethanol is often added to the dialyzer fluid to maintain blood ethanol levels. Otherwise, the ethanol would be dialyzed because of the concentration gradient. One metabolite of ethylene glycol is oxalate, which chelates calcium (hence, the hypocalcemia) and which precipitates as calcium oxalate crystals in the kidney, causing tubular necrosis. Other metabolites are aldehyde, glycolate and lactate which cause acidosis. A simplified scheme for the metabolism of ethylene glycol is shown in Figure 27. Propylene glycol is fairly nontoxic, and it is used in lotions and as a vehicle for injectable drugs. Too-rapid injection may cause cardiac depression.

A subgroup of this class of compounds is the glycol ethers. Monomethyl and monoethyl ethylene glycol are used extensively in industry, being present in latex paints and as solvents in the manufacture of lacquers, varnishes and dyes. There is some evidence that they are reproductive toxins, causing teratogenesis in experimental animals. They are not well absorbed orally, but transdermal absorption and inhalation are the important portals of entry. Human toxicity has not been well established, but there have been reports of kidney damage and bone marrow depression. Precautions are taken to limit industrial exposure of women.

Aromatic Hydrocarbons

Benzene is one of the simplest and most toxic of these cyclic, special hydrocarbons. It is highly volatile, and exposure in the workplace is mainly by inhalation. Benzene is unique in this group because of its bone marrow toxicity. Chronic exposure causes a progressive reduction in all formed elements of the blood including red cells, white cells and platelets. Aplastic anemia, resulting from almost complete destruction of the marrow, may occur and the mortality rate is high, the only treatment being bone marrow transplantation. Bone marrow depression is dose-time dependent. Leukemia also can develop. Several reports have noted a higher-than-normal incidence of acute myelogenous leukemia in workers exposed to benzene. Although leukemia has not been demonstrated in animals exposed to benzene, solid tumors have been observed.

Chromosomal abnormalities have been seen in animals exposed to benzene. The toxic agent may be a metabolite of benzene since animal studies have shown that toluene, which competes with benzene in its metabolic pathway, reduces its toxicity. A number of metabolites have been identified, including phenol.

Alkylbenzenes are a group of related compounds consisting of the aromatic ring with one or more aliphatic side chains. As a group, they lack the serious toxic effects of benzene because they are detoxified to metabolites with low toxicity, including hippuric acids which are readily excreted by the kidney. They are, however, potent CNS depressants, acting like general anesthetics. No mutagenic properties have been demonstrated. The group includes toluene (1-methylbenzene) and the xylenes (1,2-, 1,3-, and 1,4-dimethylbenzene) as well as ethylbenzene and cumene (isopropylbenzene).

Dinitrobenzene is a related substance that is used extensively in the manufacture of plastics, dyes, pigments, explosives, insecticides and in many other processes. The opportunity for an industrial exposure to toxic levels is thus fairly high. Meta, para and ortho forms exist; all are crystals in the pure form, and all volatilize with steam and are soluble in boiling water and organic solvents. They are absorbed across the skin and promote the formation of methemoglobin (MetHb), in which the normal ferrous (Fe^{++}) iron is converted (oxidized) to ferric (Fe^{+++}) iron. Methemoglobin binds irreversibly to O_2 and reduces the oxygen-carrying capacity of the red blood cells. If MetHb exceeds 1% of total hemoglobin, mild hypoxia results. If it exceeds 15%, a severe condition known as cyanosis-anemia syndrome results. Dinitrobenzene is the second most common industrial cause of methemoglobinemia. The chemical structures of some of the aromatic hydrocarbons are shown in Figure 28.

■ SOLVENT-RELATED CANCER IN THE WORKPLACE

Benzene

As noted above, benzene is capable of causing leukemia. The first case of "benzene leukemia" was observed in 1928 in a worker who was so heavily exposed that others could not work in the same environment

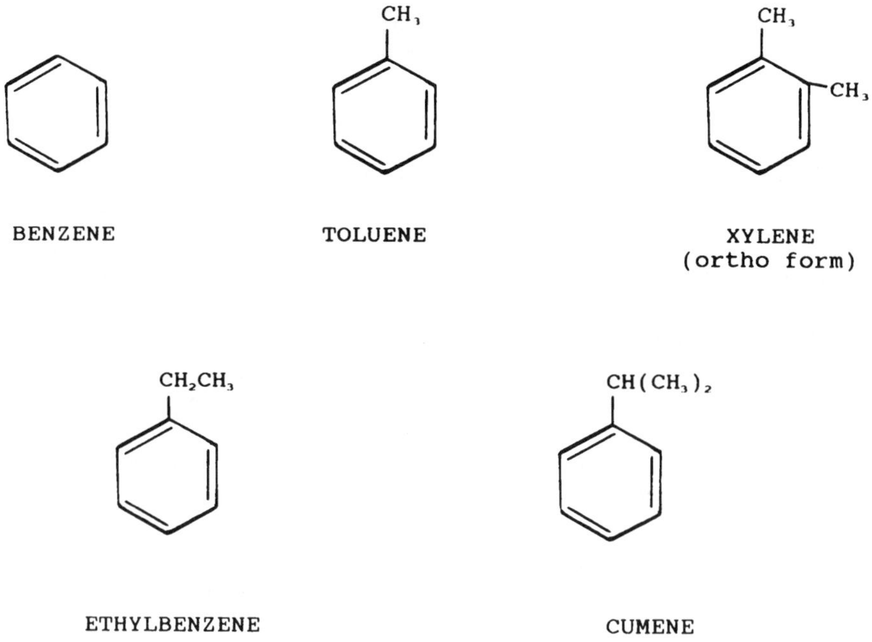

Figure 28 Chemical structures of some aromatic hydrocarbon solvents.

without becoming acutely ill, which probably saved them from a similar fate. By 1980, about 200 cases of benzene-related leukemia were reported. In Italy, which banned it in 1963, workers in the shoemaking and rotogravure industries were estimated to have a risk of leukemia 20× that of the general population. The latency period may be 15 years or more. In Japan, studies of the survivors of Hiroshima and Nagasaki revealed that the risk of leukemia was increased 2.5× in those who had worked in jobs involving benzene exposure. In the U.S., benzene was used extensively in the rubber industry until it was banned. Overall, workers in this industry had a threefold increase, and those with high exposures a fivefold increase, in their risk of leukemia.

Bis(chloromethyl) Ether (BCME)

BCME is a potent alkylating agent and carcinogen used in a variety of industrial syntheses. An American study conducted by NIOSH of 136 men exposed to high levels of BCME indicated a 10× risk of lung cancer. Exposures averaged 5 years or more, and cancer rates were highest in those exposed for 10 years or more. Sputum cytology tests indicated a 34% incidence of abnormal lung cells in exposed worker vs. 11% in those not exposed. A much larger study involving 1,800 workers in New York found a 2.5× increase in risk and in a Philadelphia chemical plant, heavily exposed workers had an 8× increased risk. The discovery that fumes of HCl and formaldehyde could react spontaneously to form BCME led to regulations prohibiting the use of these agents in the same area without special ventilation.

Dimethylformamide (DMF) and Glycol Ethers

These agents are widely used in the tanning industry for finishing hides. There have been three clusters of testicular cancer identified in this population of workers. Three cases were identified in a tannery in Fulton County, New York, between 1982 and 1984. Collection and analysis of air samples by NIOSH identified significant levels of the glycol ethers 2-ethoxyethanol, 2-ethoxyethyl acetate and 2-butoxyethanol. The workers had previously been exposed to high levels of dimethylformamide (DMF) on the job, although none was detected at time of sampling. Two other clusters of testicular cancer in workers in the tanning industry have been identified. DMF has been shown to be a testicular toxin in animal studies, but not a mutagen, and glycol ethers are known to be reproductive toxins in animals. At the present time, the precise cause of these tumors has not been identified.

Ethylene Oxide (C_2H_4O)

Although not strictly speaking a solvent, it is appropriate to consider this chemical here. It exists as a sweetish, ether-like gas at room temperature and becomes a liquid at 12°C. It is used as an industrial chemical in the manufacture of plastics, as a fumigant in agriculture and as a sterilant in health care facilities for heat-sensitive materials. All of these uses relate to the fact that this chemical is highly reactive with other organic substances, including proteins, because it contains an epoxide. The structural formula is

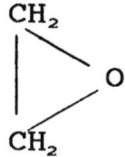

It exhibits the same toxicity as other organic solvents including CNS depression, local irritation including respiratory irritation, frostbite due to the rapid evaporation of the liquid and multi-organ toxicity following chronic exposure. It mixes with both organic solvents and water, and it floats on the latter. When exposures are minimal, as they should be when safe handling practices are observed, ethylene oxide poses little risk. Poor safety procedures or accidental exposure carry the risk of a toxic reaction. Populations at risk include workers in health care institutions, exterminators, fumigators and chemical plant workers. The question of carcinogenicity of this agent remains unanswered, but it is highly suspect because of its structure, and animal studies have shown clear evidence of increased incidences of cancers of the adrenal gland, spleen, kidney, skin, lung, stomach and brain as well as mononuclear cell leukemias. A study of workers exposed to 5–10 ppm for an average of 10.7 years did not reveal any increased frequency of cancer.

■ FACTORS INFLUENCING THE RISK OF A TOXIC REACTION

Virtually all of the factors (discussed in Chapter 1) that can influence the response of experimental animals to a toxic agent can also affect toxicity to humans. The risk of a toxic reaction is a function of the toxicity of the chemical and the duration of the exposure, hence the need to establish exposure limits like the Short-Term Exposure Limit (STEL). As noted above, toxicity may also be affected by the presence of other agents such as ethanol. N-hexane and benzene will enhance the nephrotoxicity of chlorinated hydrocarbons. This may be especially important when exposure to mixed solvents occurs, as it often does. The uptake and distribution of a toxic agent will be influenced by its air/water (blood) partition coefficient and by its water (blood)/oil (fat) partition coefficient. Body depot fat may thus serve as a reservoir for highly lipid-soluble solvents, just as it does for some anesthetics, and thus prolong their CNS-depressing effects. Age, the presence of hepatic disease, and general state of health all may influence individual risk.

■ NONOCCUPATIONAL EXPOSURES TO SOLVENTS

Solvents are present in many household products including cleaning agents, waxes and polishes, glues, paints, automotive products (cleaners, polishes, etc.) paint removers, thinners and brush cleaners, even hairsprays and of course in gasoline. Exposures may be accidental, usually involving skin contact or inhalation in an enclosed space or ingestion by small children, or deliberate, as in the ingestion or inhalation of a substance in a suicide attempt, or as a consequence of substance abuse as in gasoline sniffing. A great tragedy of our time is the high incidence of gasoline sniffing, which has resulted in several deaths, among native Canadian and American youths living on Indian reserves. Glue sniffing also can be a form of substance abuse practiced by adolescents in large cities and where access to other drugs of abuse is restricted.

A somewhat unexpected source of exposure is products used in arts and crafts. These include adhesives, paints and lacquers, and cleaning solvents. Cigarette smoke also contains volatile solvents, including benzene.

■ REVIEW QUESTIONS

For Questions 1 to 8, answer True or False.

1. Aliphatic solvents are straight or branched chain carbon-hydrogen compounds.____
2. Inhaling gasoline fumes can sensitize the heart to adrenaline.____
3. Hexane is oxidized to a hepatotoxic metabolite.____
4. Monomethyl ethylene glycol is less toxic acutely than ethylene glycol.____
5. Propylene glycol is highly toxic.____
6. Chloroform is hepatotoxic without being metabolized.____
7. Ethylene oxide is liquid at room temperature.____
8. Workers in the tanning industry may show an increased frequency of testicular cancer.____

For Questions 9 to 12, use the following code: answer A if statements 1, 2 and 3 are correct; answer B if statements 1 and 3 are correct; answer C if statements 2 and 4 are correct; answer D if statement 4 only is correct; answer E if all statements (1,2,3,4) are correct.

9. Halogenated aliphatic solvents
    1. Cause CNS depression.
    2. Are hepatotoxic.
    3. Are nephrotoxic.
    4. Never require biotransformation to exert their toxicity.
10. Methanol
    1. Is converted to formaldehyde in the body.
    2. Can cause blindness.
    3. Poisoning is treated with ethanol.
    4. May be present in fairly high concentrations in pineapple.
11. Ethylene glycol
    1. Is metabolized to a toxic substance or substances.
    2. Is toxic in the parent state.
    3. Is metabolized to oxalate which may precipitate as calcium oxalate in the renal tubules.

ORGANIC SOLVENTS AND RELATED CHEMICALS

   4. Poisoning is not treated with alcohol.
12. Benzene
   1. Depresses the CNS.
   2. May cause anemia and reduced platelet count.
   3. Is carcinogenic in animals.
   4. May cause fatal aplastic anemia.
13. Ethylene oxide
   1. Contains an epoxide bond.
   2. Is a suspected human carcinogen.
   3. Depresses the CNS.
   4. May cause multiorgan toxicity following long-term exposure.

■ ANSWERS

1. T; 2. T; 3. T; 4. T; 5. F; 6. F; 7. F; 8. T; 9. A; 10. E; 11. B; 12. E; 13. E.

■ CASE STUDY #14

Five steam press operators in a rubber plant became ill with signs and symptoms including blue discoloration of the lips and nail beds, headache, nausea, chest pain, dizziness, confusion and difficulty in concentrating. One worker suffered a seizure. Blood methemoglobin (MetHb) levels ranged from 3.8% to 41.2% (normal <1%). The product they were using was a solvent-borne adhesive.

Q. What information would you want to solicit concerning their working environment?

Q. What industrial chemicals could be responsible for the high MetHb levels?

   Several days later, the steam press was operated by a supervisor for 2 hr, so that an industrial hygienist could collect air samples. The supervisor's MetHb level was 12.5% at the end of the 2 hr. The air samples were unrevealing, so that technical assistance from the NIOSH was called in.

Q. What analysis might next be called for?

Q. What steps might you recommend in the interim?

■ CASE STUDIES #15 AND 16

#15: A 52-year-old woman was hospitalized for routine hemodialysis for chronic kidney failure. She became somnolent after dialysis and developed metabolic acidosis and shock within 12 hr. She died 24 hr after dialysis. The dialysis fluid had been made up with tap water.

#16: Two children, 4 and 7 years of age, were admitted around 7:00 pm to a small rural hospital with an acute onset of marked somnolence, vomiting, and weakness. They developed hematuria (blood in the urine) and were transferred to a nearby city hospital with a pediatric intensive care unit. Urinalysis showed calcium oxalate crystals. They were treated with intravenous fluid therapy and recovered uneventfully.

The children had attended a picnic that afternoon held in an park adjacent to a local firehall. A survey of the nearly 400 guests revealed similar but less serious symptoms in 28 of them, 2 of whom required hospitalization. Nineteen of the 28 cases, including the 2 who required hospitalization, were children under 10 years of age. The most common symptoms were sleepiness, fatigue, dizziness and unsteadiness when walking. A common denominator in those who became ill appeared to be consumption of a noncarbonated beverage made from crystals with tap water taken from the firehall. A few individuals who had not consumed the beverage, but had drunk water, also became ill. Severity of symptoms correlated with the amount of beverage or water consumed.

Q. Do you think the offending agent in these cases is
   a. A pathogenic organism?
   b. A toxic chemical?

   Give reasons for your choice.

Q. Do the calcium oxalate crystals provide a clue?

Q. What organ system is most affected?

In both of these cases, there was a direct connection between the general water supply and chilled-water air conditioning systems. Could this have been a factor in these illnesses?

■ FURTHER READING

Albert, A., *Xenobiosis*, Chapman and Hall, New York, 1987.
Cancer and the Worker, Proceedings of a Conference. New York Academy of Science, New York, 1977.
Emergency Treatment of Poisoning, Health and Welfare Canada, Publ. H49–14, 1982.
Glenn, W., Ethylene oxide: another question mark. *Occup. Health and Safety Can.*, 5, 28, 1989.
Hodgson, E. and Levi, P.E. (Eds.), *A Textbook of Modern Toxicology*, Elsevier, New York, 1987.

Klaassen, C.D., Amdur, M.O. and Doull, J. (Eds.), *Casarett and Doull's Toxicology*, Collier Macmillan, Toronto, 1986.

Methemoglobinemia due to occupational exposure to dinitrobenzene. *Morbid. Mortal. Weekly Rep.*, 37, 353, 1988.

Testicular cancer in leather workers — Fulton County, New York. *Morbid. Mortal. Weekly Rep.*, 38, 105, 1989.

8 FOOD ADDITIVES, DRUG RESIDUES AND OTHER FOOD TOXICANTS

The food that to him now is as luscious as locusts, shall be to him shortly as bitter as coloquintida.
W. Shakespeare

FOOD ADDITIVES

Food and Drug Regulations

All industrialized countries have regulations governing the ingredients and additives that can legally be present in foodstuffs. The nature of the regulations may vary from country to country, but those of Canada are fairly typical.

In Canada, regulations governing foodstuffs and food additives are part of the Food and Drug Act and are enforced by the Health Protection Branch of Health Canada. The following are some important definitions.

Food: Any substance, whether cooked, processed or raw, that is intended for human consumption including drinks, chewing gum, and any substance used in the preparation of a food, but not including cosmetics, tobacco, or any substance used as a drug. This definition includes food additives.

Food Additives: Any substance, including any source of radiation, the use of which results in, or may reasonably be expected to result in, it or its by-products becoming a part of, or affecting, the characteristics of a food. This definition does not include

1. Any nutritive material that is commonly recognized or sold as a food.

2. Vitamins, minerals, and amino acids.
3. Spices, seasonings, natural flavorings, essential oils, oleoresins, and natural extracts.
4. Accidental contaminants, such as pesticides, or drugs administered to farm livestock.
5. Food packaging materials or components thereof.

The regulations lay out which food additives are permitted, in what foods they may be used, and what are the maximum allowable amounts.

A food additive must do at least one of the following:

1. Improve nutritive value.
2. Extend shelf life.
3. Prevent spoilage during shipment.
4. Enhance appearance or palatability.
5. Assist in the preparation of the food or in the maintenance of its physical form.

It has been estimated that about one billion pounds of food additives are consumed annually in North America, or 3.5 pounds per person. The vast majority of these are harmless, but demonstrated toxicity in some experimental animals for some, and public concern about harmful effects of man-made chemicals, have created pressure on governments to tighten up regulations controlling their use and establish stricter limits on allowable levels in foodstuffs. As is the case for any xenobiotic, the use of such agents should only be undertaken on the basis of a cost/benefit analysis. If the advantage is trivial, such as enhancement of texture, then any associated risk would be unacceptable. On the other hand, prevention of spoilage or of the growth of pathogenic microorganisms might justify the acceptance of a slight risk. It is this area that generates the greatest conflict between environmental groups and growers' and manufacturers' lobbies, who may differ markedly on the definition of acceptable risk.

Food additives are used for a variety of purposes. The following are some of the major ones.

Acidifiers or acidulants provide tartness and act as preservatives by lowering pH. They may also improve viscosity.

Adjuvants for flavor facilitate the action of the principal flavoring agent.

Aerating agents (propellants, whipping agents) are used to produce a foam as in whipped toppings, etc.

Alkalies control pH, neutralize high acidity foods (tomato products, some wines) and may improve flavor.

Antibiotics are used to prevent bacterial spoilage during storage and transportation.

Antibrowning agents prevent oxidation on the surface of some foods such as lettuce that may cause brown spots.

Anticaking agents are added to powdered or crystalline products (drink mixes, powdered spices, salt, cake mixes, etc.) to prevent caking (formation of lumps).

Antimold agents are added to foods (bread, baked goods, dried fruit, cheeses, chocolate syrup) to prevent mold growth. They are also called antimycotic or antirope agents (liquid or viscous products that become moldy are described as being "ropey").

Antioxidants prevent the oxidation of fatty acids that causes rancidity, and of vitamins, which lose potency.

Antistaling agents prevent bread, etc,. from going stale.

Binders are substances used to maintain "body" and hold a product together (e.g., processed meat, snack foods).

Bleaching agents are used to whiten flour and some cheeses.

Buffers are used in many processed foods.

Chelators or sequestrants are used to bind metallic ions that can hasten oxidation of fats and shorten shelf life.

Coating agents (glazing or polishing agents) are used to coat the skins of fruits and vegetables to prevent bruising, drying or spoilage and to coat candies and tablets.

Defoaming agents (antifoaming agents, surfactants) are used to prevent excessive foaming in beverages when bottle filling.

Emulsifiers disperse fat droplets in an aqueous medium, e.g., salad dressings, milk shakes, whipped cream and toppings in pressure cans.

Extenders (fillers) are natural substances (casein, starch, soybean meal) used to add bulk to a food product.

Fixatives maintain the color of meat and processed meat.

Flavor enhancers intensify the natural flavor in soft drinks, fruit drinks, jams, gelatin desserts.

Flavors (artificial): any flavoring that does not occur in nature, even if the ingredients are all natural, is defined as an artificial flavor. When something is described as "chocolatey" rather than as chocolate, it indicates that the flavor is artificial, not natural, chocolate. This advertising ploy gets around the regulations prohibiting false advertising, and it is widely used in North America.

Food colors are added to many products, including some fruit (oranges), to restore color lost in processing or transportation. Most (90%) are synthetic. Food colors that are bound to aluminum hydroxide are known as "lakes". All synthetic food colors are highly water-soluble. Only vegetable dyes are lipid-soluble. This, of course, affects their absorption from the gastrointestinal tract.

Fumigants are toxic gases used to kill pests in harvested, dried grains and nuts.

Fungicides prevent fungal growth on the surface of some fruits.

Humectants (hydroscopic agents) retain moisture and prevent drying in some candies and in ice cream.

Maturing agents (dough conditioners): flour is better for baking if it is aged. Bleaches and other agents speed up the process.

Plasticizers (softeners) are used in chewing gum, candies, edible cheese coatings to maintain pliability.

Stabilizers (suspending agents) prevent cocoa, orange pulp, solids in ice cream from settling out.

Sweeteners (nonnutritional, artificial): there are many applications in low-calorie and diabetic diets.

The above is a partial list of the uses of food additives. It is probably unrealistic to expect that the use of such agents, many of which are synthetic chemicals, can be completely eliminated. Some of them at least are essential to allow the shipment of fresh fruit and vegetables over long distances, as is necessary if these foods are to be available in countries with a short growing season. As is so often the case, public perceptions of risk cloud the issue of artificial food additives. Although consumer advocacy groups continue to campaign for tighter controls on such agents, when saccharin was banned because animal tests had shown the development of bladder tumors in rodents fed high doses, public outcry, originating from a perceived need for this product (vanity is a powerful motivator), resulted in a partial removal of the ban (see below). The remainder of this chapter will concentrate on the more common, or more controversial, food additives.

Artificial Food Colors

The common public perception is that synthetic food dyes are inherently more toxic than natural ones. In fact, they are highly purified chemicals, most of which have received extensive toxicity testing. Moreover, they are highly water-soluble, so that absorption from the GI tract is minimal. In contrast, natural dyes are complex mixtures of compounds that are generally more lipid-soluble and therefore better absorbed. Since most natural food additives have been in use for decades, they are listed on the U.S. "Generally Regarded As Safe" (GRAS) list and have not been extensively tested. Most are lipid-soluble carotenoids (reds, oranges and yellows) and are present in carrots, squash, yams, etc., and can be regarded as harmless. The possibility of an allergy to any food component cannot be discounted, however. Natural colors tend to be more subdued and to fade more quickly than synthetic ones. Thus, synthetic colors are often preferred. (In Highland lore, "Ancient" tartans are those dyed with the original, more muted, vegetable dyes.)

Some synthetic dyes have been banned. Orange No. 1 and Red No. 3 caused diarrhea in children who consumed large amounts of candy, carbonated beverages and other confections where these colors were used extensively. More recently, Red No. 2 (amaranth) was banned because embryotoxicity was demonstrated in rats.

Considerable controversy surrounds the question of whether synthetic food dyes contribute to hyperactivity in children. Dr. Benjamin Feingold postulated that artificial colors and flavors, together with "salicylate-like" natural substances (present in apples, oranges, peaches, raisins and many berries, and in cucumbers and tomatoes) contribute to behavioral problems such as shortened attention span, easy distractability, compulsiveness and hyperactivity. Several studies have been conducted to examine this problem. In one type, groups of children were fed the "Feingold" diet or a normal diet and crossed over to the alternate diet after several weeks. This is called a double-blind crossover study. Neither the observers nor the subjects are aware of the treatment group to which they were assigned. Behavior was rated subjectively by parents and teachers. In studies at the Universities of Pittsburg and Wisconsin, 25–33% of hyperactive children showed improvement when shifted to the Feingold diet according to the parents' ratings. Teachers' ratings showed much fewer differences. Moreover, shifts from the Feingold diet to the normal one were not accompanied by behavioral changes. In other studies, children received the Feingold diet throughout the experiment, but doses of a blend of 8 certified colors were added periodically. Doses were 26–150 mg. The latter corresponds to the intake of the 90th percentile of American children. Both subjective estimates of behavior and objective tests of behavior and learning performance were used. At levels above 100 mg in a Canadian study, there was a slight deterioration in learning in 17 of 22 hyperactive children, but no change in behavior. Two of 22 were affected at 35 mg. An Australian study employing a parent questionnaire reported substantial improvement on the special diet, whereas a similar U.S. study did not.

Citrus Red No. 2 is restricted to surface use on oranges not to be processed. It has been shown to be carcinogenic in animals.

Orange B is restricted to use in the casings of sausages and hot dogs. It is related to amaranth (Red dye No. 2). The manufacturer has discontinued production because of evidence of a carcinogenic contaminant. Red No. 40 was imputed to cause cancer, but the evidence was ruled inconclusive. Yellow No. 5 (tartrazine) has been associated with allergic reactions, sometimes severe, and it is listed on the U.S. ingredients list (candies, desserts, cereals, dairy products). Cross allergenicity with aspirin is common. Canada does not have a compulsory ingredients list. This is a matter of great concern to individuals, and their parents in the case of children, who have life-threatening allergies to foodstuffs and additives.

Blue No. 1 and 2, Green No. 3, and Yellow No. 6 are considered safe, but WHO has raised questions about the adequacy of testing.

Emulsifiers

Carrageenin (Irish Moss) is extracted from several species of red marine algae. It contains a variety of calcium, sodium, potassium and ammonium salts plus a sulfated polysaccharide. It is widely used as an emulsifier and thickener in ice cream, milkshakes and chocolate drinks. It keeps milk proteins in suspension. Estimated daily intake is 15 mg per person. Only the undegraded form is permitted. Rats fed 2,000 mg/kg showed fetal deaths and young with underdeveloped bones. Increased vascular permeability and interference with complement have been shown experimentally. Although potentially serious, especially in ill people (complement is essential to the immune system), carrageenin is not well absorbed, and the FAO/WHO Committee on Food Additives has established an acceptable level of 500 mg/day. Long-term testing is probably indicated. Furcelleran is a similar substance derived from a red seaweed, and it is similarly used. Brominated vegetable oil (BVO) is also used as an emulsifier to keep flavoring oils in suspension in soft drinks. A study in 1976 indicated that the daily U.S. intake averaged less than 0.2 mg/person. These products have been in use for over 50 years, but toxicology studies in the early 1970s showed that doses of 2,500 mg/kg of cottonseed BVO caused, in rats, heart enlargement and fatty deposits in heart, liver and kidney after a few days. Doses as low as 250 mg/kg caused fat deposition in the heart. Maximum daily intake for a child probably does not exceed 0.05 mg/kg, but it may occur over a prolonged period. Corn oil BVO fed to rats and pigs at 20 mg/kg for weeks caused deposits of brominated fat in liver and other tissues. The UN joint FAO/WHO Committee on Food Additives recommended in 1971 that BVO not be used as an additive. The U.S. FDA removed it from the GRAS list, pending further safety studies by the manufacturer. These were judged to be faulty by the FDA. Its use has been discontinued.

Preservatives and Anti-Oxidants

Butylated hydroxyanisole (BHA) and butylated hydroxytoluene (BHT) are synthetics used to prevent premature rancidity in oil- and fat-containing foods. Total daily intake for a child could approach 0.5 mg/kg, which is the maximum recommended by FAO/WHO. Animal studies have consistently shown that high doses (over 500× the average human consumption) caused liver enlargement and induction of microsomal enzymes. Less than this had no effect. Recent studies have revealed evidence of carcinogenicity in offspring of mice fed BHA. These agents may act as promoters through the enzyme induction mechanism. More work is required before a decision is made. As free radical scavengers, these agents actually may have anticarcinogenic properties.

Sodium nitrate, sodium nitrite, and potassium nitrate are used as curing agents in meats such as bacon and smoked meats. They are always used in combination with salt. Nitrite also inhibits the growth of *Clostridium botulinum*, the organism responsible for botulism. Nitrate is converted to nitrite by bacterial and enzymatic action in the intestine. The average U.S. daily intake from food additives is about 11 mg. The major concern is that nitrites can combine with amines to form nitrosamines which are carcinogenic. This process is accelerated by cooking. However, it is important to note that nitrites from food additives account for less than 20% of daily intake, the rest coming from nitrates in drinking water and in vegetables such as celery, spinach, and other leafy vegetables. This is partly due to the use of nitrogen fertilizers and partly from natural sources. Saliva contains nitrate, perhaps providing over 100 mg/day to be converted to nitrites in the lower gastrointestinal tract through bacterial action. Thus, the total nitrate load is about 90 mg from natural dietary sources, 100 from saliva and only 11 from food additives. The use of nitrates and nitrites is restricted to the minimum levels required to inhibit the growth of *C. botulinum.*

A particular concern has been the poisoning of infants by nitrates in well water. There are now thousands of such cases, and many deaths, reported because the formation of methemoglobin from nitrites from nitrates impairs the oxygen-carrying capacity of the blood. A high percentage of rural dwellers in North America and Europe draw their water supply from shallow wells supplied by ground water. These are vulnerable to contamination by surface runoff and hence by nitrates from fertilizer. Typically, newborn infants in these rural areas would develop, after days or weeks, a syndrome that included cyanosis (blue baby), hypotension (nitrates and nitrites are potent vasodilators) and, eventually, coma and death. Most recovered when they were removed from the home and hospitalized. Invariably, these infants were being fed formula made with well water that contained 20–1,000 ppm of nitrates. The latency period varied according to the degree of contamination. The conversion of nitrates to nitrites occurred as a result of bacterial action either in the well or in the gastrointestinal tract of the infant, which is virtually neutral (pH 7±) and hence favorable to bacterial growth. With the decline in popularity of formula-feeding and its replacement by breastfeeding, nitrite poisoning in infants has almost disappeared in North America.

Artificial Sweeteners
Sodium saccharin is several hundred times sweeter than sucrose but leaves a bitter aftertaste. For many years, it was the only artificial

sweetener in common use. The average daily intake is about 6 mg/kg, but some individuals habituated to soft drinks may consume much more. Theodore Roosevelt first required a review of its safety in 1912. In the early 1970s, two studies, one by the FDA, reported that high doses (2,500 mg/kg) caused an increase in the incidence of bladder cancer in rats. It was not known whether this was due to an impurity. In 1977, the Canadian Health Protection Branch confirmed that the saccharin was the causative agent, but only when rats were exposed in utero to high doses. It was concluded that saccharin is a weak carcinogen and probably a cocarcinogen, but public outcry blocked its recall in the U.S. In Canada, it is available in tablet form in pharmacies, but it cannot be used as a sweetener in prepared beverages nor as a sweetener in restaurants. It is also available in Great Britain. The International Agency for Research on Cancer, an agency of the United Nations, has not deemed it necessary to place saccharin on its list of proven carcinogens.

Xylitol is a natural ingredient of many fruits and berries. It has the same caloric value as glucose but it does not affect blood glucose levels and so it can be used by diabetics. It does not cause cares because it is resistant to fermentation by plaque microorganisms. It has been used in "sugarfree" gum. Some evidence of carcinogenicity has been obtained in rats fed very high doses. Xylitol has been given i.v. to humans as a source of energy. Kidney, liver and brain disturbances have occurred and some fatalities. Use in the U.S. has been voluntarily stopped pending a review. It has been replaced by aspartame.

Sorbitol is another sugar substitute that is calorically equal to sucrose, but which will not raise blood glucose levels. Its uses are the same as for xylitol. It is also used as a humectant in jellies, baked goods and in canned bread to prevent browning. Nausea, cramps and diahrrea have occurred in some individuals.

Acesulfame potassium (an oxathiazinondioxide) was introduced in 1988 as a noncaloric sweetener. It is chemically similar to saccharin. No adverse reactions have yet been reported, but long-term safety is unknown.

Cyclamate was introduced as a sweetener after saccharin. It has the advantage of not leaving a bitter aftertaste. Both were included in the 1959 "Generally Regarded as Safe" (GRAS) list of the U.S. Food and Drug Directorate. Because cyclamate was fed along with saccharin in one study showing increased bladder tumors, it was removed from the GRAS list, and it is not available in the U.S. or Great Britain, but it is in Canada because a Canadian study had shown saccharin to be the culprit.

Aspartame is a dipeptide consisting of the amino acids aspartic acid and phenylalanine. Extensive testing has not revealed any carcinogenic

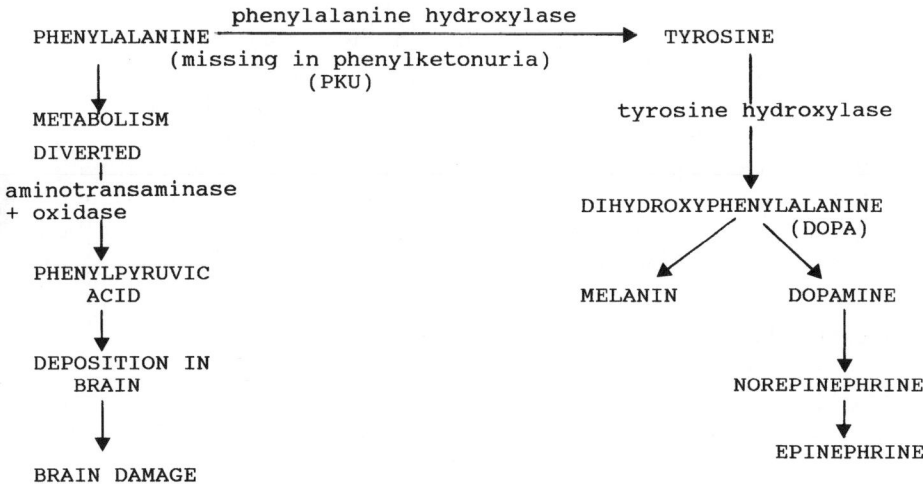

Figure 29 Phenylalanine metabolism in infants with phenylketonuria.

potential, and it has no aftertaste. It has largely replaced other sweeteners in soft drinks, gum and other dietetic foods. It should not be used by people with phenylketonuria (PKU), a hereditary defect of the enzyme phenylalanine hydroxylase, which converts phenylalanine to tyrosine (Figure 29). These people use an alternate pathway that causes the accumulation of phenylpyruvic acid. This accumulates and deposits in the brain and causes mental retardation in infants.

All newborns are tested for this and, if positive, are placed on a diet free of phenylalanine. Since tyrosine is a precursor in the synthesis of catecholamine neurotransmitters, there was some speculation that high doses could cause behavioral changes, but no evidence of this has been forthcoming in either children or adults.

A recent addition to this group is sucralose. It is a chemical modification of sucrose in which a number of H and OH groups are replaced by Cl. It cannot be broken down by digestive enzymes and it is not absorbed. Thus, it cannot enter metabolic pathways for glucose, it will not elevate blood glucose, and it provides no calories. It is rated as 600× as sweet as sucrose. It has been pronounced safe by WHO and the Joint Expert Committee on Food Additives. It has been approved in Canada and is under review in the U.S. A wide variety of products is presently in the laboratory testing stage of development, many of "natural" origin, and some of these will undoubtedly be making their appearance in the future.

Flavor Enhancers

Monosodium glutamate (MSG) has been identified as the offending agent in the "Chinese Restaurant Syndrome". Subjective symptoms of numbness and tingling of the mouth and tongue have been reported, but double-blind studies with doses up to 3 g failed to confirm an effect. It appears that certain individuals are highly sensitive. Because glutamate is an excitatory amino acid, MSG has been given in doses up to 45 g daily to mentally retarded patients, with no behavioral changes or ill effects. Animal studies have shown hypothalamic lesions, and infants under 6 months of age may be especially susceptible to MSG toxicity.

■ DRUG RESIDUES

The high-risk nature and narrow profit margin that are typical of the livestock industry have led to the extensive use of antibiotics, sulfa drugs, hormones and other pharmaceuticals to improve productivity by increasing the rate or extent of weight gain per unit of food consumed or to prevent or treat disease. Many of these are medications that are also used in human medicine; others are unique to the agricultural field, but they may have pharmacological or toxicological consequences for people as well. There are three main concerns about the possibility that traces of these substances might enter our food supply:

1. They may serve as a source of allergic sensitization.
2. Anti-infectives may contribute to the development of resistant strains of pathogenic bacteria.
3. They may exert direct toxic manifestations such as teratogenesis and carcinogenesis.

While there is ample evidence that 1 and 2 occur, the occurrence of 3 is mostly speculative.

Antibiotics and Drug Resistance

In the late 1950s, Thomas Jukes of Berkeley University reported that the antibiotic tetracycline at 50 ppm in animal feed significantly improved the rate of weight gain and the gain-to-food consumption ratio in livestock. Subsequent studies showed that this effect was not related to the prevention of disease and occurred even under optimal conditions of husbandry and hygiene. The effect was confirmed later for other antibiotics, and the mechanism remains elusive. There is by now a long list of antibiotics and other anti-infective agents that have been employed as growth promotants. Some of the more common ones are shown in Table 7.

Antibiotics are also used to prevent the outbreak of disease. Prophylactic use involves higher levels than those used for growth promotion, and a typical mixture for preventing dysentery in swine in the 1960s contained chlortetracycline 100 gm/T of feed, sulfamethazine 100 gm/T and penicillin G 50 gm/T. Concern over the potential dangers

TABLE 7
DRUGS USED AS ANIMAL GROWTH PROMOTANTS

Antibiotics
  Several tetracyclines
  Erythromycin
  The aminoglycosides neomycin and streptomycin
  Lincomycin
Antifungals
  Nystatin
Synthetic chemotherapeutics
  Sulfa drugs
  The nitrofurans furazolidone and nitrofurazone

of drug residues in food mounted over the next two decades. It began with a report in 1959 from Japan of an outbreak of *Shigella* dysentery in a hospital nursery. The outbreak was unique in that the infecting strain of the bacteria was resistant to several antibiotics and sulfa drugs, some of which had never been used in that hospital. The term "Multiple Drug Resistance" was coined for this phenomenon, and in the next few years many reports emerged of MDR in livestock, and in 1965 an outbreak of Salmonellosis in Great Britain resulted in six deaths. It was traced to the consumption of veal from calves that had been treated with several antibiotics, and the organism demonstrated MDR. In 1971 the Swann Commission in the United Kingdom recommended greater controls over the use of antibiotics in livestock. Government legislation was passed and many other countries, including Canada, followed suit.

During this period it was discovered that the pattern of resistance typical of one type of bacteria could be passed to other, unrelated genuses and species, and the term "Infectious Drug Resistance" (IDR) came into use.

For many years it was thought that IDR involved Gram negative enteric organisms exclusively. "Gram negative" refers to the histochemical staining characteristics of the bacteria (one of the major means of classifying them, it is related to the composition of their outer cell wall), and "enteric" refers to the fact that they are common inhabitants of the intestinal or enteric tract of both animals and humans. This group includes strains of *Escherichia coli, Salmonella* spp., *Shigella* spp., and *Klebsiellia* spp. It is now known, however, that most, and possibly all, bacteria are capable of developing multiple drug resistance through transference of genetic information from one cell to another.

The mechanism of IDR hinges on the fact that bacteria possess extrachromosomal units of genetic information called plasmids. These are rings of DNA that are capable of replication independent of the chromosomes and that can be passed intact from one bacterial cell to another. Genes can be inserted into these plasmids from other sources, including bacterial chromosomes, and this is dependent upon the existence of discrete sequences of 800–1,800 base pairs of amino acids called insertion sequences. When a gene for drug resistance is included between two insertion sequences, the unit is called a transposon or, more often, an R (for resistance) factor. The existence of plasmids has provided the means for genetic engineering and the bacterial synthesis of human insulin and other substances. There are several methods by which plasmids can be transferred from one bacterial cell to another. These are

1. Transformation: The lysis of a cell may release plasmids into the environment which may subsequently be absorbed by other cells. This is a highly species-specific phenomenon.
2. Transduction: This involves the participation of phage viruses which incorporate bacterial genetic information and transfer it to other cells. This also is very species-specific.
3. Conjugation: The plasmids of many bacteria possess a gene, called a fertility or "F" factor, which regulates a form of sexual reproduction. A fine tubule or "pilus" is formed between cells and intact plasmids may then be passed, along with their complement of genetic information, from one to the other. This process is not species-specific, and it is the basis for IDR, since R factors will be passed along as well. In this way, resistance genes may be shared among all enteric organisms that contact each other and a multiple resistance pattern acquired. This process is illustrated in Figure 30.

The process of conjugation is clinically important because

1. Nonpathogenic organisms may serve as a reservoir of drug resistance to be passed on to more virulent ones. There are few species barriers to transfer.
2. The process of natural selection may, when an antibiotic is used therapeutically, result in the emergence of a strain of bacteria with MDR as susceptible strains are killed or inhibited.
3. Resistance patterns, or resistant organisms, may be passed from animals to humans. Many organisms are infectious for both.
4. The process is favored by exposure to low (nontoxic) levels of antibiotics.

The question is, how serious is the problem? In the 1970s there were reports that the enteric organisms of farm workers had identical resistance patterns to those of the livestock they tended. Some consumer advocates have asked "why not just eliminate the use of antibiotics entirely?". A study in 1983 claimed that for an investment of $271 million in animal feed additives, the American consumer saved $3.5 billion in food costs. The result has been a lobbying war between the pharmaceutical and agricultural industries on one hand

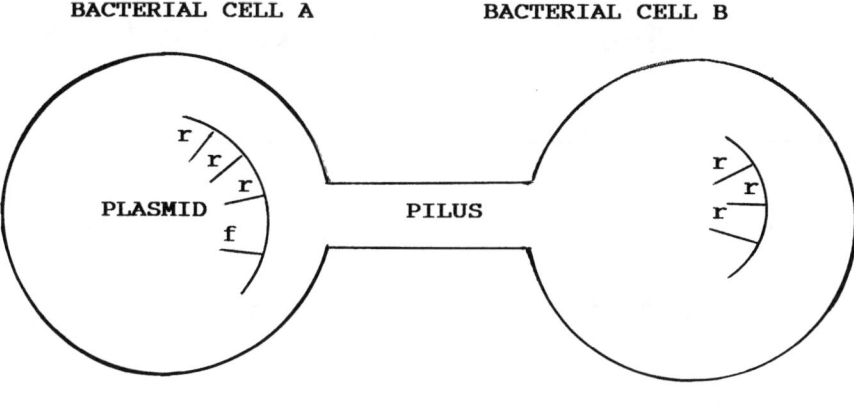

Figure 30 Bacterial conjugation and multiple (infectious) drug resistance.

and various consumer and environmental groups on the other. In the last few years, the debate has heated up as outbreaks of Salmonellosis have become more common, especially among patients of nursing homes and psychiatric hospitals, where disease resistance may be lower and personal hygiene less rigorous. In 1984, the U.S. Centers for Disease Control issued a report on 52 outbreaks of Salmonellosis between 1972 and 1983. The source of the infection was identifiable in 38 of these and in 17 (44%) it could be traced to food animals (poultry, veal, hamburger). These accounted for 69% of all outbreaks involving strains with MDR. The mortality from these strains was 4.2% vs. 0.2% from nonresistant strains. In March and April of 1985 the most massive outbreak of Salmonellosis in U.S. history occurred in Chicago, with 14,000 people eventually infected. There were only two deaths, yielding a mortality of 0.014%. This was markedly lower than the previous rates reported for either resistant or susceptible strains (the 4.2% mortality rate would have caused 588 deaths), and a followup revealed that over 16,000 people were eventually infected for a final mortality of 0.012%. Both fatalities, however, involved a tetracycline-resistant strain. The controversy continues. Thomas Jukes has criticized efforts to ban antibiotics as additives on the grounds that evidence is inconclusive, but recently Canada banned the use of chloramphenicol for any purpose in food animals. A study conducted in Ontario in the 1970s indicated that MDR was very common

TABLE 8
BACTERIAL RESISTANCE IN SEWER EFFLUENTS

Effluent	% Coliforms resistant to		
Source	Streptomycin	Chloramphenicol	Tetracycline
General hospitals	48.8	0.4	24.3
Psychiatric hospitals	9.5	0.03	0.04
Residential	0.6	0.0007	0.1

in cattle, that the most common resistance pattern was against not less than six drugs and that chloramphenicol was frequently one of these.

Avoidance of meat is no guarantee that multiple resistance patterns can also be avoided. In one study, vegetarians had a higher frequency of MDR bacteria than meat eaters. The use of manure as fertilizer for vegetables might serve as an efficient means of transfer when the vegetables are eaten raw. Meat packers also tend to acquire MDR patterns.

It must be remembered that the use of antibiotics in agriculture is not the only source of MDR strains. In one study, resistance patterns of coliform organisms in the sewage effluent from a general hospital, a psychiatric hospital, and a residential area were compared. The results are shown in Table 8.

It is evident that the incidence of MDR closely parallels the use of antibiotics in a particular setting. Thus, it is higher in the general hospital than in either the psychiatric hospital or the residential area. Moreover, resistance to chloramphenicol, which is reserved for life-threatening situations where less toxic antibiotics will not work, was very low in all situations, but the correlation still held. In 1985 chloramphenicol was withdrawn as an approved drug for use in food animals.

Resistance to antibiotics is not the only type of genetic information that can be passed in this manner. Resistance to metals and bacterial virulence also is regulated by plasmid genes. Both vertical (within a species) and horizontal (among species) transfer is now thought to be commonplace in the microbial kingdom. Even transfer between Gram negative and Gram positive organisms has been demonstrated.

No discussion of environmental hazards would be complete without passing reference to the hazard of infectious diseases. Not only has the emergence of antibiotic resistance led to clinical problems in treating old diseases such as tuberculosis, which has developed multiple drug resistance and is on the rise, but many organisms seem to be developing more virulent strains and new ones are emerging such as the AIDS virus, the *Hantavirus*, the bacteria of Lyme disease, *Legionella* and others. Perhaps this should not surprise us, given the rapidity of cell division of bacteria and the adaptability that has allowed them to survive for eons. Next to malnutrition, infectious diseases are the leading cause of death worldwide. There is an obvious relationship between the two.

Allergy

Legislation in most western countries prescribes withholding periods during which livestock may not be shipped for food following the administration of an antibiotic or other drug by any route. This is designed to prevent the entry of drug residues into the human food chain. One reason for this is the possibility that such residues could result either in allergic sensitization or in an allergic reaction in individuals who are already sensitive. Significant levels of drugs could remain at the injection site which is usually a large muscle mass that also constitutes a preferred cut of meat. Certainly, prior to the introduction of withholding times, such reactions did occur. Of the drugs used in agriculture, those with the highest allergic potential are the penicillins. Penicillin G is still available without prescription in many countries for use in livestock, and it is commonly employed for the treatment of mastitis (udder infection) in dairy cattle. It may thus be present in milk, although testing procedures are usually mandated by law. The test is not specific for a particular drug, but employs a strain of bacteria that is extremely sensitive to inhibition by almost all antibiotics. More recently enzyme-linked immunosorbant assay (ELISA) has been used.

It may be significant that 30% of patients suffering severe allergic reactions when treated with penicillin have no history of previous exposure, suggesting that sensitization occurred through environmental exposure. Although allergic reactions from contaminated food are rare and difficult to trace, they may still occur when drug residues escape detection. Sulfa drugs may also be offenders in this regard.

Diethylstilbestrol

Diethylstilbestrol or DES is one of a number of synthetic estrogens first synthesized by Charles Dodds in 1938. Since the work was conducted under the auspices of the British Medical Research Council, the chemicals were not subjected to patent restrictions, and they rapidly became widely available. As a group, they revolutionized endocrinology, made possible the development of oral contraceptives and had a major impact on the practice of gynecology. The possibility of sex without pregnancy and the consequent sociosexual revolution of the 1960s can be attributed directly to the development of these compounds and their chemical manipulation by the pharmaceutical industry.

DES mimics natural estrogens, but it is much more potent and more toxic, and it is effective orally, which natural ones are not. It is very cheap to produce, and by 1939 it was already popular in the U.K., France, Germany, Sweden and North America. It was used to treat menopausal complaints, amenorrhea, genital underdevelopment, and to suppress lactation in mothers who did not wish to nurse their infants. By 1942, agricultural use had begun on a limited scale. Early studies showed that rats fed DES gained weight more rapidly with no increase in food consumption and subsequent tests in cattle yielded similar results. Moreover, the "marbling" (diffuse fat distribution) was greater and this was considered to be a desirable flavor enhancer at the time. Athletes such as Ben Johnson were not the first to take steroids to increase weight gain!

By 1945, DES was being widely used in beef and poultry production. There were two methods of administration. One was a feed additive and the other a slow release pellet that could be injected under the skin of a steer's ear or the neck of a broiler chicken. The former method had the disadvantage of loss through spillage, and the accidental ingestion by breeding stock caused some serious fertility problems. The necks and ears containing the remains of the pellet were supposed to be discarded at slaughter, but chickens are frequently sold with the necks attached, and in 1959, a report emerged of a 40-year-old male chef in New York who developed signs of feminization including loss of facial hair, mammary development, fat deposition of the female pattern and a high voice. The problem was traced to the consumption of chicken necks taken from the restaurant and eaten as an economy measure. The use of pellet implants in poultry was banned that same year.

The carcinogenic nature of estrogens has been known for some time. Experimentally, natural estrogens were shown to produce carcinomas of the vagina, cervix and uterus in mice, and carcinogenic potency was directly related to estrogenic potency. There were sporadic reports of DES inducing cancer in mice, but in 1971 Erbst et al. reported a case of clear-cell adenocarcinoma in the vagina of a young woman who had been exposed to DES *in utero*. A DES registry was established in the U.S., and by 1984 500 cases of clear-cell adenocarcinoma had been reported. It is now estimated that females exposed to DES *in utero* have somewhat less than a 1/1,000 chance of developing it.

The use of DES in pregnancy was based on data indicating that it improved the vascularity of the uterus and promoted the synthesis of progesterone, both of which would help to maintain a pregnancy in the face of a threatened abortion. Other conditions that have been causally associated with exposure to DES *in utero* include adenosis (a self-limiting condition involving the presence of cervical cells in the wall of the uterus), infertility, carcinoma of the cervix and vagina and breast cancer. Cervical malformations occur in 18–25% of DES daughters, but most appear to correct themselves with age. Infertility is by far the most common problem experienced by DES daughters. Pregnancy may be difficult to achieve and maintain. Since the first wave of DES daughters is just now entering middle age, the possibility that increased frequencies of other forms of cancer might emerge cannot be discounted; however, the association with breast cancer is becoming more tenuous.

The fetal risks of exposure to DES appear to relate to the fact that the mechanisms that detoxify natural estrogens cannot handle DES. In rats, a glycoprotein called alpha fetoprotein binds natural estrogens but not DES. Its presence has not been confirmed in humans, but human fetuses have a very active process for sulfonating natural estrogens which may be less effective for DES. Active metabolites of DES may have a high affinity for estrogen receptors in the reproductive system, and oxidative metabolites may damage DNA. Reactive metabolites may form after attachment to the receptor by the action of peroxidase enzymes present in estrogen-dependent tissues.

Attempts by the FDA to ban the use of DES in beef cattle were blocked by court actions launched by the beef producers until 1979 when liver and kidneys from beef cattle showed significant levels of DES. Illicit use continued thereafter, and in 1980 the USDA conducted 115 prosecutions against violators. It must be emphasized that all of the health problems associated with DES in North America (barring the chef noted above) are of iatrogenic origin; i.e., they arose from medical treatment. None has been attributed to the consumption of meat from treated animals. This has not been the case elsewhere.

In the late 1970s and early 1980s, two epidemics occurred in which infants and children under 8 years of age displayed signs of abnormal sexual development such as breast development and precocious puberty. In Puerto Rico over 600 children were involved. DES was available without prescription, and meat inspection controls were poor. In Italy, high levels of DES were detected in baby foods, especially those containing veal. Several hundred infants were affected. Occurrences declined after 1979 due to tougher controls and greater public awareness.

In 1985, the DES Task Force of the U.S. Department of Health and Human Resources released its findings. These were

1. The risk of uterine carcinoma *in situ* was about twice as high in DES daughters as in nonexposed women (15.7/1,000/year vs. 7.9/1,000/year), although there were problems with the study.
2. The risk of genital herpes also was about twice as high (11.8% vs. 6.3%).

3. The risk of breast cancer was difficult to assess because of other predisposing factors.

The question of increased risk of breast cancer was addressed recently by an American multicenter epidemiological study. It compared over 3,000 women who had DES prescribed during pregnancy with 3,000 similar unexposed women. The risk of breast cancer/100,000 was 172.3 for exposed women and 134.1 for unexposed women. The rate increased markedly in both groups between 20 and 40 years post entry to the study. The results were statistically significant, and it now seems conclusive that there is a real increase in the risk of breast cancer in DES-exposed women, but probably not in their daughters.

The psychological costs to DES daughters are difficult to assess but probably quite high. Increased incidences of psychiatric problems and suicide have been reported due to feelings of helplessness and frustration. Recently, one small report suggested a higher incidence of homosexual preference in DES daughters. It is well documented that experimental exposure to sex hormones before birth influences brain development and sexual behavior. In rats and guinea pigs, alpha fetoprotein normally protects the brain from estrogens, but it does not bind to DES, and prenatal exposure resulted in abnormal patterns of sexual behavior.

There is some suggestive evidence that males exposed to DES in utero may have a higher incidence of testicular cancer and underdeveloped male genitalia, but the evidence is far from conclusive. There is conclusive experimental evidence, however, that exposure to DES *in utero* has these effects in male experimental animals.

With regard to the question of DES used as a growth promoter, the U.S. Delaney clause, which prohibits the use of anything as a food additive that has been shown to be carcinogenic in any animal, specifically exempts substances used as animal food additives, provided their residues are not detected in the final foodstuff by currently acceptable analytical techniques. The increase in sensitivity of these techniques over the past decade may render the agricultural use of DES impossible.

Concern over drug residues continues. In 1989, Canada's Health Protection Branch ordered the withdrawal of ethylenediamine dihydroiodide, a product used to treat certain infections in cattle. There was concern that the product might contribute to excessive iodine levels in people with consequences for thyroid function. It is estimated that some Canadians might consume as much as six times the recommended daily intake. Currently, a controversy is building over the use of growth hormones in meat animals. Studies indicate that they can increase lean content by at least 5% and feed efficiency by

8%. The consumer, however, remains understandably leery of their use. As long as profit margins in agriculture remain narrow, the pressure to use growth promotants will remain strong. The use of growth hormone in the U.S. has resulted in attempts to ban American beef in the European Economic Community.

■ NATURAL TOXICANTS AND CARCINOGENS IN HUMAN FOODS

Through the acquisition of folk knowledge, human beings have learned to avoid eating rhubarb leaves, daffodil bulbs and other plants that contain toxic chemicals. Under some conditions, however, foods that are normally safe may become toxic, at least for some individuals, and carcinogens may be more prevalent in human foods than previously realized.

Some Natural Toxicants

Favism

The broad bean (*Vicia fava*) may induce acute hemolytic anemia in individuals with the hereditary defect glucose-6-phosphate dehydrogenase (G6PD) deficiency. This is especially true for those individuals with the Mediterranean phenotype. It is a problem primarily for males under 5 years of age. The exact nature of the mechanism remains obscure, but there is evidence that these individuals may be deficient in hepatic glucuronide conjugase and hence unable to detoxify the offending ingredient. Susceptibility in the same individual varies from time to time, and all those with the same defect are not necessarily affected.

Toxic Oil Syndrome

In 1981, in Spain there was a remarkable epidemic that eventually affected over 20,000 people. In the first 12 months of the epidemic, there were 12,000 hospitalizations and over 300 deaths. It began in a small town on May 1, when a boy was admitted to hospital with acute respiratory failure that rapidly led to death. The epidemic peaked in June, at which time 2,000 new cases were being reported every week. The patients usually presented initially with respiratory distress, cough, low-grade fever, oxygen deficiency, pulmonary infiltrates and pleural effusions and a variety of skin rashes. Nausea and vomiting were sometimes present, as were enlarged liver, spleen and lymph nodes. Virtually all patients had elevated eosinophil counts ($>500->2,000/mm^3$, normal value, $0-500/mm^3$). The condition often progressed to severe muscle pain, muscle and nerve degeneration and even paraplegia.

An astute medical clinician traced the problem to the consumption of a cheap, unlabeled cooking oil sold as pure olive oil in the open markets in small towns and villages. The oil consisted of low-grade olive oil mixed with various seed oils including a rapeseed oil that was imported in a denatured form by mixing it with aniline to render it unfit for human consumption. Two significant clues were the facts that the condition affected low-income families almost exclusively, and nursing infants were never affected. Geographically, the epidemic was limited largely to central and northwestern Spain.

Chemical analyses of suspect oil samples, and comparison with nonsuspect samples, failed to reveal the presence of known toxicants such as heavy metals, but the suspect oil contained significant levels of aniline and fatty acid anilides, reaction products of the oil with the aniline. A dose-response relationship between the degree of contamination and the severity of signs and symptoms was also noted.

The toxicity of aniline is well known because of its heavy industrial use and although there are some similarities with the toxic oil syndrome, skin lesions, for example, aniline toxicity involves CNS symptoms (vertigo, headache, mental confusion) and blood disorders including methemoglobinemia and anemia. It was felt, therefore, that the offending agent was likely a reaction product. To date, however, the syndrome has not been reproduced in an animal model.

The toxic oil syndrome might be regarded as an historical curiosity were it not for the fact that in 1989, a similar problem emerged in the United States. Called the eosinophilia-myalgia syndrome, it affected over 1,500 people, and it appeared to be caused by consuming contaminated L-tryptophan as a food supplement. Again, the eosinophil count was usually elevated above 2,000/mm^3. Like the toxic oil syndrome, there was also inflammation of muscles and their nerve supply. There was a notable absence of acute respiratory symptoms, unlike the toxic oil syndrome.

The Centers for Disease Control determined that the product came from one supplier and that it had a contaminant, the di-tryptophan aminal of acetaldehyde, that was either the toxic agent or a marker thereof, as there was a strong association between the level of this substance and the incidence of the disease. Efforts at identifying the toxic agent and developing an animal model for the condition are continuing.

Natural Carcinogens in Foods

Bracken Fern "Fiddleheads"

Considerable evidence exists that bracken fern produces bladder cancer in cattle that eat excessive amounts when better fodder is unavailable and in rats fed large amounts of it. Because the young shoots, called fidddleheads because of their curled shape, are eaten as a delicacy in many parts of the world, including Canada and Japan, there has been concern over the potential for carcinogenic effects in humans. At one point it was suggested that the relatively high incidence of bladder cancer in Japan might be related to the consumption of bracken fern. Epidemiological studies, however, have failed to demonstrate such an association, and it is now felt that eating fiddleheads does not constitute a risk factor for cancer.

The economic significance of bracken as an agricultural toxicant has been clearly demonstrated. It is a radiomimetic substance, causing bone marrow depression in cattle and other species and thiamine deficiency in horses.

Others
Natural carcinogens and precursors have been detected in many other foods. In their 1987 review, Ames et al. list the carcinogenic TD_{50} for nitrosamines present in many foodstuffs as 0.2 µg/kg for rats and mice. In contrast, PCBs, with a similar daily dietary intake of 0.2 ug, had values of 1.7–9.6 mg/kg in these species. Other substances shown to be carcinogens in animal tests, but for which evidence of a risk to humans is weak, include allyl isothiocyanate (in kale, cabbage, broccoli, cauliflower, horseradish and mustard oil), safrole (nutmeg, cinnamon, black pepper) and benzo[a]pyrene (produced during cooking, especially charcoal broiling). By extremely conservative methods, the cancer risk for benzo[a]pyrene has been estimated at 1.5/100,000 at high levels of consumption.

It has been stated that exposure to carcinogens is an unavoidable fact of life, but that the levels in foods are so low that further reductions would not have a significant effect on cancer incidence. This "bad news" is offset by the "good news" that there are probably many more anticarcinogens in natural foods than there are carcinogens. In one study of human dietary habits, individuals who ate meat but not vegetables on a daily basis had a colon cancer risk (per 100,000 population) of 18.43, those who ate vegetables but not meat on a daily basis had a risk of 13.67, whereas those who ate both had a risk of only 3.87. Vitamins A, C and E and carotenoids have been shown to be protective against cancer, probably because of their antioxidant properties. Dietary fiber is protective, and even meat has been shown to have anticancer properties. Indoles and isothiocyanates present in cruciferous vegetables such as cabbage, cauliflower, broccoli and Brussels sprouts have been shown to be anticarcinogenic.

Carcinogenic mycotoxins abound in nature, and these will be dealt with in Chapter 10.

REVIEW QUESTIONS

For Questions 1 to 13, use the following code: answer A if statements 1, 2 and 3 are correct; answer B if statements 1 and 3 are correct; answer C if statements 2 and 4 are correct; answer D if statement 4 only is correct; answer E if all statements (1,2,3,4) are correct.

1. With regard to dethylstilbestrol (DES),
    1. It was used mainly to synchronize the estrus cycle of dairy cows.
    2. It was given to livestock as a growth promotant in feed or as subcutaneous pellets.
    3. In North America and Great Britain, its use as a feed additive for livestock resulted in significant human health problems.
    4. It was given parenterally to pregnant women to prevent impending abortion.

2. Human health problems associated with diethylstilbestrol (DES)
    1. Have been reported in some countries in infants who were fed formula containing high levels of DES.
    2. Affect women who were exposed in utero to high levels from their mother's blood.
    3. Generally involve abnormalities of the genitourinary system.
    4. Have caused widespread problems in people who eat meat containing pellet residues.

3. With regard to Salmonella infections,
    1. It is commonly transferred to humans by eating or handling undercooked meat.
    2. It is most often a problem for the general public.
    3. It sometimes involves multiple drug-resistant strains.
    4. It has a high mortality rate in normal individuals.

4. Which of the following statements is/are true?
    1. Estrogens, including DES, have been shown to be carcinogenic.
    2. Women exposed to DES *in utero* have an increased incidence of cervicouterine deformities.
    3. DES is not metabolized significantly by the human placenta or fetus.
    4. The human fetus cannot deactivate natural estrogens.

5. Regarding the use of antibiotics in agriculture,
    1. They are used exclusively for treating infections in animals.
    2. They may be used prophylactically to prevent infections in animals.
    3. There are no regulations governing their use in agriculture.

4. Very low levels added to feed may have a growth-promoting effect.
6. Multiple Drug Resistance is
   1. Characteristic of Gram negative enteric bacteria.
   2. Possible even when the organism has not contacted all of the anti-infective agents to which it is resistant.
   3. Determined by extrachromosomal DNA.
   4. Seen only in farm livestock exposed to antibiotics.
7. The process of bacterial conjugation
   1. Refers to the release of DNA from lysed cells.
   2. Requires the participation of plasmids.
   3. Refers to the transfer of DNA by phage viruses.
   4. Requires the participation of an "F" (fertility) factor.
8. Which of the following statements is/are true?
   1. Salmonella infection is rarely a problem in nursing homes.
   2. The rank order of frequency for resistant forms of intestinal infections is general hospital > psychiatric hospital > extrahospital environment.
   3. Multiple Drug Resistance occurs only in enteric organisms from animals.
   4. The frequency of MDR increases with antibiotic exposure.
9. The legal definition of food additives in most countries includes
   1. Vitamins.
   2. Food colors.
   3. Spices.
   4. Nitrates.
10. Which of the following substances used in or on foods has/have been associated with a high degree of allergic reactions?
    1. Tartrazine
    2. Carageenen
    3. Sodium metabisulfite
    4. Saccharin
11. Which of the following statements regarding nitrosamines, nitrates and nitrites is/are true?
    1. The major source of nitrosamines for people is the nitrates and nitrites used as meat preservatives.
    2. Nitrosamines have been shown to be carcinogens in animals.
    3. There are no natural food sources of nitrates and nitrites.
    4. Nitrates and nitrites inhibit the growth of *Clostridium botulinum*, the cause of ptomaine poisoning.
12. Regarding artificial sweeteners,
    1. Individuals with phenylketonuria should avoid aspartame.

2. There is no convincing evidence that aspartame in normal amounts causes behavioral problems.
3. Saccharine has been shown to cause bladder cancer in rats exposed in utero to very high levels.
4. Cyclamate or a metabolite is a proven carcinogen for humans.

13. Toxic oil syndrome
    1. Involves respiratory distress, eosinophilia and myalgia.
    2. Produced a high mortality rate in an outbreak in Spain.
    3. Resembles a disease traced to tryptophan consumption in the U.S.
    4. May be due to a reaction product of aniline and a component of a cheap, low-grade mixture of cooking oils.

■ ANSWERS

1. C; 2. A; 3. B; 4. A; 5. C; 6. A; 7. C; 8. C; 9. C; 10. B; 11. C; 12. A; 13. E.

■ CASE STUDY #17

Three men were admitted to the emergency department of a large hospital, all suffering from similar symptoms. Inquiry revealed that they had just finished dining in a nearby Chinese restaurant. All had a severe headache, cardiac palpitations, facial flushing, vertigo (dizziness), and perfuse sweating. The symptoms commenced shortly after finishing their meal. Their faces and chests were reddened, blood pressures slightly below normal and their pulses were quickened.

Q. What is the likely portal of entry of this apparent toxicant?

Q. Is this likely due to
   a. food poisoning of bacterial origin?
   b. contamination with a pesticide?
   c. a food additive?

Q. In a call to the restaurant in question, the attending physician inquired as to whether the trio had consumed mushrooms or fish. Why was this question asked?

The response to the above question was no. It was revealed that many other patrons had eaten similar food without trouble, the exception being pork chow yuk. These three were the only ones to consume this dish.

Q. How can this information be helpful?

■ FURTHER READING

Albert, A., *Xenobiosis: Foods, Drugs and Poisons in the Human Body*, Chapman and Hall, London, 1987.
Ames, B.N., Magaw, R. and Gold, L.S., Ranking possible carcinogenic hazards. *Science*, 236, 271, 1987.
Apfel, R.J. and Fisher, S.M., *To Do No Harm*. Yale University Press, New Haven, 1984.
Asulfame — a new artificial sweetener. *Med. Lett. Drug Ther.*, 30, 116, 1988.
Berkelman, R.L., Bryan, R.T. et al., Infectious disease surveillance: a crumbling foundation. *Science*, 264, 368, 1994.
Centers for Disease Control. Analysis of L-tryptophan for the etiology of eosinophilia-myalgia syndrome. *Morbid. Mortal. Weekly Rep.*, 39, 581, 1990.
Colton, T., Greenberg, E. et al., Breast cancer in mothers prescribed dithylstilbestrol in pregnancy: further followup. *JAMA*, 269, 2096, 1993.
Davies, J., Inactivation of antibiotics and the dissemination of resistance genes. *Science*, 264, 375, 1994.
Ehrhardt, A.A., Meeyer-Bahlburg, H.F.L. et al., Sexual orientation after prenatal exposure to exogenous estrogens. *Arch. Sexual Behav.*, 14, 57, 1985.
Everett, M., Growth hormones will certainly affect the meat industry. *Vet. Mag.*, April, 14–15, 1989.
Flamm, W.G., Pros and cons of quantitative risk analysis, in *Food Toxicology: A Perspective on the Relative Risks*, Taylor, S.L. and Scanlon, R.A. (Eds.), Marcel Dekker, New York, 1989, chap. 15.
Freydberg, N. and Gortner, W.A., *The Food Additives Book*. Bantam Books, Toronto, 1983.
Gans, D.A., Behavioral disorders associated with food components, in *Food Toxicology: A Perspective on the Relative Risks*, Taylor, S.L. and Scanlon, R.A. (Eds.), Marcel Dekker, New York, 1989, chap. 9.
Grasso, P., Carcinogens in food, in *Toxic Hazards in Food*, Conning, D.M. and Lansdown, A.B.G. (Eds.), Croom Helm, London, 1983, chap. 4.
Hall, R.L., Dull. B.J. et al., Comparison of the carcinogenic risks of naturally occurring and adventitious substances in food, in *Food Toxicology: A Perspective on the Relative Risks*, Taylor, S.L. and Scanlon, R.A. (Eds.), Marcel Dekker, New York, 1989, chap. 8.
Hayes, A.W. (Ed.), *Principles and Methods of Toxicology*, 2nd Edit. Raven Press, New York, 1989.
Hodgson, E. and Levy, P.I., *Modern Toxicology*, 1st Edit. Elsevier, New York, 1987.
Holmberg, S.D., Osterholm, M.T. et al., Drug-resistant salmonella from animals fed antimicrobials. *New Eng. J. Med.*, 311, 617, 1984.
Hunter, B.T., *Fact Book on Food Additives*, Keats Publishing, New Caanan, 1972.
Kilbourne, E.M., Posada de la Paz, M. et al., Toxic oil syndrome: a current clinical and epidemiologic summary, including comparisons with the eosinophilia-myalgia syndrome. *J. Am. Coll. Cardiol.*, 18, 711, 1991.
Klaassen, C.D., Amdur, M.O. and Doull, J. (Eds.), *Casarett & Doull's Toxicology*, 3rd Edit. Collier Macmillan, Toronto, 1986.

Loizzo, A., Gatti, G.L. et al., Italian baby food containing diethylstilbestrol three years later. *Lancet*, 1, 1014, 1984.

Moats, W.A. (Ed.), Agricultural use of antibiotics. *Am. Chem. Soc. Symp.*, Ser. 320, Washington, 1986.

Munro, I.C., A case study: the safety evaluation of sweeteners, in *Food Toxicology: A Perspective on the Relative Risks*, Taylor, S.L. and Scanlan, R.A. (Eds.), Marcel Dekker, New York, 1989, chap. 6.

Oberrieder, H.K. and Fryer, E.B., College students' knowledge and consumption of sorbitol. *J. Am. Dietetic Assoc.*, 91, 715, 1991.

Pariza, M.W., A perspective on diet and cancer, in *Food Toxicology: A Perspective on the Relative Risks*, Taylor, S.L. and Scanlan, R.A. (Eds.), Marcel Dekker, New York, 1989, chap. 1.

Philp, R.B., Real and potential problems associated with the use of anti-infective agents in domestic livestock. *Brief to the Ontario Ministry of Agriculture and Food Prepared for the Ontario Veterinary Association*, 1979.

Pratt, W.B. and Ferkety, R., *The Antimicrobial Drugs*. Oxford University Press, New York, 1986.

Report of the DES task force. *JAMA*, 255, 1849, 1986.

Schacham, P., Philp, R.B. and Gowdey, C.W., Antihematopoietic and carcinogenic effects of bracken fern (*Pteridium aquilinum*) in rats. *Am. J. Vet. Res.*, 31, 191, 1970.

Spika, J.S., Waterman, S.H. et al., Chloramphenicol-resistant salmonella newport traced through hamburger to dairy farms. *New Eng. J. Med.*, 316, 565, 1987.

Sun, M., Use of antibiotics in feed challenged. *Science*, 226, 144, 1984.

Thomas, A.P., Precocious development in Puerto Rico. *Lancet*, 1, 1299, 1982.

Use of EDDI in cattle. *(Canadian) Health Protection Branch Information Letter No. 757*, Feb. 10, 1989.

Williams, D.E., Dashwood, R.H. et al., Anticarcinogens and tumor promoters in foods, in *Food Toxicology: A Perspective on the Relative Risks*, Taylor, S.L. and Scanlan, R.A. (Eds.), Marcel Dekker, New York, 1989, chap. 5.

Wolff, A.H. and Oehme, F.W., Carcinogenic chemicals in food as an environmental health hazard. *J. Am. Vet. Med. Assoc.* 164, 623, 1974.

9 PESTICIDES

So naturalists observe, a flea has smaller fleas that on him prey;
And these have smaller still to bite 'em;
And so proceed ad infinitum.
Jonathan Swift

INTRODUCTION

The term pesticides refers to a large body of diverse chemicals that includes insecticides, herbicides, fungicides, rodenticides and fumigants employed to control one or more species deemed to be undesirable from the human viewpoint. They are of environmental concern for two main reasons. Although considerable progress has been made with respect to their selective toxicity, many still possess significant toxicity for humans, and many are persistent poisons, so that their long biological T1/2 allows bioaccumulation and biomagnification up the food chain (see Chapter 3). There is thus the possibility that they may enter human food supplies as well as constitute an ecological hazard. By their very nature, pesticides must have an impact on any ecosystem, since they are designed to modify it by their selective elimination of certain species. As is always the case in considering chemicals used in the service of humankind, there is a complex risk-benefit equation that must be taken into account in making decisions regarding the use of pesticides. There is no question that they have increased agricultural production when used properly, and they have, in the past, been highly effective in controlling the insect vectors of human diseases like malaria and yellow fever spread by mosquitoes, and African sleeping sickness which affects both humans and animals and which is spread by the tsetse fly. As shall be seen, however, these gains have not been without their problems.

Efforts to control agricultural pests probably evolved in parallel with cultivation techniques. Early methods included manual removal of weeds and insects, rigorous hoeing to prevent weed growth and the use of traps for animal and insect pests. The first chemical controls to be used against agricultural pests were the arsenical compounds. In 1910, Erlich discovered that arsphenamine was an effective treatment for syphilis. This was the first chemotherapeutic agent for a bacterial infection and the first example of a structure-activity relationship. It opened the door on the entire field of chemical control of both infections and pests. Paracelsus had introduced the use of inorganic arsenicals, notably arsenic trioxide (As_2O_3, white arsenic) into medicine in the 16th century, but its use was limited by its extreme toxicity. Ehrlich's discovery revived interest in these compounds, and in 1824 the Colorado potato beetle was discovered east of the Rockies and its eastward spread accelerated the search for an effective control. As_3O_3 was found to be effective and came into widespread use. Other arsenicals were developed, including Paris green (copper arsenite) which is still used as slug bait. Being a heavy metal, arsenic is persistent in the environment, the significance of which was not appreciated when it was being widely used.

Natural-source insecticides also evolved fairly early on. Certain plants have been employed as fish poisons in Southeast Asia and in South America for centuries, and in 1848 a decoction of derris root was used to control an insect infestation in a nutmeg plantation in Singapore. By 1920, large amounts were being imported into North America. The active ingredient is rotenone, and it has the advantages of low mammalian toxicity and short T1/2 in nature. Pyrethrum flowers (chrysanthemums) have been known for their insecticidal properties for centuries. Commercial manufacture began in 1828. In 1945, the U.S. imported 13.5 million pounds. By 1954, this had fallen to 6.5 million because of the widespread use of DDT, the banning of which has led to a resurgence of use of pyrethrin compounds. Nicotine sulfate (Blackleaf 40 is a 40% solution) from tobacco is used to control aphids and other insects. It has a short biological T1/2 but significant mammalian toxicity.

The mechanization of farming led to a second agricultural revolution by making possible the planting and cultivation of vast tracts of land. Pest control techniques also changed from the small scale operations of the past to include mechanized spraying from the ground and the air. This involved a marked increase in the use of pesticides, and it coincided with the introduction of the first, modern, synthetic insecticide, DDT.

Dichlorodiphenyltrichloroethane, or DDT, was first synthesized in 1874, but its insecticidal properties were not recognized until 1939. Its structural formula is shown in Figure 31. Its first major use occurred in Sicily in 1943, where it was used to halt an epidemic of tick-borne typhus.

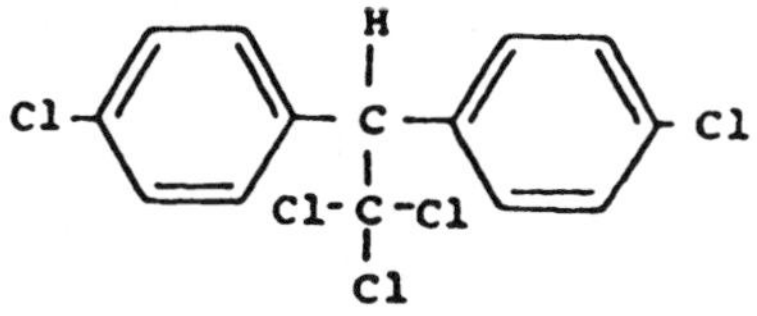

Figure 31 DDT, chemically 1,1,1-trichloro-2,2-bis (p-chlorophenyl) ethane.

Sometimes called the grandfather of all chlorinated aromatic hydrocarbons, DDT was the first of such agents to arouse environmental concern. Rachel Carson's *Silent Spring* called attention to the ecological damage caused by DDT and led to its banning in the U.S. and Canada in 1972. Prior to that, however, its use had led to the eradication of malaria in 37 countries and dramatically reduced its incidence in a further 80, providing relief to 1.5 billion people. Its effectiveness in controlling agricultural pests, coupled with its low mammalian toxicity (oral LD_{50} 113 mg/kg, dermal LD_{50} 2.5 gm/kg), resulted in extensive use in North America. U.S. production reached 50,000 metric tonnes annually. The availability of cheap, surplus aircraft after World War II resulted in the spraying of huge areas to control not only agricultural pests but human ones as well. Organochlorines, including the cyclodienes, dominated the insecticide field until the early 1960s, when organophosphates and carbamates were developed. These, plus the development of more disease-resistant hybrid crops, led to the "Green Revolution" of the 1960s, with dramatic increases in food production.

CLASSES OF INSECTICIDES

Organochlorines (Chlorinated Hydrocarbons)
As already discussed, the parent compound of this group is DDT. Its human toxicity is extremely low. In one rather heroic experiment, volunteers were fed 35 mg/day for up to 25 months without obvious ill effects. Another study of 35 male workers who had DDT levels in fat and liver 80× the American average, and who had worked in a manufacturing plant for up to 19 years, showed no ill effects. DDT is, however, a potent inducer of cytochrome P-450 hepatic microsomal enzymes and may thus affect the rate of biotransformation of other chemicals and drugs. Extremely high doses cause neurological signs and symptoms including numbness of the tongue, lips and face, dizziness, hyperexcitability, tremor and convulsions.

DDT has very high lipid solubility, and it is sequestered in body fat. Virtually everyone who was alive after 1940 has DDT in their fat. In the 1960s, significant amounts were found in people all over the world from Sri Lanka to North America. In 1970, the mean concentration in human fat was 7.88 ppm. After the ban, it fell to 4.99 in 1975. There is no evidence that chronic exposure to DDT has resulted in any health problems. In insects, DDT opens up ion channels to prevent normal axonal repolarization. Disorganized neuronal function leads to death.

Other life forms are not as resistant as humans. Fish are extremely vulnerable, and die-offs occurred after heavy rains washed DDT into streams and lakes. Deformities also occur. Predatory birds at the top of the food chain are very vulnerable as well. Reproduction is disturbed in a number of ways. DDT induces cytochrome P-450 to increase estrogen metabolism, and DDT itself has estrogenic activity which affects fertility. Ca^{2+}-ATPase is inhibited as is calcium deposition in eggshells. This effect is largely due to stable metabolites, notably DDE (dichlorodiphenyldichloroethane). Some bird species are only now recovering. The limited use of DDT against the tussock moth was reapproved in the U.S. in 1974, and its use in malarial areas has continued without interruption, so that DDT exposure on a worldwide basis still occurs.

A subgroup of the organochlorines are the cyclodienes. This group includes aldrin, dieldrin, heptachlor and chlordane. Their mechanism of insecticidal action is the same as for DDT, but their toxicity for humans is much greater because of more efficient transdermal absorption. Signs of excessive CNS excitation and convulsions occur before less serious signs appear. Several deaths, mostly in those who handle the pesticide, have occurred. These agents too are persistent in the environment. There is concern about their potential for carcinogenicity, since this has been shown in some animals. However, Ribbens reported on a study of 232 male workers who had been exposed to high levels of cyclodienes in a manufacturing plant in Holland for up to 24 years (mean 11 years). Mortality and cancer incidence were compared to the means for the Dutch male population of the same age group. The observed mortality in the group was 25, which was significantly lower than the expected mortality of 38. Nine of the deaths were from cancer, as opposed to an expected incidence of 12. These workers had been exposed to very high levels of cyclodienes in the early days of manufacture, with recorded dieldrin blood levels of up to 69 µg/l at some time in their history.

Other organochlorines include methoxychlor, lindane, toxaphene, mirex and chlordecone. Mirex and chlordecone are extremely persistent, toxic to mammals (CNS toxicity) and carcinogenic in animals. They also induce cytochrome P-450. They are no longer used in North America. Lindane shares the same toxicity but is much less

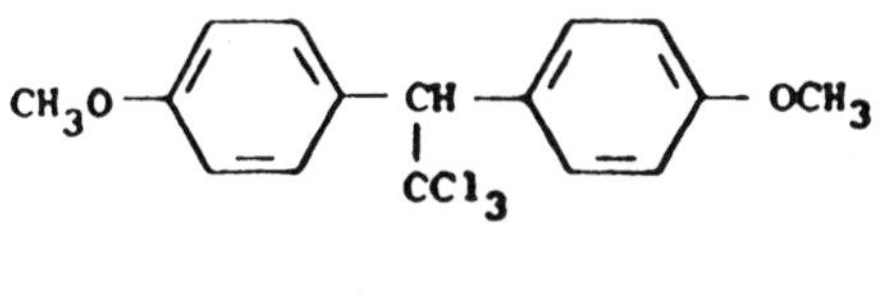

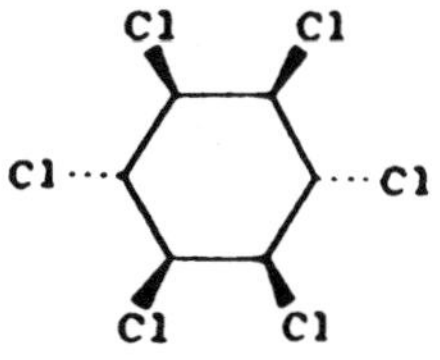

METHOXYCHLOR

LINDANE

Figure 32 Methoxychlor and lindane.

persistent, and it is used to treat head lice. Lindane (chemically 1,2,3,4,5,6-hexachlorocyclohexane) is the active isomer of benzene hexachloride. Toxaphene induces liver tumors in mice and is fairly toxic, and its use is declining. Methoxychlor is similar to DDT, but it is much less persistent and less toxic to mammals, which can metabolize it. It also is stored in fat to a much lesser degree. Its formula, along with that of lindane, is shown in Figure 32.

Organophosphates

These are the most frequent cause of human poisonings by an insecticide. The group includes parathion, dichlorvos and diazinon. They all act as irreversible inhibitors of acetylcholinesterase, so that the neurotransmitter acetylcholine is not inactivated following its release from the nerve terminal. Signs and symptoms are those of a massive cholinergic discharge and include dizziness and disorientation, profuse sweating, profuse diarrhea, constricted pupils and bradycardia (slowing of the heart), possibly with arrhythmias. Parathion has a dermal LD_{50} of 21 mg/kg and an oral LD_{50} of 13 mg/kg in male rats, but the NOEL in both rats and humans is only 0.05 mg/kg. Parathion itself is not toxic, but it is transformed in the liver to paraoxon, its oxygen analogue (see Figure 3, Chapter 1).

The following is a typical case history of organophosphate poisoning: A 52-year-old white, male farmer was admitted to a hospital emergency department following a highway accident in which his tractor collided with the rear of a motor vehicle about to make a turn. He incurred numerous lacerations and contusions and a fractured right humerus. He was restless, incoherent and required physical restraint. His pupils were bilaterally constricted; his heart rate was 55 beats/min, and he was sweating profusely. His clothing had a strong, chemical odor. His wife volunteered that he had several episodes of visual difficulty over the preceding 2 weeks. Further questioning revealed that he had been spraying organophosphate insecticides during this period (organophosphate poisoning is frequently delayed). Atropine was given intravenously in repeated small doses until the signs of cholinergic discharge abated. Another drug that

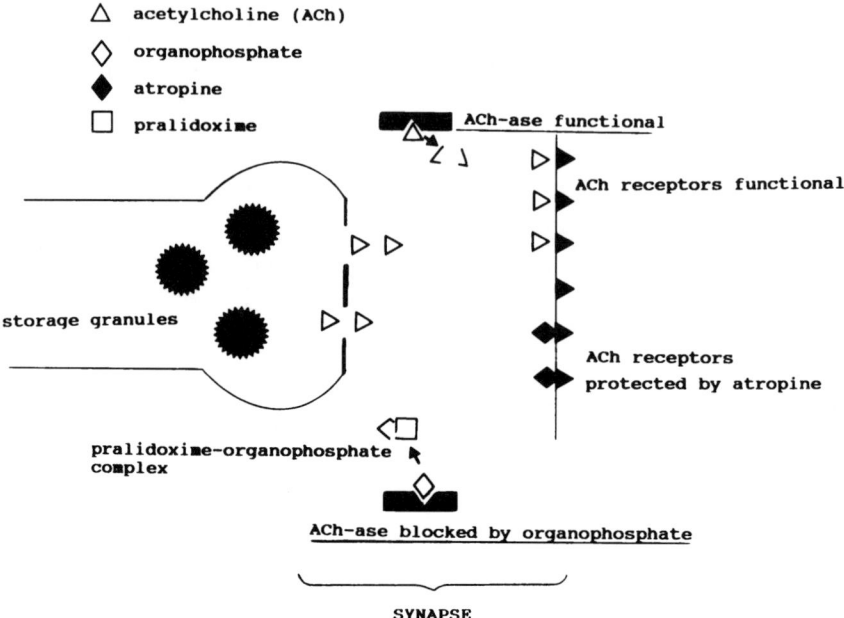

Figure 33 Sites of action of organophosphate insecticides, atropine and pralidoxime. Although the neurotransmitter site is labelled a synapse, atropine is primarily a muscarinic receptor blocking agent, acting at parasympathetic effector junctions. Acetylcholinesterase is present there, as well as in all ganglia, at the neuromuscular junction, the brain and the adrenal gland.

can be used is pralidoxime which complexes with the phosphate component of the organophosphate and releases the cholinesterase. The principal advantage of the organophosphates is their short life in the environment. The site of action of organophosphates, atropine and pralidoxime are shown in Figure 33.

Carbamate Insecticides

Carbamates (e.g., carbaryl) are also inhibitors of acetylcholinesterase, but they do not require metabolic activation and they are reversible. They are not persistent in the environment. Because they lack the phosphate group, pralidoxime cannot be used for treatment of poisoning. In fact, it is contraindicated, because it may tie up more reactive sites on the enzyme and increase the degree of inhibition. This group includes aldicarb and carbaryl. The dermal LD_{50} for aldicarb in male rats is 3.0 mg/kg. It is also fairly toxic for humans. Although these agents are generally not persistent in the environment, aldicarb may be an exception. Under certain conditions (sandy soil over aquifers), it may reach water supplies and persist for a considerable time. In Long Island, New York, it has been estimated that the levels of 6 ppb may persist for up to 20 years.

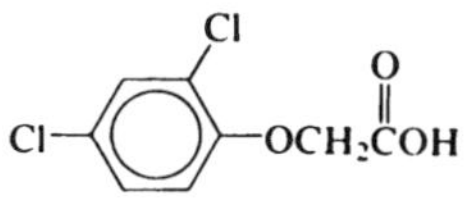

2,4-D
(2,4-dichlorphenoxyacetic acid)

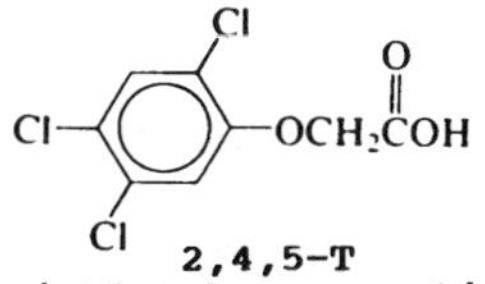

2,4,5-T
(2,4,5-trichlorphenoxyacetic acid)

Figure 34 Chemical structures of 2,4-D and 2,4,5-T.

Botanical Insecticides

The more common botanical insecticides were discussed briefly above. While it is commonly felt that natural-source insecticides are safer than synthetic ones (another example of the "nature knows best" syndrome), this is not necessarily so. Pyrethrins and rotenone have oral LD_{50} of about 600–900 mg/kg and 100–300 mg/kg, respectively. Nicotine is quite toxic, with an oral LD_{50} of 10–60 mg/kg. The main problem with pyrethrins has been the rapidity with which they are destroyed in the environment. Newer ones have been isolated with longer T1/2s to permit more effective kills.

■ HERBICIDES

Chlorphenoxy Compounds

These agents, chararacterized by 2,4-D and 2,4,5-T, act as growth hormones, forcing plant growth to outstrip the ability to provide nutrients. They are employed as a variety of salts and esters. The acute toxicity of these agents is relatively low, with LD_{50} of 300–>1,000 mg/kg reported for several species of mammals. The dog may be more sensitive (LD_{50} 100 mg/kg). Ventricular fibrillation appears to be the immediate cause of death. Acute toxicity in humans is manifested largely as chloracne.

The main concern about 2,4-D and 2,4,5-T is the likelihood of their contamination with dioxin (TCDD). The chemical structures of these compounds are shown in Figure 34. Chapter 5 discusses the toxicity of TCDD in more detail.

Dinitrophenols

Several substituted dinitrophenols are used as herbicides, the most common probably being Dinoseb (Figure 35). It has been reported to have an LD_{50} of 20–50 mg/kg in rats. Dinoseb, first registered in 1947, is out of favor because handlers may be at considerable risk for teratogenic effects, cataracts and male reproductive disturbances, even when protective clothing is worn. The U.S. EPA suspended all use in 1987. 4,6-dinitro-*o*-cresol (DNOC, Figure 35) has caused acute poisoning in humans, with signs and symptoms including nausea, vomiting, restlessness, and flushing of the skin, progressing to collapse

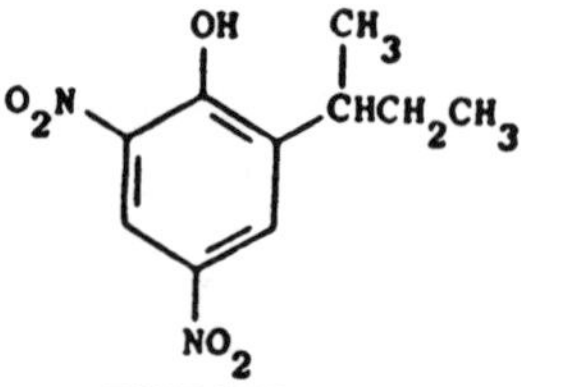

Figure 35 Chemical structures of dinoseb (2-sec-butyl-4,6-dintrophenol) and DNOC (4,6-dinitro-o-cresol).

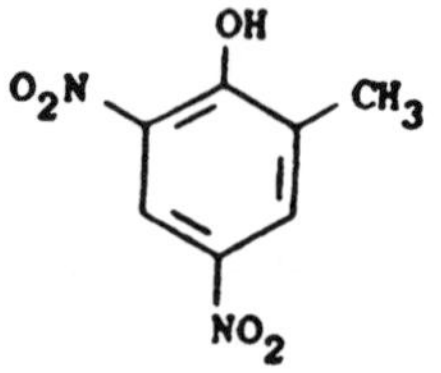

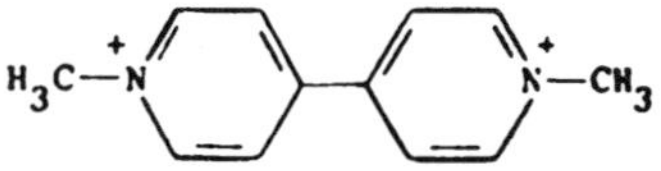

Figure 36 Chemical structures of paraquat and diquat.

and coma. Hyperthermia may occur. Death may ensue in 24–48 hr. Uncoupling of oxidative phosphorylation is probably the mechanism of toxicity. Atropine is contraindicated, because there is no anticholinesterase activity and the CNS effects of atropine may complicate the outcome. Treatment is symptomatic and includes ice baths to reduce fever, fluids intravenously and the administration of O_2.

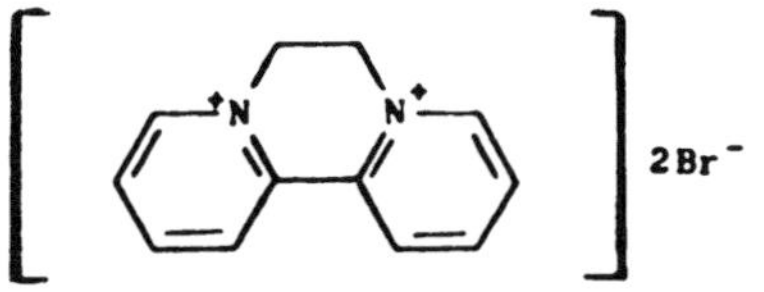

Bipyridyls

Paraquat and diquat are the most familiar members of this group (see Figure 36). Both are toxic, but their toxicity differs. The principal organ of toxicity for paraquat is the lungs, although liver and kidney also may be damaged. Respiratory failure may be delayed for several days after the ingestion of paraquat. It appears to be selectively concentrated in the lungs by an energy-dependent system. Paraquat is believed to undergo conversion to superoxide radical (O_2^-), which causes the formation of unstable lipid hydroperoxides in cell membranes. Widespread fibroblast formation occurs, and O_2 transfer to capillary blood is impaired. Treatment consists of attempts to remove or neutralize any paraquat remaining in the gastrointestinal tract by gastric lavage, cathartics and fuller's earth as an adsorbant. In complete lung failure, double lung organ transplant offers the only hope for recovery.

In contrast, diquat toxicity is centered on the liver, kidney and gastrointestinal tract. Superoxide anion formation is believed to play a role also in these organs. Poisoning with paraquat is far more common, and it has been used as an instrument of suicide on numerous occasions.

Carbamate Herbicides
Unlike the insecticide carbamates, the herbicides do not possess anticholinesterase activity. They have low acute toxicity. Dithiocarbamates are used as fungicides and have similar low acute toxicity; LD_{50} for these agents are in the gram per kilogram range for rodents.

Triazines
This group, typified by atrazine, also is characterized by low acute toxicity.

Amitrole is a herbicide somewhat related to the triazines. It has similar low acute toxicity, but it has peroxidase-inhibiting activity, and it has been associated with tumor formation in the thyroid in rats fed the chemical for 2 years.

FUNGICIDES

A wide variety of agents has been used for their fungicidal properties, some of them quite toxic. Seed grains treated with mercurials have sometimes entered the human food supply with disastrous results (see Chapter 6). Pentachlorophenol and hexachlorobenzene are halogenated hydrocarbons with the toxicity typical of that group (see Chapter 4). Thiabendazole is a fungicide of low toxicity as evidenced by the fact that it is also used as an anthelmintic in domestic animals and humans for the eradication of roundworms.

Dicarboximides
Captan and folpet are agents of some concern. Structurally similar to thalidomide, they have been shown to possess similar teratogenic properties in the chick embryo. Captan has been shown to be mutagenic, carcinogenic and immunotoxic in animals, and the EPA has judged folpet to be a probable human carcinogen with a lifetime risk of cancer of 2 per million for lettuce and small fruits and a total of 5.5 per million when all food sources are combined.

NEWER BIOLOGICAL CONTROL METHODS

The earliest form of biological control no doubt was the development of strains of plants and animals with a high degree of resistance to disease, through selective breeding. Observant farmers probably began this process soon after the domestication process began, and it continues today. Over 40 years ago, as a high school student, the author worked with Professor Waddell who developed, at the Ontario Agricultural College, the first strains of wheat to be resistant to wheat rust, a fungal infestation. Recently, a strain of American elms with a high degree of resistance to Dutch Elm Disease has been developed. Ladybugs have been bred and released to control the cottony cushion scale on oranges in California, and *Bacillus thuringiensis* is used to control forest pests.

One of the earliest, "high-tech" biological controls was developed in the 1950s and involves sterilization by radiation of millions of male insects which are then released to mate with the females. In species in which the female only mates once, this results in a high frequency of infertile unions with a resulting decline in the insect population. This method was first used successfully to control the screw-worm fly in the southern U.S. This fly lays its eggs in wounds in the skin of cattle and other livestock. The larvae then live on the flesh of the unwilling host. By 1966, the screw-worm had been successfully eradicated in the U.S. and northern Mexico. It recently resurfaced in Libya, creating a political dilemma for the U.S. Withholding technological assistance could result in massive infestations throughout Africa (the fly will also lay its eggs in wounds on humans), but the alternative was to offer help to Quaddafi. Humanitarian considerations prevailed. This form of biological control has also been used more recently to control the Mediterranean fruit fly in California.

Analogues of insect hormones have been developed that are highly specific to a given species. These hormones trigger the moulting metamorphosis in the larval stage so that the larva cannot develop normally and dies.

Pathogenic bacteria exist which can be cultured in commercial quantities and released to control specific pests. Some agents have been genetically modified for this purpose, but public concerns about "superbugs" have blocked approval of all but a few of these. Given that there are no known bacteria that are infectious for both insects and mammals (as opposed to insects being vectors for infection), this fear seems unjustified. A more legitimate concern is that beneficial or harmless species may also be attacked by the organisms (see also Chapter 13).

■ GOVERNMENT REGULATION OF PESTICIDES

Most governments have regulations regulating the use of pesticides. The Canadian regulations are fairly typical of those in place in industrialized countries. The Pest Control Products Act, administered by Agriculture Canada, regulates the introduction of new pesticides. The risk-benefit principle is applied to decisions; i.e., the degree of risk must be acceptable in light of the potential benefit to be derived from pest control with the new agent. Its relative safety and effectiveness compared to existing pesticides will influence the decision.

■ PROBLEMS ASSOCIATED WITH PESTICIDES

Development of Resistance

Insects, like microorganisms, possess the most important characteristics for the evolution of resistant strains: an extremely short reproductive cycle and the production of vast numbers of progeny. Most species of insects can go through many generations in one season, producing millions of offspring. There is thus the capacity for multiple, sequential mutations to occur and a good chance that some of these will be resistant to one or more insecticides. The development of resistance requires the presence of the appropriate insecticide to select out the resistant strain (by killing off the susceptible ones) with the means of detoxifying the chemical or excluding it from absorption. Only three years after the introduction of DDT, house flies and mosquitoes were showing signs of resistance and, by 1951, DDT, methoxychlor, chlordane heptachlor and benzene hexachloride (of which lindane is the active isomer) no longer had any effect on houseflies which proliferated abundantly. By the end of 1980, 428 species of insects and acarines (mites, ticks) were classified as resistant.

In the pre-DDT era, relatively few species developed resistance. This has been attributed to the multisite mechanisms of action of earlier pesticides (making single mutation resistance unlikely) and to their ionic nature, making detoxification by metabolism nearly impossible.

Closely related insecticides may be detoxified by the same mechanism and generally act at the same target site, so that if resistance evolves to one, either by the evolution of a detoxification process or by modification of the target molecules, cross resistance to the others will occur. This type of resistance tends to be under the control of a single gene allele or closely linked genes. Cross resistance to DDT and methoxychlor, aldrin and heptachlor has developed in this manner.

Multiple Resistance

In common with bacteria, protozoa and cancer cells, insects can develop resistance to several insecticides. This is referred to as multiple resistance. It is the result of the existence of several, independent gene alleles producing resistance to unrelated agents (e.g., organochlorines and synthetic pyrethrins, called pyrethroids) with different modes of action and different detoxification pathways. It can be a very serious problem of insect control. The mechanism of multiple resistance is obviously quite different from that of bacteria, which is discussed in Chapter 8.

Nonspecificity
Broad spectrum pesticides, as most are, make no distinction between true pests and species that are harmless or even beneficial. They therefore may disrupt the natural competition between species to permit the proliferation of one previously held in check, or as has been observed, they may kill off predator species and permit the expansion of a prey species which then becomes a pest. This has happened with spider mites.

Environmental Contamination
A greater danger than direct toxicity to humankind is probably the contamination of the environment and the subsequent bioaccumulation and biomagnification which occur with persistent pesticides. These chemicals may end up in soil, water, air or all three, depending on their characteristics (see Chapters 3 and 5). The Great Lakes are accumulating hazardous chemicals as a result of agricultural runoff and industrial discharges, frequently accidental ones. It must be stressed, however, that the greatest source of chemical contamination is residential sewage. Even after treatment, phosphates and other household chemicals may enter the water system. The water table in many areas has been contaminated with the herbicide atrazine, commonly used in corn fields.

Callous disregard for the environment can be the result of greedy individuals attempting to increase their profit margin. There have been numerous cases in Toronto of trucks dumping toxic wastes in Lake Ontario after dark, in violation of provincial and municipal laws. In the U.S., careless dumping of chlordecone in the James River by a chemical company resulted in a ban on fishing. This cyclodiene is used in ant and roach baits.

■ BALANCING THE RISKS AND THE BENEFITS

The widespread use of pesticides means that there are trace quantities present in or on almost all foodstuffs. Major advances in analytical techniques over the last 20 years mean that chemicals can now be detected at levels never before possible. Headlines proclaiming that dioxins (or PCBs, etc.) have been detected in Lake Erie fish seldom go on to say that the quantities were at the parts-per-billion level. The detection of dioxins in milk, leached from the carton paper, is an example of this. Actual levels were comparable to a drop in an Olympic-sized swimming pool.

Nonetheless, there is a growing feeling in the public that use of pesticides should be greatly curtailed because the risks are unacceptable or, what is almost as bad, unconfirmed. In recognition of this, the Ontario Ministry of Agriculture and Food has established the Foodsystems 2002 program which will attempt to reduce the use of pesticides in agriculture in Ontario by 50% by the year 2002. However, the impact of pests on food production is so great that temporary approval is sometimes granted to new agents before all of the required tests have been completed. In 1977, it was discovered that an American testing company, Industrial Bio-Test Laboratories, had misrepresented toxicological data on numerous agents. Of the 405 pesticides registered in Canada, 106 had been approved partly on the basis of Bio-Test data. In 1983, Health and Welfare Canada announced that five of these were to be withdrawn, one being the fungicide Captan, which was found to be teratogenic and carcinogenic.

Other pesticides that have been banned or which are under investigation include

- Chlordane — This cyclodiene was withdrawn in 1986 because it was considered to be an epigenetic tumor promoter. It is still registered for use against termites.
- Alachlor — This herbicide was withdrawn in 1985 because of evidence of carcinogenicity in animals. It was introduced in 1969 and widely used on corn and soya bean crops.
- Cyhexatin — Dow Chemical voluntarily withdrew this insecticide because of evidence of teratogenicity (hydrocephaly) in rabbits.

■ TOXICITY OF PESTICIDES FOR HUMANS

Pesticide applicators are at risk primarily from inhalation and dermal contact with pesticides. Protective clothing and equipment are important means of reducing risk. Nonoccupational poisonings occur largely from oral ingestion of contaminated food, although dermal exposure has resulted in poisonings in infants (the pentachlorophenol treatment of hospital linens), and inhalation exposure from spray drift can occur. Household pesticides can cause poisoning by all three routes.

There have been a few isolated fatalities in North America from acute pesticide poisoning, but elsewhere in the world, many cases of mass poisonings have occurred. Consumption of seed grains treated with hexachlorobenzene and organic mercury have resulted in mass outbreaks of poisoning in Turkey, West Pakistan, Iraq and Guatemala. Accidental contamination of foodstuffs like flour, sugar and grain with parathion and others has occurred in several places around the world.

Effects of long-term exposure to very low levels of pesticides on human health remain conjectural. This does not preclude the possibility that as detection methods and evaluation techniques become even more sophisticated, such effects will be identified. There is also the possibility that contaminants may emerge as a greater risk than the pesticide itself, as was the case with TCDD. There is no doubt that persistent poisons can have a catastrophic effect on the environment, and this is probably the most compelling argument for limiting their use.

There is some hope that nature may be developing her own protective processes against pesticides. There is some new evidence that fields sprayed with the same chemicals year after year may develop a population of bacteria that break down the pesticides and may even adapt to the point that they utilize them as a food source. Tests on prairie soils indicated that the organism *Rhodococcus* breaks down thiocarbamate insecticides in the test tube within 2 hours. It may be possible through gene splicing to develop plants that protect themselves against pesticide residues.

■ REVIEW QUESTIONS

For Questions 1 to 8, use the following code: answer A if statements 1, 2 and 3 are correct; answer B if statements 1 and 3 are correct; answer C if statements 2 and 4 are correct; answer D if only statement 4 is correct; answer E if all statements (1,2,3,4) are correct.

1. Which of the following statements is/are true?
   1. Lindane is commonly used for the control of head lice.
   2. Toxicological testing is usually performed only on the active ingredient of an insecticide.
   3. Lindane is an organochlorine.
   4. A cause-and-effect relationship for progressive chronic disease resulting from prolonged pesticide use is well established.

2. Which of the following is/are true?
   1. The herbicide atrazine has contaminated the water table in many areas.
   2. Eggshell strength is adversely affected by contact with pyrethroids by the female bird.
   3. The fungicide captan is a teratogen and carcinogen in experimental animals.
   4. Natural pesticides are always safer than synthetic ones.

3. The main reason for the carcinogenicity in animals of the herbicides 2,4,-D and 2,4,5,-T is/are
   1. Their action as growth hormones.
   2. Their mutagenicity.

3. Their photosensitizing properties.
4. The presence of dioxin (TCDD) as a contaminant.

4. The mechanism of TCDD carcinogenicity in humans may involve
   1. Its pleiotropic response to the Ah locus.
   2. Its conjugation with glutathione.
   3. Its lack of gene restriction.
   4. Its ability to cause chloracne.

5. The insecticide parathion
   1. Is an organochlorine.
   2. Is biotransformed by a mixed function oxidase to its toxic form.
   3. Biomagnifies in the environment.
   4. Causes symptoms that can be treated with atropine.

6. Which of the following statements concerning DDT is/are incorrect?
   1. It has a low acute LD_{50}.
   2. It is very persistent in the environment.
   3. It is very water-soluble.
   4. It is still used for mosquito control in many places.

7. Which of the following statements is/are true?
   1. Organophosphates are very persistent in the environment.
   2. Organophosphates are inhibitors of acetylcholinesterase.
   3. Carbamate insecticide poisoning can be treated with pralidoxime.
   4. Carbamate insecticides act as reversible inhibitors of acetylcholinesterase.

8. Which of the following statements is/are true about insect resistance to insecticides?
   1. Over 400 species of insects, mites, etc. showed resistance to pesticides by 1980.
   2. Resistance to preorganic insecticides is especially prevalent.
   3. Cross resistance to closely related chemicals may occur.
   4. Resistance is more likely to occur if an insecticide has several sites of action.

ANSWERS

1. A; 2. B; 3. D; 4. A; 5. C; 6. E; 7. C; 8. B.

CASE STUDY #18

A 43-year-old male crop duster was admitted to the emergency department of a rural hospital following an accident in which the aircraft he was attempting to land on a grass strip hit hard, collapsed the undercarriage and nosed over. The pilot suffered numerous lacerations and bruises but no serious injuries. He was restless and incoherent, and he had to be physically restrained. A rapid breath alcohol test was performed and it was negative. His pupils were constricted, his heart rate was slowed, and he was sweating perfusely. His ground assistant volunteered the information that the pilot had complained of visual disturbances on several occasions during the previous few days. His clothing smelled strongly of a chemical.

Q. Some of these symptoms could be due to a head injury or to chemical intoxication. What facts point to the latter?

Q. What information would you want to seek from his assistant?

The ground crew revealed that the pilot had been spraying crops with parathion during the preceding two weeks.

Q. What class of pesticide is this?

Q. What drugs would be indicated for treatment?

Q. How would treatment differ if a carbamate insecticide such as Sevin had been used?

Q. What blood test might assist in confirming the diagnosis?

■ CASE STUDY #19

During the months of June and August of 1993, 26 men, 19–72 years of age, were admitted to three different local hospitals with an array of symptoms that included nausea, vomiting, dizziness, visual disturbances, muscle weakness, abdominal pain, headache, sweating and excessive salivation. The men all worked in apple orchards, 19 different ones in all.

Q. What do these symptoms suggest?

Q. What inquiries would you want to make of these men?

Q. What inquiries would you want to make of the orchard operators?

■ FURTHER READING

Albert, A., *Xenobiosis*, Chapman and Hall, New York, 1987.
Carson, R., *Silent Spring*, Fawcett Crest Books, 1962.
Claus, E.P. and Tyler, V.E., Jr., *Pharmacognosy*, Lea and Febiger, Philadelphia, 1967.

Flanders, R.V., Potential for biological control in urban environments, in *Advances in Urban Pest Management*, Bennett, G.W. and Owens, J.M. (Eds.), Van Nostrand Reinhold, New York, 1986, chap. 6.

Garner, R.J., *Veterinary Toxicology*, Bailliere, Tindall, London, 1957.

Georghiou, G.P. and Saito, T. (Eds.), *Pest Resistance to Pesticides*, Plenum Press, New York, 1983.

Hall, S.H. and Dull, B.J., Comparison of the carcinogenic risks of naturally occurring and adventitious substances in food, in *Food Toxicology: A Perspective on the Relative Risks*, Taylor, S.L. and Scanlan, R.A. (Eds.), Marcel Dekker, New York, 1989, chap. 8.

Klaassen, C.D., Nonmetallic environmental toxicants: air pollutants, solvents and vapors, and pesticides, in *Goodman and Gilman's The Pharmacological Basis of Therapeutics*, Gilman, A.G., Rall, T.W., Nies, A.S. and Taylor, P. (Eds.), Pergamon Press, New York, 1990, chap. 67.

Murphy, S.D., Toxic effects of pesticides, in *Casarett and Doull's Toxicology*, 3rd Edit. Klaassen, C.D. and Amdur, M.O. (Eds.), Macmillan, New York, 1986, chap. 18.

McEwen, F.L. and Stephenson, G.R., *The Use and Significance of Pesticides in the Environment*, Wiley Interscience, 1979.

Palca, J., Libya gets unwelcome visitor from the west. *Science*, 249, 117, 1990.

Ribbens, P.H., Mortality study of industrial workers exposed to aldrin, dieldrin and endrin. *Int. Arch. Occup. Environ. Health*, 56, 74, 1985.

Schneider, M.-J., *Persistent Poisons*, New York Academy of Science, New York, 1979.

Williams D.E., Dashwood, R.H. et al., Anticarcinogens and tumor promoters in foods, in *Food Toxicology*, Taylor, S.L. and Scanlan, R.A. (Eds.), Marcel Dekker, New York, 1989, chap. 5.

10 MYCOTOXINS AND OTHER TOXINS FROM UNICELLULAR ORGANISMS

Anything green, that grew out of the mold,
Was a wonderful drug to our fathers of old.

INTRODUCTION

Mycotoxins are a group of chemically diverse and complex substances present in a wide variety of filamentous fungi (molds). Some mycotoxins may have a survival advantage by virtue of their toxicity to competing organisms in the microenvironment. The biological function of others is unclear, but many have significant biological activity, and several are toxic to mammals. They therefore have significant public health and economic implications. It is estimated that 25% of the world's annual food crops are contaminated by mycotoxins.

Some are of interest to pharmacologists and toxicologists because they have served as research tools to study cell function and to identify various types of neurotransmitters and blocking agents. Poisonous mushrooms are solid (not filamentous) spore-forming fungi that also constitute a health hazard and which have historic, pharmacological significance, but they shall be considered separately in the following chapter.

SOME HUMAN HEALTH PROBLEMS DUE TO MYCOTOXINS

Ergotism

Reports of toxicity from molds are as old as recorded history. The oldest, recorded source of fungus poisoning is ergotism. Ergot is the common name for the fungus *Claviceps purpurpea* that affects cereal grains, especially rye, and that produces a number of very potent pharmacologically active agents that cause toxic reactions when people eat bread made from contaminated flour. Periodic epidemics of ergotism have occurred throughout history, and in medieval times these were often attributed to supernatural causes. The earliest reference to ergot seems to have been on an Assyrian tablet circa 6000 B.C., which refers to "a noxious pustule in the ear of grain". The Parsees, an ancient religious community in India, referred in their writings to noxious grasses that caused women to abort and to die shortly thereafter. The ancient Greeks escaped the scourge of ergotism because they never developed a taste for rye bread, which they referred to by a phrase which translates roughly as "that filthy Macedonian muck". Since rye bread did not reach Europe until after the decline of the Roman Empire, few if any references to ergotism exist in Roman writings.

There was an early association of ergot with St. Anthony of Egypt, who lived some time between 250–350 B.C. He is considered to be the founder of Christian monastic life and spent long sojourns in the desert, experiencing visions and hallucinations frequently involving attacks by Satan in various guises (wild beasts, soldiers, women) who either physically attacked him or tempted him. Contemporary witnesses claimed that he behaved like an individual being physically abused. Because the signs and symptoms of ergot poisoning resembled his attacks, sufferers in the Middle Ages attached religious significance to them, and they came to be known as "St. Anthony's Fire" (or sometimes "Holy Fire"). In 1100, the Monastery of the Hospitallers of St. Anthony was established at La Motte in France, and it became the site of pilgrimages by those afflicted with this malady. The signs and symptoms included intense, burning pain in the extremities followed by a blackened, necrotic appearance, hence, the association with fire. Epidemics of madness in the Middle Ages may also have been the result of ergotism.

Ergot was employed as an abortifacient long before it was known to be the cause of St. Anthony's Fire, and this use continued well into the 19th century when it was finally abandoned because of its highly toxic nature. One of the components of ergot is still employed to stop postpartum hemorrhage because it constricts blood vessels and contracts the uterus. To summarize, the symptoms of ergotism include drowsiness, nausea, vomiting, muscle twitch, staggering, gangrene, hallucinations and abortion. Ergot contains a veritable potpourri of pharmacological agents that were characterized by Sir Henry Dale early in this century and that contributed to the development of many new drugs.

The active components of ergot, the ergot alkaloids, are all derivatives of lysergic acid. They bear some molecular resemblance to adrenaline (epinephrine), dopamine and serotonin, all central neurotransmitters. Methysergide is the precursor of LSD. Two derivatives still have medicinal application. Ergometrine (ergonovine) is the one used to control postpartum bleeding. Ergotamine is a powerful vasoconstrictor responsible for the gangrene of ergotism. In very small doses, it may be used to treat migraine.

Ergotism may also affect farm livestock, causing similar signs and, in addition, loss of milk production in cattle and necrotic combs, feet and beaks in poultry.

Aleukia

Aleukia (literally "absence of white cells") occurs when millet and other grains that have become moldy are consumed. The toxin is a tricothecene from species of *Fusaria,* and it damages bone marrow. Because of severe food shortages in Russia during World War II, the eating of moldy grain resulted in several epidemics of aleukia, including a very large one in 1944. The mortality rate of those with marrow damage was 60%. Other symptoms included hemorrhages in the skin and mucous membranes. The condition also is referred to as alimentary toxic aleukia.

Aflatoxins

Aflatoxins are a family of heterocyclic, oxygen-containing compounds secreted by the molds *Aspergillus flavus* and *A. parasiticus.* These molds grow abundantly on many kinds of plants in very hot conditions. Peanuts are especially prone to infection. Taste is not affected, and the toxins are heat stable. Aflatoxin B_1 is the most frequently encountered member of the group, and it is a potent carcinogen in experimental animals (rodents, birds and fish). Rats fed a diet containing 15 ppb of aflatoxin B_1 develop hepatitis that is often followed by cancer of the liver. Epidemiological studies of populations in Uganda, Kenya and Thailand have shown a close correlation between the incidence of liver cancer and the consumption of food containing aflatoxins. Thus, this natural carcinogen, at least as potent as TCDD in animal studies, is strongly implicated as a cause of human cancer, unlike the latter synthetic agent, that has received much more media coverage. Other aflatoxins found together are B_2, G_1 and G_2. Acute toxicity also can occur in humans. In 1974, an outbreak in India resulted in about 100 deaths, and in Kenya 12 died in an outbreak in the early 1980s.

The aflatoxin story began in 1960, with a serious and mysterious outbreak of a disease in turkeys in Great Britain. Turkey poults developed loss of appetite, feeble fluttering, lethargy, and they frequently died in a few days. Necropsies revealed hemorrhage and necrosis of the liver and kidney. The disease was dubbed "X" disease. Outbreaks also occurred in ducklings in Europe, the U.K. and Africa. The common denominator was groundnut (peanut) meal used as feed and subsequently shown to be contaminated with *A. flavus*. Livestock (mammals) may also be affected. Death has occurred in humans consuming an estimated 6 mg/day of aflatoxin B_1.

In North America, there is concern about the long-term effects of consuming low levels of aflatoxins, and monitoring systems are in place. Fortunately, the mold is not adapted to the colder northern climate so that Canadian-grown peanuts are free of the toxins. The majority of peanuts and peanut butter are still imported from subtropical climes, however. Peanuts are not the only source of these toxins. Because of drought conditions in the U.S. in the summer of 1988, up to 30% of the corn crop may have been contaminated. The Quaker Oats Company was turning away almost one truckload in five at its Cedar Rapids plant in the fall of 1988. The company tests six samples from each truck. Inspections of the 340,000 tons of corn crossing the Canada-U.S. border annually were stepped up in 1988–1989 by Agriculture Canada.

Fumonisins

Produced by *Fusarium moniliforme* and *F. proliferatum*, these mycotoxins are ubiquitous in many parts of the world, including South Africa, where they were first identified, and many states bordering the Great Lakes. Of particular concern is the recently identified fumonisin B_2 (FB_2). It is felt by many experts that it will become the most significant mycotoxin for human health. FB_2 has been shown to be carcinogenic for animals (a promoter and initiator of liver cancer in rats), and there is a high degree of correlation between the incidence of esophageal cancer in humans and the presence of FB_2 in corn in specific areas of South Africa. The fungus infects corn, millet, sorghum and rice around the world. Fumonisins are toxic for many species of animals, especially horses, and outbreaks have caused numerous deaths in horses and swine in Texas, Iowa and Arizona. Severely infected corn cobs may contain up to 900 mg/kg. Symptomatology in various species is as follows:

> Swine — Vomiting, convulsions, sudden death, abortion, pulmonary edema (porcine pulmonary edema syndrome). Symptoms may occur at levels above 20–50 mg/kg of FB_1.
> Poultry — Ataxia, paralysis, sudden death, stunted growth. Toxicosis occurs at levels of contamination of 10–25 mg/kg FB_1.
> Cattle — Poor weight gain, liver damage.

Horses — Equine leukoencephalomalacia (ELEM) is the disease caused by fumonisins. Brain degeneration occurs with focal necrosis. Symptoms may involve blindness, wild behavior, liver damage, staggering, ataxia, etc., and may occur at levels of 10 mg FB_1/kg for over 40 days.

FB_1, FB_2 and hydrolyzed FB_1 have been shown to be specific inhibitors of *de novo* sphingolipid synthesis and sphingolipid turnover. It has been hypothesized that this interferes with normal cell function and hence could account for the toxicity of fumonisins.

The fumonisins are themselves analogues of sphingosine, with structural similarities to the phorbol esters, which are known carcinogens (see Chapter 11).

Other Mycotoxic Hazards to Human Health

Several mycotoxins may be potential health hazards by virtue of direct toxic effects or because of carcinogenic or teratogenic properties, although it must be emphasized that hard evidence linking these to human health problems is scanty. Ochratoxin A is formed by *Aspergillus* and *Penicillium* species. It has been shown to be embryotoxic and teratogenic in several laboratory species of mammals (pig, dog, mouse, rat) and birds, and it is therefore viewed as a potential human teratogen. Acute effects in animals (including swine and poultry) include renal and hepatic destruction. Both humoral and cellular immune systems are adversely affected. Contamination of bread and cereals has been documented in parts of central Europe (Yugoslavia, Bulgaria, Poland, Germany), and levels have been detected in human milk, urine, blood and kidneys.

Patulin is potentially a carcinogenic toxin produced by several species of fungi including some *Penicillium* spp. It is a highly potent inhibitor of RNA polymerase, having a strong affinity for sulfhydryl groups. It therefore inhibits many enzymes. Teratogenicity has not been demonstrated in mammals, but embryolethality occurs at higher doses (2 mg/kg i.p.). A common source of patulin is *Penicillium expansum*, a common spoilage microorganism in apples. Apple juice can sometimes contain significant amounts of patulin. Acute toxicity in rodents is manifested largely as gastrointestinal symptoms, including hemmorrhage. Carcinogenicity studies in rats were negative, but clastogenic activity has been shown in some systems.

T-2 toxin is produced by various *Fusaria* and is both potent and common. It inhibits protein and DNA synthesis, and it is therefore potentially teratogenic and carcinogenic. Acute effects include loss of appetite and vomiting.

Fungi tend to produce mixtures of toxins, so that exposure to a single agent is unlikely to occur. There is some experimental evidence that ochratoxin A potentiates the teratogenic effects of T-2. It must be emphasized that any mycotoxin that has been shown to be toxic in several mammalian species (including various farm livestock) must be regarded as potentially toxic for humans.

■ ECONOMIC IMPACT OF MYCOTOXINS

In addition to their direct effects on human health, mycotoxins have a tremendous impact on agriculture through spoilage of field crops that are thus rendered useless for animal or human consumption (the loss of 20% of the corn crop noted above is one example) or because they cause poor weight gain and outright illness in livestock that consume contaminated feeds. Losses are difficult to estimate, but they undoubtedly run to many millions and possibly billions of dollars in North America. The presence of trace quantities in meat, dairy products and eggs constitutes a further, if largely unconfirmed, health hazard to people. In the Great Lakes basin, various species of *Fusarium* are the most common offenders, especially in eastern Canada. They may produce a host of toxins with potent pharmacological actions.

Fusarium Life Cycle

F. graminearum is the fungus responsible for maize ear rot in corn and head blight in wheat. Its life cycle is typical of all *Fusaria*. Spores survive in crop debris from the previous season (stubble, stalks, seeds) to reinfect the next year's crop. Intensive farming practices that involve planting susceptible species in the same fields year after year thus favor the spread of infection. Spores may also be spread by birds such as starlings and red-winged blackbirds which puncture the corn kernels to eat the milk. Insects like the picnic or corn-sap beetle seek out damaged kernels and also may spread spores.

Certain weather conditions favor the spread of infection. Fungus growth is favored by warmth (15–35°C) and by surface wetness for more than 48 hr. After the infection is established, weather is not critical to the production of the toxin. Mold growth will continue throughout the season and even afterwards, if not properly dried or if storage conditions are poor (too damp, too warm, poor air circulation). Mold growth can even occur during feed preparation and in poorly cleaned feed troughs. Late harvest may allow the growth of another fungus, *F. sporotrichiodes*, which produces the toxins T-2, HT-2 and diacetoxyscirpenol.

Contaminated grains thus will contain complex mixtures of toxins and metabolites. The following is a list of the other important ones and their effects.

Zearalonone

This estrogen-like toxin causes (in swine) swollen, red vulva, vaginal and rectal prolapse, vulval enlargement in piglets and fertility problems. Developmental defects and lethality have been shown in some laboratory species.

Vomitoxin (Deoxynivalenol or DON)

This trichothecene causes decreased feed intake and reduced weight gain in pigs at about 2 mg/kg of feed and vomiting and refusal of feed at very high concentrations (>20 mg/kg feed). DON will be used as a "prototype" mycotoxin to illustrate agricultural problems associated with these agents (see below).

Other Trichothecenes

T-2 and HT-2 toxins and diacetoxyscirpenol are more toxic than DON and cause reduced feed intake, vomiting, irritation of the skin and g.i. tract, neurotoxicity, teratogenicity, impaired immune function and hemorrhage. Adverse effects seen in farm animals are generally caused by mixtures of these toxins rather than by single ones. Blending of several grains in the preparation of feed may further contribute to the toxic diversity of the mixture. Potentiation of effects may occur. Thus, DON at the subthreshold level of 1 mg/kg plus low (ppb) concentrations of T-2 and other, unidentified toxins may cause severe toxic manifestations in a sensitive species like swine. The chemical structures of some of these toxins are shown in Figure 37.

■ DETOXIFICATION OF GRAINS

Because of the diverse chemical properties of the mycotoxins, physical and chemical procedures that are effective against one toxin may have little or no influence on the toxicity of others. Thus, there is no single process that can be utilized. The most important control factor must be the avoidance of conditions favoring fungal growth at all stages of food production.

Harvesting and milling. Infected kernels may represent less than 5% of all the grain. They may be broken or shrivelled and in wheat may take on a "tombstone" appearance. In corn, the tips of cobs may have shrivelled, highly infected kernels containing up to 3,000 mg/kg of DON. Grain dust may be very contaminated. Screening and blowing will remove much of the dust, particles and withered kernels.

Wet milling of corn has been shown to remove about two/thirds of T-2 toxin, but milling had little effect on the DON content of flour from hard wheat, nor did baking the flour into bread. Some

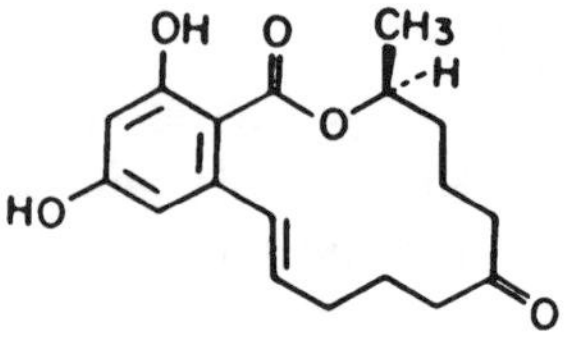

Zearalenone

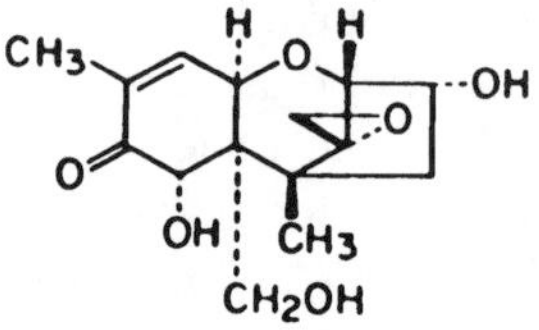

Deoxynivalenol (DON, vomitoxin)

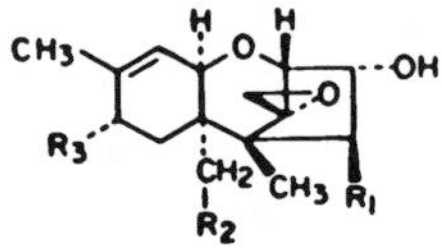

T-2: CH_3COO at R_1 and R_2, $(CH_3)_2CHCH_2COO$ at R_3

HT-2: OH at R_1, CH_3COO at R_2, $(CH_3)_2CHCH_2COO$ at R_3

Diacetoxyscirpenol (DAS): CH_3COO at R_1 and R_2, H at R_3

Figure 37 Structure of several mycotoxins.

milling procedures may actually increase the DON content of the finished product. In mild infections, washing and roasting may significantly reduce toxin levels.

Chemical treatments. Laboratory tests have shown that moist ozone, ammonia, microwaving and convection heating reduce DON concentrations in moldy grain. Aqueous sodium bisulfite plus heat effected a complete detoxification. Studies have shown that this technique resulted in normal feed intake and weight gains when contaminated corn was treated and fed to swine.

Binding agents. The addition of binding agents such as bentonite, anionic and cationic resins and vermiculite-hydrobiotite were tested on the toxicity of T-2 in rats. Bentonite prevented T-2 toxicosis by blocking intestinal absorption. Polyvinylpyrrolidone or ammonium carbonate had no effect on DON toxicity in swine. Alfalfa fiber has been shown to partially overcome the growth-depressing effect of zearalenone in rats but not the estrogenic effects in swine.

Other techniques. Dilution of contaminated feed with clean feed will improve palatability and feed consumption. More concentrated diets with respect to calories, protein, etc. may overcome the effects of a moderate reduction in feed intake. Experimentally, antibodies against zearalenone have been raised in swine and shown promise in protecting against its toxic effects.

Species Differences in DON Kinetics

Swine appear to be much more sensitive to the anorexic and weight loss effects of DON than ruminants (cattle, sheep) or poultry, which are very tolerant. In one study, laying hens actually preferred a diet containing 5 ppm of DON over clean feed. These differences are due in part to differences in absorption and in part to differences in biotransformation and elimination. Studies with radiolabelled DON indicated that sheep absorbed 9% or less of an orally administered dose; in turkeys 20% or less was absorbed, whereas pigs absorbed up to 85% of a single oral dose.

Intravenous administration of radiolabelled DON in sheep revealed an initial distribution phase (T1/2 = 18 min) followed by an elimination phase (T1/2 = 66 min). A glucuronide conjugate was formed and comprised 15–20% of plasma levels. In turkeys there was an extremely rapid distribution phase (T1/2 = 3.6 min), a rapid elimination phase (T1/2 = 46 min) and the formation of a conjugate (probably glucuronide) comprising up to 10% of the total dose. Again, swine showed a much different picture. There was a very rapid distribution phase (T1/2 = 5.8 min), a secondary slower distribution phase (T1/2 = 96.7 min) and a very prolonged terminal elimination phase (up to 510 min). There was no evidence of significant biotransformation in swine. Thus, it took seven to ten times longer for swine to clear the toxin than for the other two species.

Toxicity is a function of the concentration of toxin reaching the target organ, which in turn is affected by the rate of absorption at the portal of entry, the extent of distribution to nontarget sites (i.e., where no toxic effects occur), the rate and extent of biotransformation to nontoxic metabolites (or to toxic ones, as the case may be) and the rate of elimination in urine and feces. The effect of species differences in some of these factors was introduced in Chapter 2. Volume of distribution (Vd) and clearance data provide some information regarding the fate of the absorbed toxin. The apparent Vd is a mathematical calculation of the volume of diluent required to dilute an administered dose of a substance (usually intravenously) to the observed concentration. Vd = M/C where M = mass (amount of substance) and C = concentration of substance. Calculations of Vd for DON yielded values of 0.167 L/kg for sheep, vs. 1.3 L/kg for swine, suggesting that in the former DON was confined mainly to the extracellular fluid, whereas in the latter it was taken up by tissues. Initial systemic clearances were not all that different, being 1.37 ml/min/kg for sheep and 1.81 ml/min/kg for swine. An interpretation of this data suggests that DON is initially rapidly distributed to tissues and then slowly released back into the plasma, yielding the slow, terminal elimination phase. Turkeys also had a very large Vd (2.33 L/kg), but they also had an extremely rapid clearance (35.0 ml/min/kg), indicating that DON was rapidly distributed to tissue compartments but not held there.

Thus, the extreme sensitivity of swine to DON is the result of (1) high oral bioavailability, (2) wide distribution to tissues, (3) slow elimination from the body, and (4) minimal detoxification through biotransformation.

Regarding species differences in capacity to biotransform xenobiotics, evolutionary factors have played a role. Given the plethora of toxic substances in the plant kingdom, pure herbivores would have been exposed to the greatest risk of poisoning, as compared to carnivores or omnivores, and would therefore have needed to evolve nonspecific detoxifying enzymes. The toxin most likely encountered by pure carnivores, however, would have been botulinum toxin, and their need would have been to evolve a method of eliminating it quickly before this potent neurotoxin could paralyze movement and respiration. Thus, canines and felines are very poor metabolizers of drugs (e.g., aspirin can be extremely toxic to cats and dogs), but they have exquisitely sensitive vomiting mechanisms to eliminate bad meat before the toxin is absorbed. Conversely, herbivores are efficient metabolizers but generally lack good vomiting reflexes. Emesis does not occur in equines, ruminants or rodents. Omnivores like humans and swine fall in between. The importance of diet in the evolution of detoxifying systems applies even to primates. Despite their closeness to us on the phylogenetic tree, fruit- and plant-eating primates are generally more efficient metabolizers of xenobiotics than we are. In further support of this theory, it can be pointed out that the T1/2 for amphetamine is about 86 min in both the rabbit and the horse but is 390 min for the cat and 300 min for humans.

The question of mycotoxin residues in human food sources remains largely unanswered, but studies indicate that DON residues are not a problem. Feeding very high levels of DON to dairy cattle resulted in only trace quantities in milk, and when fed to poultry, no appreciable tissue residues were measured. Again, this is the consequence of the species pharmacokinetic characteristics.

■ OTHER TOXINS IN UNICELLULAR MEMBERS OF THE PLANT KINGDOM

Many soil organisms produce substances that are toxic to others, and a continual state of chemical warfare exists in the microenvironment. These organisms provide the source of all of our antibacterial and antifungal antibiotics, many of which have significant mammalian toxicity. Many others were tested and discarded because of their high toxicity. Some antibiotics are teratogenic and are used in the treatment of cancer because of their effects on cell reproduction (e.g., actinomycin D, doxorubicin, adriamycin). Some organisms may be responsible directly for poisonings in humans.

The blue-green algae called *Cyanobacteria* produce pentapeptides that are hepatotoxic and that have caused numerous deaths when they contaminate drinking water.

Some strains of that ubiquitous organism *Staphylococcus aureus* are capable of producing a protein enterotoxin, 1 µg of which can induce vomiting, severe colic, and profuse diarrhea. The bacteria are usually introduced to foods from infected handlers, and they proliferate in warmth and especially in creamy foods (cream pies, salad dressings, etc.).

The most potent toxin known is the protein toxin from *Clostridium botulinum*, 1 µg of which may be fatal to a human. Lab animals may show symptoms at 10^{-6} µg/kg. Botulinum toxin blocks the release of acetylcholine from peripheral nerve endings.

■ REVIEW QUESTIONS

For Questions 1 to 10, use the following code: answer A if statements 1, 2 and 3 are correct; answer B if statements 1 and 3 are correct; answer C if statements 2 and 4 are correct; answer D if statement 4 only is correct; answer E if all statements (1,2,3,4) are correct.

1. Regarding mycotoxins, which of the following statements is/are true?
    1. *Fusarium* species produce numerous mycotoxins which are hazardous for many species.
    2. Deoxynivalenol (DON) causes fertility problems in swine.
    3. DON causes vomiting and poor weight gain in swine.
    4. Other trichothecenes are less toxic than DON.

2. Which of the following statements is/are true?
    1. All varieties of wheat and corn are susceptible to fungal infections in varying degrees.
    2. Rapid drying of grain minimizes the risk of post-harvest mold growth.
    3. Mycotoxin toxicity in livestock usually results from mixed fungal contamination.
    4. There is no practical way of detoxifying corn and wheat contaminated with mycotoxins.

3. Which of the following statements is/are true?
    1. There is no evidence that fumonisins are carcinogenic for humans.
    2. Zearalenone has estrogen-like activity.
    3. Aflatoxins are not toxic for poultry.
    4. Ochratoxin causes renal damage in swine and poultry.

4. The symptoms of ergot poisoning
    1. Are similar for humans and animals.
    2. Include nausea, dizziness, burning pain in the extremities and abortion.
    3. Arise largely from consuming rye contaminated with ergot alkaloids.
    4. Are caused by ergot alkaloids produced by *Fusarium moniliforme*.

5. Which of the following statements is/are true regarding Fumonisin B_2 (FB_2)?
    1. It strongly suspected of being a carcinogen for humans.
    2. It produces brain damage in horses.
    3. Horses are very susceptible to FB_2 toxicity.
    4. In swine, the signs of toxicity are similar to those of DON.

6. Regarding ergot alkaloids,
    1. They are derivatives of lysergic acid.
    2. Chemically they resemble catecholamine neurotransmitters.
    3. They are chemically related to lysergic acid diethylamide (LSD).
    4. They may produce hallucinations.

7. Which of the following statements is/are true regarding ergot alkaloids?
    1. Ergometrine (ergonovine) causes contractions of the gravid uterus.
    2. Ergometrine has never had any medical application.
    3. Ergotamine in low doses has been used to treat migraine headaches.
    4. Ergotamine is nontoxic except at very high doses.

8. Regarding aflatoxins,
    1. The mold that produces them is very hardy and survives in cold climates.
    2. They are oxygenated heterocyclic compounds.
    3. They cause hepatic damage but are not carcinogenic.
    4. Epidemiological studies have shown a high degree of correlation between liver cancer and the consumption of foods contaminated with aflatoxin B_1.

9. Regarding detoxification of grains contaminated with mycotoxins,
    1. A technique that works on one mycotoxin may not work on others.

2. DON concentration may be reduced by heating.
3. Binding agents may reduce toxicity of T-2.
4. Antibodies against zearalenone have shown promise in protecting swine against estrogenic effects.

10. Regarding DON toxicity,
    1. Swine are more susceptible than poultry or ruminants.
    2. Turkeys form a conjugate and rapidly eliminate DON.
    3. Sheep form a glucuronide conjugate.
    4. A high volume of distribution (Vd) is generally associated with higher toxicity in that species.

11. Match the statement with the correct toxin or effect.
    a. Cyanobacteria cause this.
    b. Blocks acetylcholine release from peripheral nerve endings.
    c. Adriamycin.
    d. *Staphylococcus aureus.*

___ i   Produces a protein enterotoxin.

___ ii. Produces hepatotoxic pentapeptides.

___ iii. A teratogen used in cancer chemotherapy.

___ iv. *Botulinum* toxin.

ANSWERS

1. B; 2. A; 3. C; 4. A; 5. A; 6. E; 7. B; 8. C; 9. E; 10. A; 11. i. d, ii. a, iii. c, iv. b.

FURTHER READING

Albert, A., *Xenobiosis: Food, Drugs and Poisons in the Human Body*, Chapman and Hall, New York, 1987.

Foster, B.C., Trenholm, H.L. et al., Evaluation of different sources of deoxynivalenol (vomitoxin) fed to swine. *Can. J. Anim. Sci.*, 66, 1149, 1986.

Fumonisins — a current perspective and view to the future (various authors). *Mycopathologia*, 117, 1, 1992.

Keeler, R.F. and Tu, A.T., Plant and animal toxins, in *Handbook of Natural Toxins*, Vol. 1. Marcel Dekker, New York, 1983.

Long, G.G. and Dickman, M.A., Characterization of effects of zearalenone in swine during early pregnancy. *Am. J. Vet. Res.*, 47, 184, 1986.

Miller, J.D., Young. J.C. and Trenholm, H.L., Fusarium toxins in field corn I. Time course of fungal growth and production of deoxynivalenol and other mycotoxins. *Can. J. Bot.*, 61, 3080, 1983.

Philp, R.B., *An Introduction to Comparative Pharmacology*, Department of Pharmacology, U.W.O., London, 1977.

Rheeder, J.P., Marasas, W.F.O. et al., *Fusarium moniliforme* and fumonisins in corn in relation to human cancer in Transkei. *Phytopathology*, 82, 352, 1992.

Riley, R.T., Norred, W.P. and Bacon, C.W., Fungal toxins in foods: recent concerns. *Ann. Rev. Nutr.*, 13, 167, 1993.

Scott, P.M., Trenholm, H.L. and Sutton, M.D., *Mycotoxins: A Canadian Perspective*, Natural Resource Council of Canada, No. 22848, Ottawa, 1985.

Seamon, W.L., Epidemiology and control of mycotoxigenic *Fusaria* on cereal grains. *Can. J. Plant. Pathol.*, 4, 196, 1982.

Shier, W.T., Sphingosine analogs: an emerging new class of toxins that includes the fumonisins. *J. Toxicol.*, 11, 241, 1992.

Theil, P.G., Marasas, W.F.O. et al., The implications of naturally occurring levels of fumonisins in corn for human and animal health. *Mycopathologia*, 117, 3, 1992.

Trenholm, H.L., Hamilton, R.M.G. et al., Feeding trials with vomitoxin (deoxynivalenol)-contaminated wheat: effects on swine, poultry and dairy cattle. *J. Am. Vet. Assoc.*, 184, 527, 1985.

Trenholm, H.L., Prelusky, D.B. et al., Reducing mycotoxins in animal feeds. *Agriculture Canada Publication 1827/E*, 1988.

World Health Organization, Aflatoxins, in *Evaluation of Certain Food Additives and Contaminants. 31st Report of the Joint FAO/WHO Expert Committee on Food Additives.* WHO, Geneva, 1987, 33–37.

World Health Organization, Ochratoxin A, in *Evaluation of Certain Food Additives and Contaminants. 37th Report of the Joint FAO/WHO Expert Committee on Food Additives,* WHO, Geneva, 1991, 29–31.

World Health Organization, Patulin, in *Evaluation of Certain Food Additives and Contaminants. 35th Report of the Joint FAO/WHO Expert Committee on Food Additives,* WHO, Geneva, 1990, 29–30.

11 ANIMAL AND PLANT POISONS

He was a bold man that first ate an oyster.
Jonathan Swift

INTRODUCTION

Chemical warfare is widely practiced in the animal and plant kingdoms. Just as microorganisms produce antibiotics that inhibit the growth and reproduction of competing organisms, more complex plants synthesize chemicals that render them unpalatable to potential predators or that are truly toxic, thus selecting for individuals that avoid them or producing aversive reactions that limit consumption to once or twice only. Occasionally, however, these plants accidentally enter the food chain of livestock or humans, directly or indirectly, leading to a toxic reaction.

Similarly, toxins and venoms are used in the animal kingdom for defense and prey capture. Toxins are consumed and operate much like those of plants. They may also be secreted by special cells or glands in the skin, so that toxicity may occur simply by taking the intended victim into the mouth. Toxic reactions have occurred among students indulging in the fad of "toad-licking". These toxins tend to be low molecular weight peptides with neurotoxicity. Some fish actually swim in a "cloud" of toxin secreted by these skin glands. Their toxins are usually steroid glycosides and choline esters. A neurotoxin secreted in this way by a species of flounder of the Red Sea is so powerful that a shark attempting to bite it is incapable of closing its jaws and instantly convulses. It has been proposed as a shark repellant for downed fliers and divers. These skin toxins also serve as antibacterial agents to prevent infection, as the slime secreted by the skin of amphibians is an ideal culture medium for bacteria.

Venoms are injected in some way and may be employed both for defense and for prey capture. Humans and animals sometimes accidentally become the victims of envenomation or intoxication by poisons of animal origin. The term toxin is used also to refer to the individual components of venoms.

Numerous texts have been written on these subjects, and this chapter can do no more than skim the field and discuss some of the more important, or more interesting, examples. Emphasis will be placed on agents that have become important to the biological sciences, either as drugs or as research tools.

TOXIC AND VENOMOUS ANIMALS

Toxic and Venomous Marine Animals

Venoms and toxins are distinguished by the manner in which they are inflicted upon the victim. A venom is a substance kept in a special poison gland and administered by a complex injection apparatus or by lacerating spines. This type of system may be employed in defense or in prey capture. A toxin is a poison usually ingested as an accidental component of tissues or organs. Toxins likely evolved as a species (rather than an individual) protection. Predators with a preference for that prey would be selected out if the toxin is fatal. Conditioned avoidance could occur in survivors.

The term for poisoning by the muscle tissue of scalefish, as opposed to shellfish, is ichthyosarcotoxism. It includes toxins that are accumulated up the food chain from plankton as well as toxins that are synthesized by the fish itself. In the former situation, poisoning is usually by a mixture of toxins.

Scalefish Toxins

Ciguatoxin. The condition ciguatera is caused by this toxin, which probably concentrates up the food chain from a photosynthetic dinoflagellate to reach toxic levels in large marine fish, such as grouper, snapper, amberjack, barracuda, parrot fish, etc. These species are either territorial predators of smaller coral reef species, or they are coral browsers. Toxic symptoms are both gastrointestinal (nausea, vomiting, cramps, diarrhea) and neurological (numbness, tingling of lips, throat and tongue, dizziness, headache). Ciguatoxin is thought to increase membrane permeability to sodium, and it is responsible for the neurological signs and symptoms. Ciguatoxin is very toxic to mice which are the test species for detecting its presence. It is a large, colorless, heat-stable lipid molecule, the structure of which has not been elucidated. Other, unidentified agents are likely involved in ciguatera poisoning. Ciguatera is by far the most common cause of scalefish poisoning. Any large, warm-water species is a potential source of the toxin.

Barracuda also sometimes contain a related neurotoxin that is very toxic to cats. In some parts of the West Indies, it is customary to feed some of the meat from a large barracuda to a local cat before humans eat it.

Tetrodotoxin. The term comes from *Tetraodontidae*, meaning four teeth, the name of the genus that produces the toxin. This toxin is present in puffer fish, which are called "fugu" in Japan. It is a very potent blocker of fast sodium channels and therefore inhibits nerve conduction in a manner like local anesthetics. It causes parathesias, paralysis, anesthesia and loss of speech, but consciousness is retained. Prognosis is improved if vomiting occurs early after ingestion. In Japan there are about 150 cases of poisoning annually with a 50% mortality. Fugu is considered a delicacy in Japan, and special chefs are licensed to prepare it. The toxin is concentrated in the liver and gonads of the fish. Connoisseurs of fugu prefer that just enough toxin remains to cause a slight tingling of the lips when it is consumed. In Japan, this custom has spawned a somewhat macabre poetry form: "Last night he and I ate fugu; Today I helped carry his coffin".

Tetrodotoxin is also found in some newts and frogs. The blue-ringed octopus, a small octopus that inhabits tidal pools and shallow waters around Australia and other central Indo-Pacific waters, produces a toxin, administered by a bite, that is believed to be identical to tetrodotoxin. The bite of a larger specimen can be fatal in minutes.

Tetrodotoxin is of interest as a research tool because of its potent sodium channel blocking activity. Its occurrence in such diverse species can be explained by the fact that it is not synthesized by the animal itself, but rather by certain species of *Vibrio* bacteria which exist in a symbiotic relationship with the host.

Scrombroid poisoning. The name "scrombroid" comes from *Scrombridae*, referring to dark-muscled fish like tuna and mackerel.

This condition occurs from eating fish rich in histidine. Improper refrigeration results in the conversion of histidine to histamine by surface bacteria. The signs and symptoms resemble a histamine reaction. They may onset in minutes to hours and include dizziness, headache, diarrhea, facial flushing, tachycardia, pruritus and wheezing. Fish contaminated in this way are often described as having a spicy or peppery taste. Levels of histamine in contaminated fish may exceed 100 mg/100 g. The FDA has set 50 mg/100 g as the hazard level. Histamine is not destroyed by cooking. There is some evidence that another product of decomposition, saurine, may contribute to the symptomatology.

Icthyotoxin. Many species of fish, including fresh-water varieties, sometimes contain a toxin in the gonads that can cause severe gastrointestinal symptoms.

Shellfish Toxins

Saxitoxin. The cause of paralytic shellfish poisoning, this toxin is produced by dinoflagellates (single-celled animals that are the cause of red tides), and it concentrates in bivalves such as oysters, clams and mussels. The mechanism and symptomatology are the same as for tetrodotoxin. The minimum lethal dose is about 8 µg/kg for both. "Never eat oysters in a month without an R" is an old adage that still is good advice in the Northern Hemisphere as dinoflagellates bloom in the summer months.

Domoic acid. In 1987, there was an outbreak of mussel poisoning in Canada caused by mussels from the waters around Prince Edward Island. The toxin was domoic acid, which concentrates from a seaweed called chondria. It is also found in a diatom, and it is rare in North Atlantic waters, being more common in Japan. The toxin was identified by the Canadian National Research Council Atlantic Laboratory. There were three fatalities in elderly nursing home residents in Quebec, and over 100 individuals were affected. Several have been left with short-term memory deficit. It was suggested that marine pollution might have changed the environment to favor the growth of the offending diatom. Domoic acid is responsible for a condition called amnesic shellfish poisoning.

Okadaic acid. Also produced by a dinoflagellate, okadaic acid is responsible for a condition known as diarrhetic shellfish poisoning.

There is some evidence that there has been a dramatic increase in toxic algal blooms, the so-called red tides. The first confirmed outbreak of diarrhetic shellfish poisoning in North America occurred in 1990, and it was traced to dinoflagellates in Canadian waters. Brown pelicans eating anchovies off California were dying of domoic acid poisoning in 1991, saxitoxin has been found in Alaska crabs, and in 1987–1988 shellfishing off North Carolina was shut down because of a red tide of a dinoflagellate that produces a neurotoxin, brevitoxin.

Stinging Fishes

In venomous fish such as the stonefish, lionfish, scorpionfish and stingray, spines are located ahead of the dorsal fin, on the tail or around the mouth. A heat-labile protein causes intense pain and cardiac shock. Heat above 50°C may afford some relief. Stonefish and stingrays are most often stepped upon because they conceal themselves in the sand bottom. Fatalities have occurred from envenomation by these fish, generally as a result of cardiovascular shock, with AV block and bradycardia. Other symptoms include numbness, inflammation and edema at the site of injury, severe pain in surrounding tissues, delirium, nausea, vomiting and sweating. Stingray venom is a large, heat-labile, protein (molecular weight > 100,000) with neurotoxic as well as cardiotoxic properties. There is an antivenin for stonefish venom.

Mollusc Venoms

Conotoxins. These are found in marine cone snails. All are strongly basic peptides highly crosslinked by disulfide bonds. Alpha conotoxins are nondepolarizing, neuromuscular blocking agents like curare, and they therefore cause paralysis. They are 13–15 amino acid peptides. Mu conotoxins are 22 amino acid peptides that block sodium channels, thus acting like tetrodotoxin and saxitoxin. Omega conotoxins block presynaptic, voltage-dependent calcium channels. Recent research has identified a number of subtypes with specificity for various channel subtypes, making them useful research tools. Cone snails are univalve gastropods with a complex envenomation apparatus they use for prey capture.

Coelenterate Toxins

These are found in jellyfish and corals. Fire coral produces a protein venom that causes intense local burning when touched. It feels like a cigarette burn. Physalis (Portuguese man-o-war, bluebottle, mauve stinger) causes signs and symptoms like fire coral. Red streaks, called straps, occur where tentacles touch skin. Allergic reactions can occur, also generalized symptoms such as fever, nausea, cardiac and respiratory distress. Bluebottles washed up on beaches remain toxic until completely dried out. First aid consists of washing the affected area with vinegar, dilute acetic acid, or meat tenderizer (papain), all of which denature protein. Even urine may help relieve the pain. This species is common around the world.

The sea wasp (box jellyfish) is the most venomous marine animal known. It inhabits the Indo-Pacific region and accounts for many deaths annually. Contact with tentacles causes intense, agonizing pain, coma and cardiac shock. Mortality is 25% and children, the elderly and heart patients are most vulnerable. First aid is denaturation as above, removal of tentacles using forceps, gloves, a knife and fork or any means to avoid touching the tentacles, and cardiopulmonary resuscitation (CPR). Local anesthetic spray such as is used for sunburn may be helpful. An antivenin is available.

The "Irukandji" is a tiny, four-tentacled, Indo-Pacific jellyfish causing similar symptoms plus massive sympathetic discharge.

Echinoderm Venoms

Sea urchins and sea anemones such as the long-spined or black sea urchin and crown-of-thorns sea urchin possess toxins. Injury usually occurs to the feet and lower limbs when a diver or swimmer steps on these bottom-dwellers. Spines are driven into the flesh and break off. The toxin produces local pain and burning similar to that of the bluebottle. It is heat-labile, so immersing the affected part in water as hot as the person can stand may help. Unlike the usual first aid for envenomations, movement and trauma may actually help by breaking up the spines so they can be absorbed more quickly. Vinegar or even urine may also help.

Sea anemones also produce very potent toxins that act when they are ingested. Best characterized of these are the equinatoxins (EqTs I, II and III) from *Actinia equina*. In rats they cause coronary vasospasm, cardiac arrest and other cardiorespiratory toxicity. Hemolysis also occurs, as does degranulation of blood platelets and white cells.

Toxic and Venomous Land Animals

Venomous Snakes

Many snakes are venomous, but most have rear fangs that are designed to paralyze prey after it has been taken into the oral cavity. These are incapable of inflicting a venomous bite. The four genuses of poisonous snakes that are a danger to humans are the *Viperinae, Crotalinae, Elapidae* and *Hydrophidae*.

Viperinae. These old world vipers have hollow, needle-like fangs that are set in short, movable maxilla which rotate to bring the fangs into the biting position as the mouth is opened. The head is large and triangular. True vipers are found in Europe, Africa, Asia and Australia and include the European adder (the only poisonous snake in Great Britain), Russell's viper and the Australian death adder.

Crotalinae. These "pit vipers" are similar to the old world vipers, but they also have a deep, infrared-sensitive pit between the eye and the nostril that is used for tracking prey by body heat. This genus includes all rattlesnakes, the water moccasin, cottonmouth and copperhead. All are found in North America. Canada has the Western Diamondback and, in Ontario, the Massasauga rattlesnake.

Elapidae. These are distributed worldwide and account for the most poisonous and feared snakes of the tropics and subtropics. This group includes the cobras, the boomslang, and many Australian species (taipan, tiger snake, brown snake, etc.). There are two species in the U.S., the Eastern (Florida) and Arizona coral snakes.

Hydrophidae. These are the sea snakes. They are restricted to the warmer waters of the Pacific and Indian Oceans. Although they are highly venomous, they are not aggressive. Most bites occur to commercial fishermen because the snakes become trapped in the nets. For some years, sporadic discussions have occurred concerning the feasibility of digging a sea-level canal across the Isthmus of

Panama to connect the Atlantic and Pacific Oceans. One environmental concern which has been raised is that such a canal could introduce the yellow-bellied sea snake to the Gulf of Mexico and the Caribbean Ocean where conditions are favorable for their proliferation.

Snake venoms. These are complex mixtures of proteins, many of which are proteolytic enzymes. In general, venoms of the *Elapidae* and the *Hydrophidae* tend to be neurotoxic with myonecrosis (breakdown of muscle tissue) occurring at the bite wound, whereas venoms of the vipers and crotalids are generalized coagulants with local anticoagulant activity to spread the venom, causing much local damage (pain, necrosis, bleeding). Neurotoxicity is less prominent.

Signs and symptoms of neurotoxic venoms include progressive paralysis, muscle spasm, respiratory distress or failure, muscle ache (myalgia), kidney failure with myoglobinuria (the appearance of myoglobin in the urine) and cardiac failure. These symptoms apply to coral snake bites in North America. The venom of the Banded Krait contains alpha bungarotoxin which is an irreversible blocker of acetylcholine receptors, and beta bungarotoxin which causes a massive release of neurotransmitter vesicles. Alpha bungarotoxin is used as a research tool in biomedical research.

Signs and symptoms of viper and crotalid bites (including the Massasauga rattlesnake) include immediate, intense burning at the site of envenomation (like a bee sting), followed by numbness, swelling, shock, and hematuria. Necrosis and possibly gangrene may occur later at the bite wound. Subcutaneous hemorrhages may be present and severe cases may show signs of neurotoxicity. Table 9 lists some of the major components that have been identified in snake venoms. The venom of Russell's viper contains an activator of clotting Factor X, and it is used in coagulation research.

Phospholipase A_2 complexes in rattlesnake venoms are neurotoxic as well as inducing tissue damage. Myotoxic components have exhibited mitogenic multiple effects on growing cultured myocytes.

Elapidae and *Viperinae* both possess a-neurotoxins that act postsynaptically to prevent acetylcholine from binding to its receptor, as well as neurotoxins that affect transmitter release presynaptically. B-neurotoxins have phospholipase A_2 activity. The mambas and other African snakes have voltage-dependent K^+ channel blockers (dendrotoxins), noncompetitive inhibitors of acetylcholine (fasciculins), muscarinic toxins, and L-type calcium channel blockers (caliseptins). All are small proteins containing about 60 amino acids and three or four sulfides. Some components of snake venoms are listed in Table 9.

First aid. Regardless of the type of snake bite, the most important first aid measure is a tension bandage applied to the entire affected limb with the same tension one would use to bandage a sprain.

TABLE 9
SOME COMPONENTS OF SNAKE VENOMS

Component	Elapids, hydrophids	Vipers, crotalids
Nicotinic blocker (neuromuscular blockade)	+	+ −
Cholinesterase (neurotoxic)	+	−
Coagulant protease (increases clotting)	−	+
Anticoagulant protease (decreases clotting locally)	+ −	+
ATP (shock, hemolysis)	+	+
Phosphatase (shock, hemolysis)	+	−
Phospholipase A (histamine and SRSA release[a])	+	−
Bradykininogen (forms bradykinin, causes pain at site)	−	+
Hyaluronidase (breaks down connective tissue, spreads venom)	+	+

[a] SRSA = slow-releasing substance of anaphylaxis (leukotrienes).

Rest and reassurance are important. Transportation to a medical center possessing antivenin should occur as soon as possible. Modern hospitals should have the antivenin, or polyvalent antivenin, appropriate to their area in stock. Forced exercise and alcoholic beverages are definitely contraindicated.

There are a few venomous lizards, such as the Gila monster which inhabits the American Southwest. They are not generally as dangerous as venomous snakes. The venom lacks neurotoxins but contains coagulants and enzymes as well as serotonin. The reaction tends to be more local. Lethal doses in animals lead to cardiorespiratory collapse.

Venomous Arthropods

Members of the order *Hymenoptera* (bees, wasps, ants) of the class *Insecta* and of the class *Arachnida* (spiders, scorpions) have venoms that contain substances commonly involved in the mammalian pain response. The Old World scorpion *Leiurus quinquestriatus* produces charybdotoxin which affects calcium-activated K^+ channels as does the bee venom apamin, whereas the Mexican scorpion *Centroides noxius* produces noxiustoxin which blocks voltage-dependent K^+ channels (see Table 10). Mast cell degranulating peptide in bee venom also blocks these channels.

TABLE 10
COMPONENTS OF HYMENOPTERA VENOMS

	Histamine releasers	Bradykinin	Serotonin	K⁺ channel blockers
Bees				Apamin (Ca^{2+} activated) Mast cell degranulating peptide (voltage activated)
Ants	+	+		
Wasps	+	+	+	
Scorpions	+	+	+	
Old World				Charybdotoxin (Ca^{2+} activated)
Mexican				Noxiustoxin (voltage activated)

Allergic reactions may occur to any of these insect venoms and may be life-threatening if severe or if multiple stings occur. The African honey bee is not dangerous because it is more venomous, but because it is extremely aggressive so that multiple stings occur.

Neurotoxins tend to predominate in spider venoms. Virtually all spiders are venomous, but only a few have an envenomation apparatus suitable for piercing human skin. So-called "widow" spiders (*Latrodectus* spp.) have a worldwide distribution in temperate and tropical climes. The Black Widow, also known as the death spider, hourglass spider and by many other names, inhabits all of Ontario south of Sudbury, but it is rarely encountered since the demise of the outdoor privy. The venom is extremely toxic, but very little is injected so fatalities are rare. The toxin is complex. At least seven "latrotoxins" have been identified; five are specific for insects, one for crustacea and one for vertebrates. The latter is alpha latrotoxin (α-LTX).

It is a medium-sized protein now available in pure form (see below) that induces the release of most if not all neurotransmitters. It bears similarity to β-bungarotoxin (found in the Banded Krait). Both cause a massive release of cholinergic vesicles. Fasciculations and board-like rigidity of the muscles of the trunk occur rapidly followed by muscle cramps, pain in the muscles and respiratory distress. The presence of a specific receptor on vertebrate neurons for α-LTX has been identified, and this is the likely explanation for the species specificity of the toxin. Recovery from Black Widow spider bite takes about 12 hr.

The Brown Recluse (*Loxosceles* spp.) known also as the violin spider, has been working its way north from Mexico and Florida and has reached New York state and probably southern Ontario. The bite is painless. Local necrosis develops and expands over the next week due to local blood clotting and microthrombosis. Occasional deaths have occurred from hemolytic anemia and kidney failure. The venom is complex and includes coagulants, enzymes and a complement inhibitor.

The Australian funnel-web and red-backed spiders are claimed to be especially venomous. Two funnel-web toxins have been identified, a polyamine (FTX) and omega agatoxin. Both block high voltage-dependent P Ca^{2+} channels, and they are now used as research tools for this purpose.

■ TOXIC PLANTS AND MUSHROOMS

Folk knowledge, passed orally from generation to generation, usually determines what we can eat and what we cannot eat, and even what parts of a plant are edible. Thus, we make pies from rhubarb stalks, but we never make salad from the leaves, which are toxic. Nor do we eat the bulbs or stalks of the many ornamental flowers and shrubs in our gardens. Seldom do we consider the fact that these decisions are toxicologically based. The number of plants that are potentially harmful is too voluminous for extensive coverage. The subject has formed the basis of many texts. The following list contains examples of major groups of toxicants chosen because they are common or because of their pharmacological significance.

Vesicants

Many plants contain oxalates that can cause corrosive burns to the mouth, esophagus and stomach. Symptoms also include vomiting and diarrhea. Since oxalates are anticoagulant, bleeding may also occur. Dieffenbachia contains calcium oxalate which is a vesicant. Rhubarb leaves contain a variety of oxalates and may be fatal if consumed in quantity due to their renal toxicity (see also ethylene glycol). May apple, buttercup and philodendron contain vesicants. Philodendron sometimes causes poisoning in cats which chew the leaves. Poison ivy, poison oak and poison sumac all contain urushiol, which is a phenolic vesicant.

Cardiac Glycosides

White and purple foxglove (*Digitalis lanata* and *D. purpura*) are the commercial source of medicinal digitalis. Many garden plants contain similar active components, including lily of the valley, star of Bethlehem and oleander. Symptoms are those of digitalis overdose: nausea, vomiting, visual disturbances (a green halo seen around objects) and cardiac arrhythmias.

Astringents and Gastrointestinal Irritants (Pyrogallol Tannins)

Acorns, geraniums, sumac berries, hemlock bark, rhubarb leaves, and horse chestnut all contain these. North American natives learned to remove tannins from acorns by steeping the crushed nuts repeatedly in fresh water. This will not work for horse chestnuts, however.

Autonomic Agents

Deadly and wooded nightshades contain atropine (hyoscyamine) and scopolamine (hyoscine). The signs and symptoms are those of blockade of muscarinic cholinergic receptors plus CNS symptoms (disorientation, hallucinations). *Amanita muscaria* is the common poisonous mushroom. Signs of poisoning are those of massive cholinergic discharge due to the toxin muscarine, plus CNS effects like those of nightshade due to the presence of anticholinergic agents. CNS depression and cardiac failure may follow. Symptoms may be delayed 12–24 hr.

Dissolvers of Microtubules

Colchicine from the autumn crocus is used in the treatment of gouty arthritis and as an experimental tool. It dissolves microtubules to prevent mitosis and also phagocytosis. Overdose causes severe diarrhea. Vincristine and vinblastine from the periwinkle plant share this property and are used to treat childhood leukemias. Podophylotoxin from the may apple also dissolves microtubules.

Phorbol Esters (e.g., Phorbol Myristate Acetate, PMA)

These are components of croton oil from spurge plants. These substances are cocarcinogens. They directly activate protein kinase C (substituting for diacyl glycerol) independently of extracellular calcium, and they are used as experimental tools for this reason. They act as drastic purgatives. The site of action of PMA is shown in Figure 38. It affects an important control mechanism for intracellular regulation.

Many other carcinogens, cocarcinogens and anticarcinogens exist in plants. Safrol is a liver carcinogen found in some spices (nutmeg, cinnamon) and in oil of anise (licorice flavoring). The use of anise oil and of oil of sassafras has been banned. Some tannins are liver carcinogens, and the polyaromatic hydrocarbon (PAH) benzo-[a]-pyrene is a potent carcinogen that occurs in green vegetables, unrefined vegetable oils, coconut oil and chicory. Benzanthracenes are other PAHs that occur in vegetables. Numerous others exist.

Solanine and Chaconine

These are glycogenic alkaloids that occur in potatoes and tomatoes (members of the *Solanaceae* family) when they are exposed to

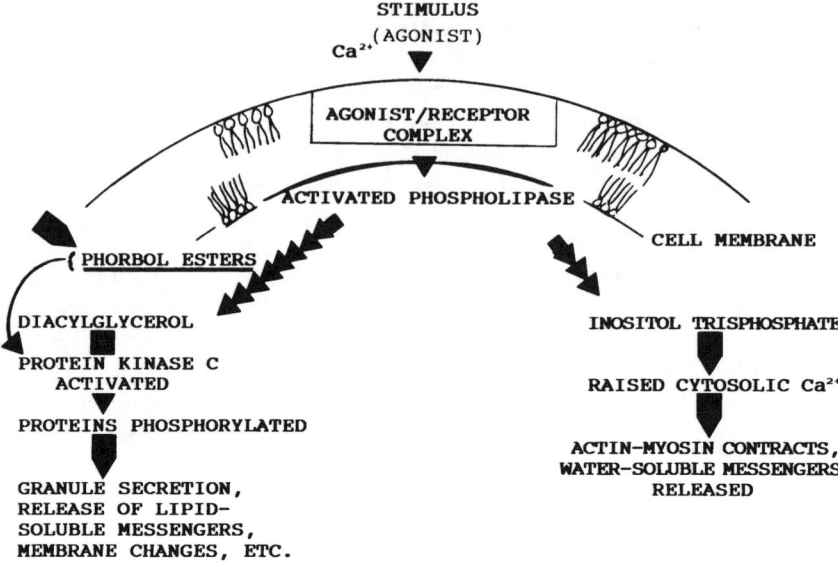

Figure 38 Site of action of phorbol esters as activators of protein kinase C.

excessive sunlight, blight, sprouting and prolonged storage. The compounds have cholinergic activity and have been shown to be teratogenic. Symptoms of poisoning include vomiting, diarrhea, cramps, dizziness, visual disturbances and other CNS manifestations.

Cyanogenic Glycosides

Substances such as amygdalin in almonds, dhurrin in sorghum, linamarin and lotaustralin in cassava and lima beans and prunasin in stone fruit (cherries, peaches, chokecherries) are cyanogenic glycosides capable of forming hydrogen cyanide (HCN) by beta glucuronidases from the plants when cells break down or from the microflora of the gastrointestinal tract. Cyanide poisoning can occur in ruminant animals from eating vegetation high in cyanogenic glycosides or in humans who have consumed improperly stored or prepared foods such as lima beans or cassava. In humans, CN can be formed from organic nitriles by cytochrome P450-dependent monooxygenases and from organic thiocyantes by glutathione S-transferases.

Detoxification of Hydrogen cyanide. Hydrogen cyanide (HCN) is detoxified by conversion to thiocyanate, which requires sulfur-containing amino acids and vitamin B_{12}. Deficiencies of these increase the risk of toxicity. The metabolic detoxification system is overwhelmed, and hydrogen cyanide interferes with electron transport in the cytochrome $a-a_3$ complex, with resulting tissue hypoxia. This leads to rapid failure of the CNS and death. Treatment is the

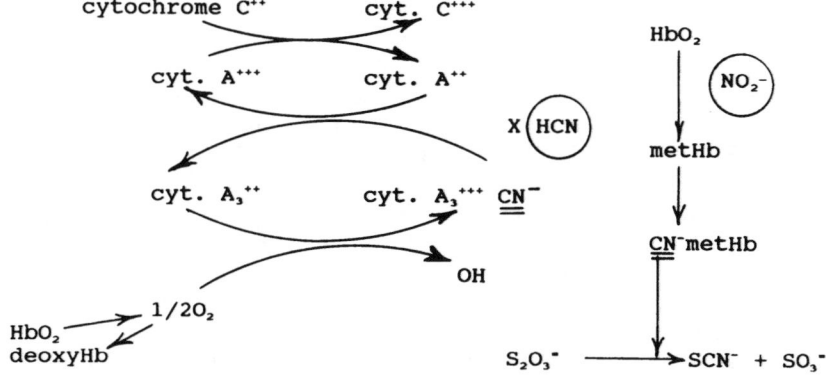

Figure 39 Site of action of HCN and detoxification by nitrites.

administration of intravenous nitrites which form methemoglobin. Methemoglobin has a high affinity for HCN and binds it to protect the cytochrome and allow time for biotransformation to occur. These events are summarized in Figure 39.

Convulsants

Water hemlock, the poison of Socrates, typifies this group. The toxin is cicutoxin (from the plant's Latin name *Cicuta maculata*). This plant resembles parsnips, smells like turnips, tastes sweet and is the most toxic indigenous plant in North America. The toxin is present in all parts of the plant but is concentrated in the root. It is most toxic in springtime. Mild intoxication produces nausea, abdominal pain, epigastric distress and vomiting in 15–90 min. Early vomiting may be protective. Severe poisoning produces profuse salivation, sweating, bronchial secretion, respiratory distress and cyanosis. Convulsions occur and *status epilepticus* precedes death. Mortality rates are of the order of 30%. There is no known antidote. Fatal poisonings in children have occurred from using toy whistles made from the stem.

In the period 1978–1989, 58 persons in the U.S. are known to have died from ingesting toxic plants mistaken for edible wild fruit or vegetables. Water hemlock was responsible for at least five of these.

Many chemicals of animal and plant origin are useful as research tools in physiology and pharmacology. A partial list includes the following:

Tetrodotoxin from puffer fish — It blocks fast sodium channels and is used to study nerve conduction (saxitoxin from shellfish also).

Alpha conotoxin from cone snail — This is a nondepolarizing neuromuscular blocking agent.

Mu conotoxin from cone snails — This acts like tetrodotoxin.

Omega conotoxin from cone snails — This is a specific inhibitor of presynaptic, voltage-dependent Ca^{2+} channels.

Russell's viper venom — This activates Factor X in the clotting system and is used in certain clotting tests and coagulation research.

Alpha bungarotoxin from the Banded Krait — This is an irreversible blocker of acetylcholine receptors.

Beta bungarotoxin from the Banded Krait and the Black Widow spider — This causes massive release of peripheral neurotransmitter vesicles.

Apamin from bees — This is a blocker of K^+ channels.

Charybdotoxin from the scorpion — This is also a potent K^+ channel blocker.

Digoxin from foxglove — This blocks Na^+/K^+ ATPase and is used to treat congestive heart failure.

Atropine from nightshade — A muscarinic blocker, it has many uses.

Tannins from a variety of plants are used in astringent lotions.

Cytochalasins from fungi — These fix cell membranes and microtubules *in vivo*.

Phorbol myristate from croton oil — An activator of protein kinase C, it is used to study calcium intracellularly.

Colchicine from autumn crocus — It dissolves microtubules and arrests mitosis. It is also used to treat acute attacks of gouty arthritis.

Capsaicin from chili peppers — It is used as a counterirritant in liniments and ointments to provide heat through vasodilitation.

Vincristine and vinblastine from periwinkle — Arrest mitosis; these are used as anticancer drugs.

Recent research into the nature and chemical composition of polypeptide venoms has led to their availability in pure form as research tools. Some examples are as follows:

From the Eastern Green Mamba (*Dendroaspis angusticeps*) — α-dendrotoxin; it blocks certain voltage-gated K^+ channels. β-dendrotoxin blocks certain voltage-gated K^+ channels in synaptosomes and smooth muscle cells.

From the Australian Taipan (*Oxyuranus s. scutellatus*) — Taicatoxin; selectively blocks high threshold voltage-gated Ca^{2+} channels in heart cells.

From the Black Widow spider (*Latrodectus tradecemguttatus*) — α-latrotoxin; a 130,000 Dalton protein, it is the principal toxic component of the venom, causing massive exocytotic secretion of neurotransmitter vesicles both centrally and peripherally.

Numerous other examples exist.

■ REVIEW QUESTIONS

Answer the following questions True or False.

1. Crotalid venoms are predominantly anticoagulant._____

2. Hyaluronidase in snake venoms helps to disseminate the poison at the site of the bite.____

3. Omega conotoxin blocks presynaptic, voltage-gated calcium channels.____

4. Ciguatoxin does not biomagnify up the food chain.____

5. Beta bungarotoxin is found only in the venom of the Banded Krait.____

6. Potassium channel blockers are found in the venom of the Brown Recluse spider.____

7. Urushiol is an astringent found in the horse chestnut.____

8. Hyoscyamine is the same as scopolamine.____

9. Alpha bungarotoxin is an irreversible blocker of acetylcholinesterase.____

10. Cytochalasin fixes microtubles *in situ.*____

11. The venom of vipers is predominantly neurotoxic.____

12. Tetrodotoxin is a potassium channel blocker.____

13. Saxitoxin is synthesized by dinoflagellates.____

14. Lily of the valley contains cardiac glycosides.____

15. The bite of the Brown Recluse spider is extremely painful.____

For the following questions, match the statements with appropriate response from the list below.

    a. Vinegar
    b. Traumatizing the area
    c. A tension bandage over the entire affected limb
    d. Phorbol myristate acetate
    e. Tetrodotoxin
    f. Domoic acid
    g. Ciguatoxin
    h. Okadaic acid

16. Causes generalized paralysis due to fast sodium channel blockade.____

17. The cause of amnesiac shellfish poisoning.____

18. Directly activates phosophokinase C.____

19. General first aid for any snakebite.____

20. First aid for the sting of a bluebottle (Physallis).____

21. May reduce the pain of an imbedded sea urchin spine._____

22. The cause of diarrhetic shellfish poisoning._____

23. May cause gastrointestinal and neurological symptoms when large marine scalefish are eaten._____

ANSWERS

1. T; 2. T; 3. T; 4. F; 5. F; 6. F; 7. F; 8. F; 9. F; 10. T; 11. F; 12. F; 13. T; 14. T; 15. F; 16. e; 17. f; 18. d; 19. c; 20. a; 21. b; 22. h; 23. g.

CASE STUDY #20

In 1988, several patrons of a restaurant experienced signs and symptoms of illness including nausea, headache, dizziness, facial flushing, and diarrhea. The symptoms onset about 5–60 min after the meal (median 38 min) and persisted for about 9 hr. Only these six patrons (four males, two females) experienced problems even though an estimated 50–60 had partaken of the same buffet lunch.

Q. What questions would you wish to ask of the affected and the unaffected patrons?

Q. What possible causes of this reaction could there be?

Several of the affected individuals noted upon questioning that a particular fish dish had a "Cajun" or peppery flavor.

Q. Does this help to identify the problem?

CASE STUDY #21

In June of 1990, six fishermen aboard a private fishing boat off the Nantucket coast of Massachusetts developed symptoms that included numbness and tingling of the mouth, tongue, throat and face, vomiting, loss of sensation in the extremities, periorbital edema and 24 hr later, lower back pain (in all six). The initial symptoms persisted for about 14 hr, the back pain for 2–3 days.

Q. What organ system is primarily affected?

Q. What information would you want to obtain from the victims?

It emerged that all six men had comsumed blue mussels at the same meal. The blue mussels had been harvested in deep water about 115 miles off shore. The mussels had been boiled for about 90 min and were consumed with boiled rice, baked fish and a salad. There appeared to be a correlation between the severity of the symptoms and the number of mussels consumed.

Q. What is the likely cause of the poisoning?

Q. What fish could have been responsible for the same array of symptoms?

■ CASE STUDY #22

Over a period of 72 hr in August, eight seasonal tobacco workers were admitted to a regional hospital with a variety of signs and symptoms that included weakness, nausea, vomiting, dizziness, abdominal cramps, headache and difficulty in breathing. They had all been working in the fields in the morning following an evening of steady rain. The average time of onset of the symptoms was 10 hr after commencing work. All patients were males, 18–32 years of age. All required hospitalization for 1 or 2 days.

Q. Is this likely an occupational disease, a food poisoning from something in the breakfast meal, or an infection?

Q. What occupational hazards might these workers encounter?

Q. What lab tests might help in the differential diagnosis?

■ CASE STUDY #23

On August 9, eight persons were admitted to the emergency department of a Florida hospital with one or more of these symptoms: cramps, nausea, vomiting, diarrhea, chills and sweats. All reported having eaten amberjack, a predatory scalefish, at a local restaurant within the preceding 9 hr (mean time to symptoms 5 hr). Three of the victims required hospitalization. These symptoms persisted for 12–24 hr. Within 48–72 hr, most of these patients developed pruritus and parathesias of the hands and feet and muscle weakness.

Subsequent investigation uncovered 14 similar cases, all of whom had eaten amberjack at one of several local restaurants. These received the fish from the same supplier in Key West.

Q. What organ systems are involved in this intoxication?

Q. Does the evidence point to restaurant kitchens as the source of the toxin?

Q. What potential causes of this problem must be considered?

Q. Which is your choice?

■ CASE STUDY #24

In the fall of 1992, two young men were foraging in the Maine woods for wild ginseng. Several plants were collected. The younger man, aged 23, took three bites from the root and his 39-year-old brother took one bite from the same root. Within 30 min, the younger man vomited and began to convulse. They walked out of the woods and received emergency rescue within 45 min of the onset of symptoms. At this point the man was unresponsive, cyanotic and had tachycardia, dilated pupils and perfuse salivation. He had several clonic-tonic convulsions, developed ventricular fibrillation and was dead on arrival at the local hospital despite resuscitative attempts. The older brother was not showing symptoms at this time and was given gastric lavage and activated charcoal. Some time later, he developed delirium and seizures. He recovered with symptomatic treatment.

Q. What is the likely source of this problem?

Q. What organ systems are involved?

Q. What is the plant and the toxin that would cause this array of symptoms?

> *Massasauga rattlesnake, eat brown bread.*
> *Massasauga rattlesnake, fall down dead.*
> *If you catch a caterpillar, give it apple juice.*
> *But if you catch a rattlesnake, TURN IT LOOSE!*
> Old Ontario skipping rhyme

■ FURTHER READING

Culotta, E., Red menace in the world's oceans. *Science*, 257, 1476, 1992.

Dickey, R.W., Fryxell, G.A. et al., Detection of the marine toxins okadaic and domoic acid in shellfish and phytoplankton in the Gulf of Mexico. *Toxicon*, 30, 355, 1992.

Dreisbach, R.H., *Handbook of Poisoning,* 11th Edit. Lange Medical Publishers Los Altos, 1983.

Dunson, W.A., *The Biology of Snakes*, University Park Press, Baltimore, 1979.

Edmonds, C., Lowry, C. and Pennefather, J., Dangerous marine creatures, in *Diving and Subaquatic Medicine,* 3rd Edit. Butterworth-Heineman, Oxford, 1992, chap. 24.

Evaluation of certain food additives and naturally occurring toxicants. *39th Rep. Joint FAO/WHO Expert Comm. on Food Additives*, 828, 30, 1992.

Gilman, A.G., Goodman, L.S., Rall, T.W. and Murad, F. (Eds.), *Goodman and Gilman's The Pharmacological Basis of Therapeutics,* 7th Edit. Collier Macmillan, Toronto, 1985.

Hall, R.L. and Dull, B.J., Comparison of the carcinogenic risks of naturally occurring and adventitious substances in food, in *Food Toxicology: A Perspective on the Relative Risks,* Taylor, S.L. and Scanlan, R.A. (Eds.), Marcel Dekker, New York, 1989.

Hashimoto, Y., *Marine Toxins and Other Bioactive Marine Metabolites,* Japan Science Society Press, Tokyo, 1979.

Lampe, K.F., Toxic effects of plant toxins, in *Casarett and Doull's Toxicology: The Basic Science of Poisons,* Klaassen, C.D., Amdur, M.O and Doull, J. (Eds.), Macmillan, New York, 1986, chap. 23.

Noble, R.C., Death on the half-shell: hazards of eating shellfish. *Pers. Biol. Med.,* 33, 313, 1990.

Olivera, B.M., Gray, W.R. et al., Peptide neurotoxins from fish-hunting cone snails. *Science,* 230, 1338, 1985.

Phelps, T., *Poisonous Snakes,* Poole Press, Dorset, 1981.

Russell, F.E., Toxic effects of animal toxins, in *Casarett and Doull's Toxicology; The Basic Science of Poisons,* Klaassen, C.D., Amdur, M.O. and Doull, J. (Eds.), Macmillan, New York, 1986, chap. 22.

Scrombroid fish poisoning. *Morbid. Mortal. Weekly Rep.,* 38, 140, 1989.

Suput, D. and Zorec, R. (Eds.), Toxins and exocytosis. (proceedings of a symposium). *Ann. NY Acad. Sci.,* 710, 1994.

Sutherland, S.K., *Venomous Creatures of Australia,* Oxford University Press, Melbourne, 1985.

Trease, G.E. and Evans, W.C. (Eds.), *Pharmacognosy,* Bailliere Tindal, London, 1971.

Water hemlock poisoning — Maine, 1992. *Morbid. Mortal. Weekly Rep.,* 43, 229, 1994.

12 RADIATION HAZARDS

■ INTRODUCTION

The electromagnetic spectrum encompasses all forms of radiant energy. Table 11 lists the various components and their wavelengths in decreasing order. Ionizing radiation, that portion of the spectrum that can cause serious cell damage, includes all wavelengths of 1,000 Å or less (Figure 40). Ionizing radiation, by stripping electrons from molecules as it passes through tissues, produces ionized species of everything from H_2O to macromolecules like DNA. These ionized species are unstable and reactive and can produce dramatic disruptions in cell function including mutation. Figure 40 illustrates the radiation spectrum and the location of the cutoff for risk. There is still controversy regarding the degree of risk.

Radiophobia, an illogical fear of radiation hazards, has led to considerable controversy over the extent of the environmental risks of ionizing radiation. Consider the following conflicting statements:

Dr. K.Z. Morgan, a pioneer in health physics: "It is incontestable that radiation risks are greater than published."

C. Rasmussen, nuclear engineer at MIT: "There is a lot of evidence that low doses of radiation not only don't cause harm but may in fact do some good. After all, humankind evolved in a world of natural low-level radiation." (About 82% of our total radiation exposure comes from natural sources.)

R. Guimond, EPA: "We can't avoid living in a sea of radiation."

In some cultures, deliberate exposure to natural-source radiation is done in the belief that it has curative powers. In Japan, exposure to radon is courted in "radon spas" where natural hot springs occur.

TABLE 11
TYPES OF RADIATION AND THEIR WAVELENGTHS

Type of radiation	Wavelength
Radio waves	30 km to 3 cm
Microwaves	3 cm to 10 mm
Thermal (heat)	.078 to .001 mm
Infrared (includes thermal portion)	.5 mm to 10,000 Å
Visible light	7,800–4,000 Å
Ultraviolet	4,000–1,850 Å
Extreme ultraviolet	1,850–150 Å
Soft X-rays	1,000–5 Å
X-rays	5–.06 Å
Gamma rays	1.4–.01 Å
Cosmic rays	to 1/10,000 Å

ELECTROMAGNETIC SPECTRUM
IONIZATION OF TISSUES OCCURS AT WAVELENGTHS AT OR BELOW 1000 °A.
1cm = 100,000,000 °A (1 °A = 10^{-8} cm).
1000°A = 10^{-5} cm.

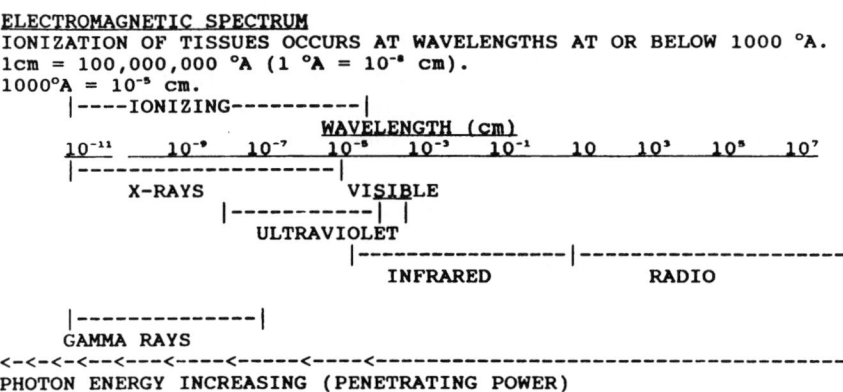

Figure 40 Electromagnetic spectrum and radiation hazards.

■ SOURCES AND TYPES OF RADIATION

The Cause of Radiation
Elements that exist in an unstable form are continually decaying to more stable ones. In the process, they give off energy in several ways. Ionizing radiation arises when an unstable nucleus gives off energy. An unstable nucleus is called a radionuclide. In contrast, X-irradiation is a form of cosmic ray and also occurs when a suitable target such as tungsten is bombarded with electrons. It does not arise from nuclear decay.

Sources
Natural sources of radiation include cosmic rays from space, solar rays that intensify during solar storms (sun spots), radiation emanating from rocks and ground water, and radiation coming from within

our own bodies, mainly from decay of radioactive potassium in muscle. Of considerable concern at present is the exposure to radon gas, a radioactive decay product of radium, a common radioactive element in soil and rock. This will be discussed in more detail below.

Anthropogenic sources of radiation include medical X-rays and radioisotopes, ion sensors in smoke detectors, uranium used to provide the gleam in dentures, mantles in camping lanterns, radioactive wastes, nuclear accidents (e.g., Chernobyl) and careless handling of nuclear materials such as the Cesium 137 that was discarded in a dump in Brazil and which ultimately killed 4 people and contaminated 249. Until fairly recently, radium was used to hand-paint luminous watch dials. Workers used to "point" their brushes by running them between their lips, a practice that led to cases of radiation sickness.

Types of Radioactive Energy Resulting from Nuclear Decay

1. A nucleus can eject two protons and two neutrons to lose mass and convert itself into another element. The ejected components constitute an alpha particle. An alpha particle is slow-moving and cannot penetrate skin, but can cause dangerous ionization if ingested. An alpha particle is actually identical to the nucleus of helium.
2. The neutron of a nucleus can lose an electron to become a positron. The lost, negatively charged particle is called a beta particle.
3. Even after emitting alpha or beta particles, a nucleus may remain in an agitated state. It may rid itself of excess energy by giving off gamma rays. These are short, intense bursts of electromagnetic energy with no electrical charge. They can penetrate lead and concrete and can cause extensive tissue damage by ionization.
4. Neutrons are ejected from the nucleus during nuclear chain reactions. They collide and combine with the nuclei of other atoms and induce radioactivity of the above types. This is the primary source of radioactivity following a nuclear explosion.

■ MEASUREMENTS OF RADIATION

There are two, different types of measurements of ionizing radiation. One is concerned with the level of energy actually emitted from the source, and the other is concerned with the amount of tissue damage that can be produced by a particular form of radiation. The field is unfortunately further confused by a more recent shift to the international (SI) system. The equivalent values are shown in Table 12. They are not easily interchangeable.

Measures of Energy

1. As the nucleus of an atom decays, it gives off a burst of energy (the ionizing radiation) called a "disintegration". The number of disintegrations

TABLE 12
EQUIVALENT VALUES, OLD AND NEW SYSTEMS

Old	New (Si)
1 curie (Ci)	37 gigabecquerels (Gbq)
1 rem (rem)	10 millisieverts (mSv)
100 rem	1 Sievert (Sv)
1 rad (rad)	10 milligrays (mGy)
100 rad	1 gray (Gy)
1 roentgen (r)	258 millicoulombs/kg (mC/kg)

per unit time varies with the nature of the source. Various counting instruments (scintillation counters, etc.) measure disintegrations per minute (dpm). The basic unit of measuring radiation energy is the curie (abbr. Ci), and it is the number of disintegrations (3.7×10^{10}) occurring in 1 sec in 1 gm of Radium 226. Radioisotopes used in science are usually provided in milli Ci (mCi) amounts. The new unit is the becquerel (Bq). A Bq represents one disintigration per second. 1 Ci = 37×10^9 Bq (37 gigabecquerels). Table 12 gives equivalent values for the old and new systems.

2. The first unit of radiation was the roentgen (abbreviated r). It has a complicated definition based on the amount of X-rays or gamma rays required to cause a standard degree of ionization in air.

Measures of Damage

1. The earlier measure of radiation damage is the rem, which stands for roentgen-equivalent-man. It is that amount of ionizing radiation of any type that produces in humans the same biological effect as 1 r. The new international unit is the sievert (Sv). 1 sievert = 100 rem.
2. The rad is the amount of radiation absorbed by 1 gm of tissue. The new international unit is the gray (Gy). 1 gray = 100 rad.

A rough scale of toxicity is as follows: 10,000 rem is rapidly fatal because of damage to the CNS. Whole body exposure to 300 rem is about the LD_{50}. Between 100 and 300 rem, radiation damage occurs. The assumption is made that the risk associated with radiation is linear all the way to zero. When it comes to assessing carcinogenic potential, however, accuracy is extremely difficult. Below 10 rem, effects are unclear due to confounding factors like smoking, pollution and diet. Over 300 agents have been shown to be carcinogenic in animal tests. Residents of Denver have lower cancer death rates than those of New Orleans despite higher radiation exposures because of increased levels of cosmic radiation at their high altitude.

SOME MAJOR NUCLEAR DISASTERS OF HISTORIC IMPORTANCE

Hiroshima

The group of people from whom the most reliable data have been gathered concerning radiation hazards are the survivors of the atom bomb dropped on Hiroshima at 8:15 a.m., August 6, 1945. In 1947, the Radiation Effects Research Foundation was established. Exhaustive studies have shown that the heavily exposed people, called the "hibakusha", had a 29% greater chance of dying from cancer than normal. Excess numbers of leukemia cases began appearing in the late 1940s and peaked in the early 1950s, but by the early 1970s, they had dropped to levels near those of unexposed Japanese. Now the surviving hibakusha have longer life expectancies than the overall population, perhaps because of closer medical supervision. One of the most feared hazards of radiation is that of congenitally deformed infants because of radiation-induced genetic defects in the mother. While such defects have been demonstrated experimentally, the Hiroshima study compared 8,000 children of hibakusha with 8,000 of unexposed Japanese and found an incidence of chromosome damage in 5/1,000 of the former and 6/1,000 of the latter. Protective mechanisms may be functioning in humans (e.g., spontaneous early abortion). Fetuses exposed in utero are a different story. Dozens of mentally retarded infants were born in the areas around Hiroshima and Nagasaki (target of the second atomic bomb) in the months following the blasts. Abortions were also numerous. The fetus appears to be most vulnerable between 8 and 15 weeks. One recent development, however, is that a special panel formed by the U.S. National Research Council recently released a report following a reassessment of the Hiroshima and Nagasaki data and concluded that the levels of exposure were much lower than previously calculated. Original estimates were based on tests at the Nevada nuclear test site using much flimsier buildings than were actually present in those cities. As a consequence, the established safe limits have had to be revised downward. This is mainly a concern for persons exposed to radiation on the job, but it has revived the controversy about whether there is any safe level of exposure. Ironically, evidence is now surfacing that scientists and technicians who worked on the atomic bomb project during the war are showing up with elevated incidences of cancer which, when adjusted for exposure level, may be even greater than those of the hibakusha. The new safe exposure limit is 20 mSv/yr averaged over 5 years with no more than 50 in any 1 year. The old level was 50 mSv.

Chernobyl

The most recent and highly publicized nuclear disaster was Chernobyl in April of 1986 (a much worse but largely concealed disaster occurred in Russia in 1958). Twenty percent of the plant's radioactive iodine escaped along with 10–20% of its radioactive cesium and other isotopes. 135,000 people lived in a 30-km radius of the power plant. There were 30 deaths and 237 cases of severe radiation injury. Two thousand children have been born to women who were living in the accident zone at the time of the disaster. No abnormalities have yet been detected in them. An examination of about 700,000 people over a wider area has so far not revealed any physical problems. Russian scientists estimate an increase in the cancer rate of 0.04% over the next 20 years. In western Europe, exposed to the drift of radioactive dust, it is estimated that, over the next 50 years, 1,000 additional cancer deaths will occur. Normally, there would be 30,000,000 cancer deaths in this period, so the increase is 0.003%. Aside from those people directly exposed to the effects of the explosion, the greatest risk of exposure seems to come from eating contaminated food. Thousands of reindeer had to be destroyed in northern Scandinavia because they had grazed on contaminated pasture.

Dr. Marvin Goldwin, chief of the Joint U.S.-Soviet Medical Team, made the following points in a recent report.

1. Everyone in the Northern Hemisphere received a small dose of radiation. The degree of exposure of those people at highest risk cannot be accurately identified.
2. Radioactive iodine posed an early risk to the thyroid glands of exposed people.
3. As of 1991, there was no detectable increase in the incidence of cancer, but leukemia may yet show up, and solid tumors may not show up for 10 years.

Chernobyl provides another example of how psychological damage can often exceed physical damage in an environmental disaster. Thousands of people received a radiation exposure which exceeded the maximum recommended lifetime allowance. Since radiation levels in their locales have fallen to low levels similar to those of surrounding areas, it makes no medical or scientific sense to relocate them. Stress and fear, however, create an understandable desire in these people to be moved out of the area, and their wishes are presently being acted upon.

Three Mile Island

The partial core meltdown of the Unit 2 reactor at TMI in March 1979 was largely responsible for bringing the nuclear power program in the U.S. to a halt. Of the "defense in depth" safety features, all but the outer water shield failed. Some authorities claim that even

RADIATION HAZARDS 249

a complete meltdown would not have breached this defense. Despite concerns of nearby residents, only 15 Ci of radioactivity were actually released. The news media exploited the event with sensational reports of "deadly clouds of radioactive gas" and made much of the potential for explosion of a large bubble of hydrogen in the reactor. In fact, there was none because no oxygen was present. The group at greatest risk from radiation were the workers who were involved in the cleanup. U.S. federal regulations limit the maximum exposure of workers in the nuclear industry to 12 rem per year. Workers in the "hot" areas receive about 1 millirem per hour, the equivalent of one chest X-ray. 1987 totals for such workers were about 710 millirem (see Chapter 2 for more on Three Mile Island).

The Hanford Release
In contrast, massive amounts of ^{131}I were deliberately released (for purposes still classified as top secret) from the military nuclear facility at Hanford, Washington state, in the 1940s and 1950s. Some residents may have received as much as 2,295 rem! Again, the greatest source of exposure may have been the consumption of contaminated meat and vegetables. Multimillion-dollar studies have recently been commissioned to seek answers to the degree of risk and to assign responsibility. Obviously, the potential for civil action is considerable.

The consistent element present in all of these disasters, including the discarding of Cesium 137 in Brazil, has been human error, misjudgment and negligence. At Hanford, charges of unsafe practices and antiquated, dangerous equipment are still being made. Conversely, the real danger to the public has routinely been overblown, and events have been exploited by the media and by antinuclear groups. In the minds of many, nuclear reactors equate with nuclear bombs.

■ RADON GAS: THE NATURAL RADIATION

As noted above, 82% of the ionizing radiation to which North Americans are exposed comes from natural sources. It has been estimated that there is enough natural radioactive material in the human body that, if it were a laboratory animal, it would have to be disposed of as hazardous waste! The average annual, natural background exposure in Great Britain is about 1 mSv (0.1 rem or 100 mrem). In Canada, it is somewhat higher because the Canadian Shield (the band of granite rock that spans midnorthern Ontario, Quebec and Manitoba) is rich in uranium deposits. Radon gas is by far the biggest potential health hazard from natural radiation. It is the decay product of uranium, and it seeps up through faults in the substrata of soil and may

TABLE 13
RADON CONCENTRATIONS IN CANADIAN HOMES IN PCI/LITER AIR

City	# Homes tested	# >4.5 pCi/L	%
St. Lawrence P.Q.	432	63	14.6
Sherbrooke P.Q.	905	64	7.1
St. John N.B.	866	51	5.9
Fredricton N.B.	455	26	5.7
Thunder Bay Ont.	627	29	4.6
Charlottetown P.E.I.	813	35	4.3
Sudbury Ont.	722	29	3.8
St. John's NFL.	585	17	2.9
Montreal P.Q.	600	9	1.5
Quebec P.Q.	584	9	1.5
Toronto	751	1	0.1
Vancouver	823	0	0

Compiled from data reported by McGregor, R.G., et al., *Health Physics*, 39, 285–299, 1980.

leak into houses through cracks in basement walls, drains, etc. The advent of airtight houses has increased the risk by trapping radon gas in the house. In the British study, it was calculated that 1,000,000 people were exposed to radon at levels of 5–15 mSv annually. In contrast, only 5,100 workers in their nuclear industry were exposed to levels as high or higher. This is the equivalent of 50–150 chest X-rays annually! The distribution of radon homes is extremely random, with highly contaminated homes located right beside radon-free ones. The federal government conducted a survey of Canadian homes and found that Winnipeg had the highest radon levels of any Canadian city, with high levels also found in parts of Saskatchewan and northern Ontario and Quebec (Table 13). Toronto has low levels.

A map of radon risk areas in the United States, published by *National Geographic*, shows a band running slightly east of north through the middle of Ohio, and another running east-west through New York state. A U.S. federal study surveyed 20,000 homes in 17 states and found that 25% had potentially hazardous levels of radon. Radon was described as the largest environmental radiation health hazard in America. Debate over the degree of risk plagues this area as it does others. The EPA study measured radon levels in basements where they would be highest. Calculations of risk have estimated the lifetime risk of dying from radon-induced cancer at 0.4% for exposed individuals, which is the same as your chances of dying in a fire or a fall. If radon were a manmade carcinogen, it would unquestionably be banned, and most certainly would be the target of antinuclear activists. Nevertheless, there is still controversy regarding the degree of risk or perhaps more correctly, the degree of exposure. A British study calculated that 6–12% of all myeloid leukemias may be attributed to radon exposure, with levels rising to 23–43% in Cornwall, where the highest exposures occur. Worldwide, their calculations suggested 13–25% of all myeloid leukemias in all age groups could be due to radon.

In 1984, a construction worker at the Limerick nuclear generating station near Reading, Pennsylvania, consistently set off radiation alarms despite the fact that he had never worked in a "hot" area. An examination of his home subsequently revealed radon levels of 2,600 pCi/L, the highest ever recorded. North Dakota, with 63% of homes showing levels of 4 pCi/L or greater, leads the states in radon exposure, followed by Minnesota (46%), Colorado (39%), Pennsylvania (37%) and Wyoming and Indiana (26%). After cigarette smoking, radon is probably the most common cause of lung cancer. Radon-222 (^{222}Rn) and ^{220}Rn are the only gaseous decay products of uranium. Their half-life is 3.8 days, and they decay to particles (not gases) including radioactive poloniums, which actually are the alpha-emitting toxins causing cell damage. Alpha particles, unlike gamma rays, can only cause cell damage for a radius of about 70 μm, hence the risk of lung cancer. The increased incidence of leukemias is hard to explain on this basis.

■ TISSUE SENSITIVITY TO RADIATION

In general, tissues with a high rate of turnover are more susceptible to the effects of ionizing radiation. Thus, thyroid, lung, breast, stomach, colon and bone marrow have high sensitivity; brain, lymph tissue, esophagus, liver, pancreas, small intestine and ovaries are intermediate; and skin, gall bladder, spleen, kidneys and dense bone are low. This order of sensitivity roughly parallels the frequency of primary cancer in these tissues. If molecular disruption is sufficient, the cell will die. Since hair follicles and gastrointestinal mucosa have high turnovers, radiation sickness involves hair loss and severe diarrhea. Because bone marrow cells also have a high turnover, repair of DNA may not be complete before replication occurs and the daughter cells may be malignant. This is why leukemia is the most common cancer associated with radiation injury.

Questions regarding the safety of the nuclear industry continue to emerge. In August of 1989, two workers at the Pickering Nuclear Plant in Ontario were mistakenly given unshielded practice equipment to change a new type of fuel rod just recently introduced. They received what were widely reported as the highest levels of radiation ever encountered by workers in the Canadian nuclear industry — 5.6 and 12.2 rem. The annual allowable limit set by the Atomic Energy Commission is 2 rem. The radiation exposure from an average chest X-ray is 15 mrem, which has been calculated to (theoretically) cause one additional cancer per 100,000,000 population. In other words, if the entire population of North America were to receive one chest X-ray, one could expect three additional cancer cases as a result. These workers received about 700× this amount, which would cause an additional cancer per 143,000 people. This risk will be lessened if they are removed to areas where there is no possibility of additional exposure for at least 1 year. The safety maxim that should apply in all situations involving exposure to radiation is ALARA (as little as reasonably achievable). It should be noted that, once again, human error was responsible for this accident.

In Great Britain, a disturbing report was released early in 1990 to the effect that offspring of nuclear plant workers had an increased incidence of birth defects. Since these children are not directly exposed to radioactive material, the conclusion, if the data are correct, is that exposure of the parents (mostly men) caused genetic damage. These results contradict earlier studies, and it has been pointed out that clusters of birth defects occur geographically in the absence of nuclear generators (or other identifiable causes), but the data will be of concern until they can be explained or disproved.

Public pressure has resulted in the cancellation of 50 new nuclear power plants in the U.S. As a result, there is greater reliance on coal-fired generators. About 200 coal miners die each year in mine accidents and an equal number from "black lung disease" (pneumoconiosis). Recent (1992) mine accidents include 26 killed in Nova Scotia, 400 in Turkey and 38 in Russia. Such events rarely cause a ripple of concern among opponents of nuclear energy. There is also the problem of acid rain resulting from sulfur pollution of the atmosphere by coal-fired generators. Has the public traded a potential but high-profile risk for a real and greater one that is less visible? Table 14 compares various sources of radiation encountered by Americans.

■ MICROWAVES

Microwaves are the shortest waves in the radio portion of the electromagnetic spectrum (1 mm–30 cm). They are at very high frequencies (1,000–300,000 megacycles/sec), and they are used in radar, for long distance transmission of phone signals and TV signals, and, of course, in microwave ovens.

TABLE 14

AVERAGE ANNUAL U.S. DOSES

Natural sources — <100 mrem.
Medical and dental X-rays — 78 mrem.
Radio-isotopes — 14 mrem.
Weapons testing — 4–5 mrem. Much higher at down-wind locations in Nevada, Utah, etc.
Nuclear industry — <1 mrem.
Building materials (brick and masonry) — 3–4 mrem.
Total average annual exposure about 200 mrem.
One chest X-ray gives an exposure of 10–15 mrem.

Because of its high dielectric constant, water dissipates energy as heat when exposed to microwaves. At high enough energies, thermal damage may occur in living tissues. Recent concern has been expressed over possible carcinogenic effects of microwaves given off by cellular phones. The energy level is so low, however (<5 watts), that no thermal effects can be detected. Little information is available concerning nonthermal effects of microwaves, but evidence for carcinogenicity is scanty and anecdotal.

■ ULTRAVIOLET RADIATION

Ultraviolet radiation occupies the electromagnetic spectrum between 400 and 4 nanometers: UVa — 400–320 nanometers; UVb — 320–280 nanometers; UVc — below 280 nanometers. An increase of 1–2% of UVb radiation is associated with an increase of 2–4% in skin cancer. Uvb is in the ionizing radiation range and can therefore damage DNA, leading to mutations and cancer. The effect on melanocytes appears to be more complicated.

Melanomas often appear first at sites not directly exposed to sunlight. Tropical and subtropical areas have much higher incidences of skin cancer than temperate zones. In North America, it has been claimed that the incidence of skin cancer has increased by 400% in recent years, presumably because of the destruction of the ozone layer.

Medical Uses of UV Radiation

1. Photophoresis — Blood is removed from a patient and exposed to uv light, then returned to the patient. The technique is useful for treating mycosis fungoides, a complication of skin cancer, and it is promising for some leukemias.
2. UVa is used in conjunction with a photosensitive drug called psorelen to treat the skin lesions of psoriasis. The technique is called puva (psorelen–UVa).

EXTRA-LOW FREQUENCY (ELF) ELECTROMAGNETIC RADIATION

ELF waves are nonionizing radiation waves with extremely long wavelengths (several hundred kilometers) and very low frequencies (<300 Hz). Exposure to artificial ELF fields occurs near high-tension electrical lines and much lower levels emanate from household appliances. Electric blankets especially are thought to be potent sources of ELF exposure because of the close proximity and prolonged contact they entail. Other types of low frequency waves include microwaves, emissions from TVs, and radiofrequency fields. High-tension lines create both electrical and magnetic fields and, although insulation will shield the former, the latter have great penetrating powers and will penetrate the human body. The unit of measure of magnetic fields is the gauss, which has a very complicated definition. Electromagnetic fields are also measured in volts/meter (V/m). The field immediately below a 400,000 V transmission line is 10,000 V/m, dropping to 500–1,000 V/m at a distance of 100 m.

Numerous studies have shown deleterious effects from low levels of ELF radiation including deformities of chick embryos, behavioral alterations and physiological changes. Mice and snails have shown increased sensitivity to heat following exposure to ELF. Experimental findings have not always been consistent, however. Calcium uptake by cells is inhibited in some systems. Most of these studies employed exposures ranging from 30 milligauss (the level emitted by some electric blankets) to 1 gauss and exposure times ranging from 30 min to several days or weeks (in the case of the chick embryo studies). The field strength of the earth is about 0.3–0.6 gauss.

The greatest controversy centers on the interpretation of epidemiological data. Several studies have been conducted that purport to show increased incidences of cancer, especially leukemia, recurrent headache and depression, congenital deformities and other health problems in people living near high-voltage power lines or in high-exposure working environments. These studies have been reviewed by several epidemiologists and have been faulted on several counts, including failure to account for other risk factors, failure to balance control and test groups and the use of inappropriate statistics. In particular, the use of the Proportional Mortality Rate (PMR) has been criticized. This is the practice of expressing mortality rates due to certain causes as a percentage of the total. With this method, a decrease in one cause, (e.g., traffic accidents) leads to an apparent increase in the other causes, even though none has actually occurred or even when a decrease has occurred. Most reviewers concede that there is some evidence of a marginal increase in the risk of leukemia in electrical workers, but qualify this by stating that the question remains open. This has not discouraged Paul Brodeur, a staff writer for the *New Yorker*, from recently publishing a book on the subject entitled *Currents of Death*, to follow up on his previous success with *The Zapping of America* on the subject of the "hazards" of video display terminals.

Two studies in 1988 provided further conflicting evidence. One, conducted in Washington state, found no correlation between the incidence of acute lymphocytic leukemia and exposure to electromagnetic radiation from high power lines located within 140 ft of homes whether compared to actual measures within the homes or to computed values based on the wiring configurations. The other, conducted in Colorado, looked at cancer incidences in children 14 years old and under and found no association between total cancer incidence and exposure levels but did find a modest association for lymphomas and sarcomas. The situation was not clarified at a 1992 meeting in San Diego at which a Swedish group, using more accurate data regarding exposure levels than has been previously available, showed a correlation between exposure and cancer incidence. Conversely, a literature study commissioned by the U.S. government found no convincing evidence of an increased cancer risk for exposure to ELF or electromagnetic fields (EMF).

The results of a massive study of some 224,000 utility workers in Quebec, Ontario and France were released in the spring of 1994. This was a case-control study in which 4,151 new cancer cases identified during the study period were matched with 6,106 cancer-free controls from the same population to eliminate confounding factors such as smoking, etc. Past exposure to electromagnetic fields (EMF) was documented. There was no association observed between the overall cancer incidence and past EMF exposure. This was true for lymphoma, melanoma, lung, stomach and colon cancer. There was a statistically significant association between cumulative exposure and a rare form of adult leukemia, acute nonlymphoid leukemia and its subtype, acute myeloid leukemia. Risk was increased by a factor of 2.41, but a causal relationship was not demonstrated. The level of signifance, moreover, was at the 95% confidence limit, which is usually considered to be the minimum acceptable. There was a nonsignificant association of cumulative exposure to the highest levels of EMF and astrocytoma, a rare brain cancer. Given the size of this study, it seems unlikely that any more conclusive evidence soon will be forthcoming.

One rather peculiar theory attempts to associate geographic areas of high electromagnetic fields with deep underground tension between tectonic plates of the earth. One such area is in Canada, and the suggestion (largely unfounded) is that these areas are associated with luminous atmospheric phenomena, UFO reports and increased incidence of cancer, notably brain tumors!

IRRADIATION OF FOODSTUFFS

The use of ionizing radiation to preserve foodstuffs by killing spoilage microorganisms was proposed as early as 1905 when British patent No. 1609 was issued to "J. Appleby, miller, and A.J. Banks, analytical chemist". They noted that sterilization might be accomplished in the complete absence of foreign chemicals. It was not until the postwar arrival of the "atomic age", however, that sources of ionizing radiation were sufficiently abundant to make this procedure practical and economical. In the immediate post-war era, there was considerable public enthusiasm for the development of nuclear technology for peaceful purposes, and there was good acceptance of the notion of irradiated foods such as milk and vegetables. By 1965, the U.S., Canada, and the former USSR had approved the marketing of certain foods treated with low-dose irradiation, but no manufacturers had taken advantage of this situation. By this time the health concern over strontium 90 levels in milk was widely known, the antinuclear movement was in full swing, and radiophobia was growing. Acceptance of irradiated foods was in rapid decline, and it became replaced with concerns over potential health hazards. These concerns tended to fall into three areas:

1. Foods would be made radioactive and those who consumed them might develop radiation sickness.
2. Foods might be altered in some way, rendering them toxic, even carcinogenic.
3. Microorganisms could mutate to new and horrific pathogenic forms.

The first objection is easily dismissed. Foods are not rendered radioactive by low-dose ionizing radiation. Extensive toxicological testing, including tests of carcinogenicity, over several decades led a United Nations Joint Expert Committee on Irradiated Foods to conclude, in 1980, that "... the irradiation of any food commodity up to an overall average dose of 10 kGy presents no toxicological hazard; hence, toxicological testing of foods so treated is no longer required". The second objection is therefore readily disposed of. The third concern stems from experimental evidence that pure, dilute solutions of glucose become mutagenic for *Salmonella typhimurium* after irradiation, and that polyunsaturated fats, in an oxygen atmosphere, form lipid peroxides when irradiated. Neither change has ever been noted in complex food materials. Nor is there any evidence of a direct mutagenic effect on microorganisms by ionizing radiation under the conditions in which it is employed. Nevertheless, resistance to irradiation varies greatly from bacterium to bacterium with spore formers like *Clostridium botulinum* being very resistant. Ionizing radiation is thus likely more useful to prevent spoilage than to eliminate potential pathogenic organisms. Public acceptance of irradiated foods remains a problem.

■ REVIEW QUESTIONS

For the following questions, answer A if statements 1, 2, and 3 are correct; B if statements 1 and 3 are correct; C if statements 2 and 4 are correct; D if only statement 4 is correct; and E if all statements (1,2,3,4) are correct.

1.
   1. Gamma rays have a longer wavelength than microwaves.
   2. Ionizing radiation includes all wavelengths of 1,000 Å or less.
   3. Only a small fraction of radiation comes from natural sources.
   4. Ionizing radiation strips electrons from molecules as it passes through tissues.

2.
   1. Alpha particles consist of two neutrons and two protons.
   2. An alpha particle is identical to the helium nucleus.
   3. Gamma rays have no charge.
   4. Beta particles have no charge.

3.
   1. A roentgen defines the amount of ionization in air caused by any radiation.
   2. rem stands for "reactive emission material".
   3. Predicting the carcinogenic potential for radiation is very difficult below 10 rem.
   4. ALARA stands for "always let active rediation alone".

4.
   1. The main source of radon in the environment is nuclear power plants.
   2. Alpha-emitting daughters of radon bind to dust particles that lodge in the lung.
   3. Strontium 90 poses no threat to human health.
   4. Radon daughters are likely the second most common cause of lung cancer after smoking.

5.
   1. In some geographic areas, the public may get higher radiation exposures than most nuclear power plant workers.
   2. The thyroid gland is very sensitive to radiation damage.
   3. Neutrons given off during radioactive decay may collide with other atoms to induce alpha, beta and gamma radiation.
   4. Smoke detectors contain a radioactive component in the ion detector.

6. Which of the following statements is/are true?
   1. Alpha particles cause tissue damage only over a very short distance (70 µm).

2. Tissues with a rapid turnover are most vulnerable to radiation damage.
3. Tissues sensitive to radiation damage include hair follicles, bone marrow and gastrointestinal mucosa.
4. Safe limits for radiation exposure have been reduced recently because of new evidence that low exposures cause cancer in mice.

7. 1. Extra-low frequency (ELF) electromagnetic radiation is emitted by high-voltage power lines.
   2. Evidence that ELF radiation causes cancer is inconclusive.
   3. ELF radiation has been shown to affect many biological systems adversely in single-celled organisms and chick embryos.
   4. ELF radiation is given off by electric blankets.

8. With respect to irradiated foods,
   1. Fruit is sometimes irradiated to prevent spoilage.
   2. Low-level irradiation has been shown conclusively to induce mutation in several species of bacteria.
   3. Lipid peroxides have been shown to form from polyunsaturated fats *in vitro* but not *in vivo*.
   4. Irradiated food becomes radioactive.

■ ANSWERS

1. C; 2. A; 3. B; 4. C; 5. E; 6. A; 7. E; 8. B.

■ FURTHER READING

Bowie, C. and Bowie, S.H., Radon and health. *Lancet,* 1, 409, 1991.
Cobb, C.E., Living with radiation. *Nat. Geogr.,* 175(Apr.), 403, 1989.
Cole, L., Much ado about radon. *The Sciences,* Jan./Feb., 19, 1990.
Diehl, J.F., *Safety of Irradiated Foods,* Marcel Dekker, New York, 1990.
Dowson, D.I. and Lewith, G.T., Overhead high-voltage cables and recurrent headache and depressions. *The Practitioner,* 232, 435, 1988.
Editorial. Living under pylons. *Br. Med. J.,* 297, 804, 1988.
Eijgenraam, F., Chernobyl's cloud: A lighter shade of gray. *Science,* 250, 1245, 1991.
Henshaw, D.L., Eatough, J.P. and Richardson, R.B., Radon as a causative factor in induction of myeloid leukemia and other cancers. *Lancet,* 335, 1008, 1990.
Hobbs, C.E. and McClellan, R.O., Toxic effects of radiation and radioactive materials, in *Casarett and Doull's Toxicology: The Basic Science of Poisons,* 3rd Edit. Klaassen, C.D., Amdur, M.O. and Doull, J. (Eds.), Macmillan, New York, 1986, chap. 21.

Jerrard, H.G. and McNeill, D.B. (Eds.), *Dictionary of Scientific Units,* 6th Edit. Chapman & Hall, 1992.

Langone, J., Tracking the radon threat. *Time,* 62, Sept. 1988.

McGregor, R.G., Vasudev, P. et al. Background concentrations of radon and radon daughters in Canadian homes. *Health Physics,* 39, 285, 1980.

Michaelson, S.M., Influence of power frequency electric and magnetic fields on human health, in *Environmental Sciences,* Sterrett, F.S. (Ed.), New York Academy of Science, 502, 55, 1987.

Pool, R., Electromagnetic fields: the biological evidence. *Science,* 249, 1378, 1990.

Report of the working group on electric and magnetic ELF fields. *Electric and Magnetic Fields and Your Health,* Publ. #H46–2/89–140E, Mineral Supply and Services Canada, 1990.

Rutkowski, C.A. and Del Bigio, M.R., UFOs and cancer? *Can. Med. Assoc. J.,* 140, 1258, 1989.

Savitz, D.A., Wachtel, H. et al., Case-control study of childhood cancer and exposure to 60 Hz magnetic fields. *Am. J. Epidem.,* 128, 21, 1988.

Severson, R.K., Acute nonlymphocytic leukemia and residential exposure to power frequency magnetic fields. *Am. J. Epidem.,* 128, 10, 1988.

Shore, R.E., Electromagnetic radiations and cancer. Cause and prevention. *Cancer,* 62, 1747, 1988.

Smith, H., ICRP publ. 50. Lung cancer risk from indoor exposures to radon daughters. *J. Can. Assoc. Radiol.,* 39, 144, 1988.

Stone, R., Can a father's exposure lead to illness in his children? *Science,* 258, 31, 1992.

Stone, R., Polarized debate: EMFs and cancer. *Science,* 258, 1724, 1992.

Tar-Ching, A.W., Living under pylons. *Br. Med. J.,* 297, 1469, 1988.

Theriault, G., Goldberg, M. et al., Cancer risks associated with occupational exposure to magnetic fields among electric utility workers in Ontario and Quebec, Canada and France: 1970–1989. *Am. J. Epidem.* 139, 550, 1994.

Tronnes, D.H. and Seip, H.M., Health risks caused by indoor radon exposure, in *Risk Assessment of Chemicals in the Environment,* Richardson, M.L. (Ed.), Royal Society of Chemistry, London, 1988, chap. 16.

Wertheimer, N. and Leeper, E., Magnetic field exposure related to cancer subtypes, in *Environmental Sciences,* Sterrett, F.S. (Ed.), New York Academy of Science, 502, 43, 1987.

Woolf, A.D., Radon. *Clin. Tox. Rev.,* 13, Aug. 1991.

13 GAIA AND CHAOS: HOW THINGS ARE CONNECTED

Life itself is a religious experience.
James Lovelock

THE GAIA HYPOTHESIS

Formulated in 1965 by the independent British biologist James E. Lovelock and elaborated by Lynn Margulis, distinguished biology professor at the University of Massachusetts, it proposes that certain kinds of life on the planet grow, change and die in ways that lead to the persistence of other life forms. In some circles, this has been interpreted as meaning that life on Earth forms a single, complex continuum, one ecosystem throughout time and space. The Earth, according to this view, can thus be considered as a single organism and its various components as cells in that organism. The name is taken from the Greek earth goddess Gaea. Although Lovelock intended his first book to be taken as a scientific treatise, there was a considerable amount of mysticism and spiritual significance attached to it by segments of the public, and this tended to turn serious scientists away from the theory for a long time. As information accumulated about the role of rainforests in consuming CO_2 and producing O_2, and of wetlands in purifying water, and of ocean phenomena such as El Niño in affecting climate, the idea of Earth as an integrated biosystem gained credibility, especially as it became evident that human disruption of components of it, such as the ozone layer, could have serious consequences for life on Earth.

To a microbiologist like Margulis, the Gaia hypothesis made perfect sense as it seemed, when stripped of its earth goddess mystique, to be simply another, perhaps more complex, example of symbiosis, so commonly encountered in the world of microorganisms. As she pointed out during an address to other microbiologists, "Without the few pounds of bacteria in each of our guts, no one would ever digest food, and without the nitrogen-fixing bacteria in the soil, no food would ever grow in the first place".

Both Lovelock and Margulis take considerable pains to point out that the Earth-as-single-organism view is not what Gaia is all about. To quote Margulis, "...the surface temperature, chemistry of the reactive gaseous components, the oxidation-reduction state and the acidity-alkalinity of the Earth's atmosphere and surface sediments are actively (homeorrhetically) maintained by the metabolism, behavior, growth and reproduction of organisms (organized into communities) on its surface. Gaia is not an individual; it is an ecosystem".

One example of how one organism can affect others is 2 billion years old. At that time, a family of organisms called cyanobacteria were dominant on the earth. They were photosynthetic, and in the process of consuming CO_2 they produced large quantities of O_2 which was poisonous to most other species. Those with resistance to O_2 survived and gradually evolved into the aerobic organisms that came to dominate the planet's species. There has never been as radical a change in the Earth's population since.

The resurgence of interest in the Gaia hypothesis as a result of environmental concerns has had both good and bad consequences. On the plus side, there is increased awareness of the interconnectedness of life on Earth and the possibility that a disruption in one part of an ecosystem can have far-reaching consequences. Less desirable is the proliferation of fuzzy-minded philosophies (we are all one with the universe, etc.) which has led to such new age phenomena as Shirley MacLaine, with her crystals and prior lives (why was everyone a princess or a warrior in a previous life, but never a toilet cleaner?). Beliefs such as these tend to erode interest in, and trust of, science, and this may be reflected in declining enrollments in science programs at a time when society needs to be improving and increasing its science and technology.

■ CHAOS THEORY

In the introduction to his book *Chaos: From Theory to Applications*, Tsonis points out that simplicity and regularity are associated with predictability, whereas complexity and irregularity are virtually synonymous with unpredictability. Chaos, in the language of mathematics, is defined as "randomness" generated by simple deterministic systems.

The word randomness is presented in quotation marks to suggest that the random nature may be apparent and that the determinism may persist, albeit in forms difficult to recognize.

Chaos theory has its roots in bifurcation theory first formulated by Poincare, but it was developed by Edward Lorenz who showed that nonlinear differential equations exhibited final states that were nonperiodic, i.e., apparently random. During investigations using computer simulations of computer networks, it was discovered that, under some conditions, routings became random and chaotic instead of following the orderly sequence which the system had been designed to use. For example, if a request was placed to use a specialized computer in location A (e.g., designed for theoretical mathematics) during a slack period (lunch hour), the request would be honored and several minutes of use might be available. If, however, the request clashed with several other simultaneous ones, it might be rerouted to location B along with other requests and disrupt use of this facility with resulting further rerouting to tertiary locations and the subsequent production of ripples throughout the system. Graphic plots of usage reveal oscillating patterns that are neither organized nor completely random. They have been described as "organized complexity".

An important feature of chaos theory is the existence of so-called low-level attractors. Using the example of a free-swinging pendulum, where x_1 = position and x_2 = velocity, the tendency to return to the equilibrium state, where both x_1 and x_2 = 0 is defined as the trajectory and the equilibrium state is the attractor. In a system in which the effect of friction is offset by a mainspring, a disturbance in the motion of the pendulum will eventually be overcome and it will return to its periodic state. In this case, the cycle is the attractor. When chaotic processes are plotted on phase-space (three-dimensional) graphs, patterns are produced that are not truly random, as they would be if the process was completely disorganized. These patterns are called "strange attractors".

Chaos theory is now being applied to such diverse fields as the physics of fluid mixing and weather forecasting. In the latter regard, computer models suggest literally that a butterfly beating its wings in China can influence weather patterns in North America. While this may seem farfetched in the real world, it illustrates once again the interconnected nature of nature, and it helps explain why accurate weather forecasting is a difficult goal to achieve over a time span of more than a few days. There are a number of models used for weather forecasting. They differ in the physics and other parameters. Chaos theory may provide a means of exploring which model is most appropriate under given conditions.

SOME EXAMPLES OF INTERCONNECTED SYSTEMS

Students of ecology should be highly familiar with the concept of symbiosis and how important is in the maintenance of an ecosystem, but for those whose education has been centered largely on human health, a few examples of how biological events can be interconnected might be illustrative.

A Vicious Circle
In Chapter 11 the subject of fungal infections of cereal grains was discussed and how these caused economic losses measured in billions of dollars in Canada alone. Such economic loss to farmers means fewer dollars to spend on consumer goods. This leads to a slowdown in the manufacturing sector, leading to higher unemployment, leading to further declines in the purchase of consumer goods (a vicious circle) and, eventually, contributes to a recession. This is an economic vicious circle, but a biological one may also occur. Figure 41 illustrates how this might work.

Domino Effects of Global Warming
Even a small increment in mean annual temperature, including a cyclical one, could have profound effects on the biosphere. The current "hotspot" for fungal infestations of grain in Canada is southwestern Ontario. Warming would move the demarcation line northward, so that *Aspirgillus flavus*, the source of the carcinogenic mycotoxin aflatoxin, which cannot survive cold winters, could change its distribution.

Certain insects could also move north. The Africanized honey bee is already in the southwestern U.S. and it would follow a warming trend northward. Mediterranean fruit flies, screw-worm flies, even malarial mosquitoes, could follow. Insects are vectors of many diseases of animals and humans. These would spread along with their hosts. Venomous insects also could move north. The Brown Recluse spider has moved from Florida to Pennsylvania and will undoubtedly cross the Great Lakes at some time if it has not already done so. There is some evidence that ecological disturbances caused by anthropogenic activity may result in viral infections jumping species barriers, especially from animals to humans. Recently, the Marburg virus and the Ebola virus emerged as life-threatening infections of humans and are believed to have originated in monkeys. Every variety of influenza antigen has been identified in ducks and other water fowl. The plague bacillus, *Yersinia pestis*, periodically jumps from rats to humans, carried by fleas. Slash-and-burn agriculture, practiced in Africa for eons, creates an environment favorable to the Anopheles mosquito that carries malaria. Lyme disease has already

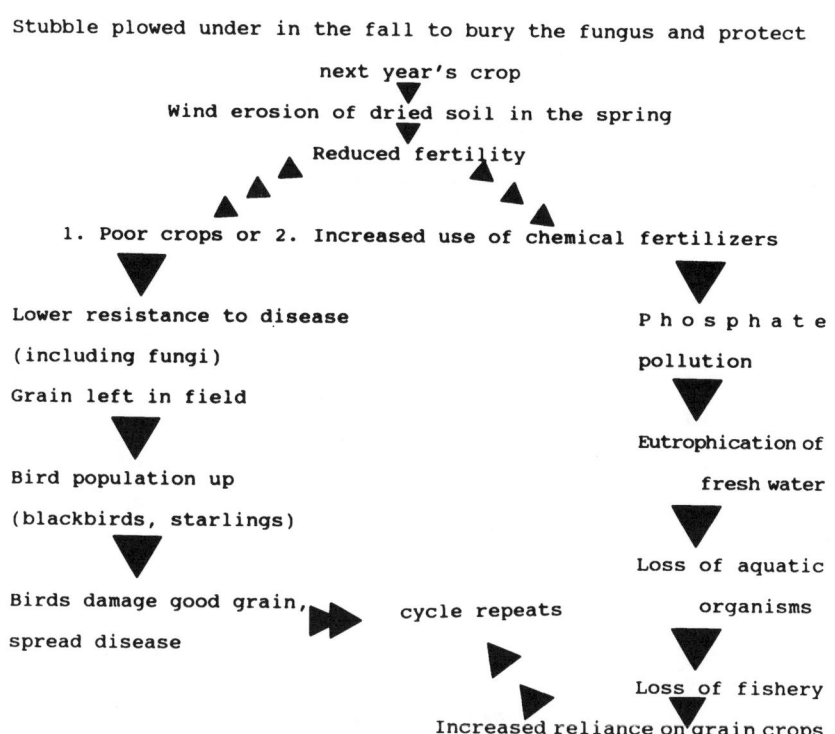

Figure 41 A vicious circle of fungal infection.

reached southern Ontario from its original identification site in New England. A recent outbreak of a viral infection has recently (1993) killed 12 people in New Mexico. The virus (Hantavirus) was identified as belonging to the Hantaan group, which is spread in the feces and urine of rodents. It was first identified during the Korean war as the cause of hemmorhagic fever in soldiers. The condition observed in the American Southwest is called hantavirus pulmonary syndrome (HPS). An exceptionally high yield of piñon nuts, a staple diet for rodents, led to an explosion in the rodent population and a resulting increase in the exposure of humans to their droppings. It has since been identified as a cause of human infection in many states, including Florida. The deer mouse appears to be the most common carrier, but other rodents such as the cotton rat have also been identified as carriers. On a more cheerful note, a longer growing season could mean increased crop yields and more tillable soil.

A Feedback Loop

An elegant example of how an ecosystem can self-regulate is the manner in which water temperature is regulated. It has been shown recently that bioregulation of water temperature occurs in small bodies

```
        PISCIVOROUS FISH INCREASE
            (eg through stocking)
                    EAT
                     ▼
              ZOOPLANKTIVORES
                  (decline)
                    EAT
                     ▼
                ZOOPLANKTON
                  (decline)
                    EAT
                     ▼
                   ALGAE
                  (decline)
                     ▼
            WATER TURBIDITY LOWERED
                     ▼
          LIGHT PENETRATION INCREASED
                     ▼
                WATER WARMED
                     ▼
                ALGAE BLOOM
                     ▼
           LIGHT PENETRATION LESSENED
                     ▼
                WATER COOLED


                 WATER LAYERS

                   SURFACE
-------------------------------------------------
EPILIMNION- freely circulating, warmed by sun
-------------------------------------------------
METALIMNION- layer of rapidly-declining temperature
-------------------------------------------------
-----------------------THERMOCLINE----------------
HYPOLIMNION- layer of fairly constant temperature (3-4° C)
                    BOTTOM
```

Figure 42 Thermoregulation in a small body of fresh water. If fish-eating predators (piscivores) proliferate, they deplete the zooplankton that consume algae. The algae bloom, increasing turbidity and hence lowering temperature which, in turn, inhibits algal growth, allowing more light penetration and a warming effect. A decline in piscivores has the opposite effect.

of fresh water in a manner reminiscent of negative feedback control systems in mammals. The system is shown in Figure 42. Algae are photosynthetic, consuming CO_2 and releasing SO_2 and O_2.

■ FOOD PRODUCTION AND THE ENVIRONMENT

This text is primarily about the relationship between the environment and human health, and it is undeniable that starvation and

malnutrition are the greatest killers of humankind and that
late to the ability of the Earth to feed its population. Some co
ation of this question is thus not inappropriate.

Meat vs. Grain

It is often stated that a vegetarian diet is environmentally friendlier than a diet containing meat because one can produce more food by growing plants than by grazing animals. Besides, animals produce methane, which is a greenhouse gas. This argument is frequently put forth by animal rights activists to support their philosophical position, which also draws heavily on the mystical side of Gaia. This conventional "wisdom" has even appeared in the popular press in articles written by nutritionists. But does it stand up to scrutiny? Consider the following:

1. Livestock can be, and are, grazed on grasslands unsuitable for cultivation and in colder climates with a very short growing season. This occurs in several locations in North America and northern Europe. In Australia and New Zealand, sheep are grazed extensively on land that is harsh and inhospitable to cultivation.
2. Before the North American prairies were plowed up to plant grain, they supported an estimated 50,000,000 bison and millions of pronghorn antelope, elk and caribou (at least seasonally) without damage to the soil. Cultivation coupled with drought brought the dustbowl of the 1930s. In southern Ontario, a dry spring and high winds may result in thick clouds of brown dust coating autos, houses and grassland as topsoil is blown from surrounding farms that were fall-plowed to eliminate the spores of the *Fusarium* mold.
3. Even in semiarid areas such as the American Southwest, studies have shown that the footprints of wild ungulates and cattle form little traps for seeds and water. Grass tufts develop in these that help stabilize the soil. It must be kept in mind that there is a vast difference between intelligent grazing and overgrazing. Much of the evidence against livestock grazing is taken from underdeveloped countries where overgrazing occurs extensively and agricultural technology lags. Another bit of evidence is taken from the Amazon basin, where deforestation for cattle ranching destroys the soil and the rainforest. This is quite true, but deforestation also occurs for paper production (some years ago, Japanese entrepreneurs floated an entire pulp and paper mill up the Amazon) and for crops such as sugar cane. All are equally destructive because the culprit is the deforestation.
4. What about that methane? Best estimates are that the rice paddies of the world produce as much methane as all of the animals combined, both domestic and wild (about 100 megatons/yr). Since rice is a staple grain for much of the world, shifting reliance to it from meat would not prevent as much methane pollution as one might think. Besides, if we increase the wildlife population by protective measures (a laudable goal), we will also be increasing methane

production. Is there good methane and bad methane? How much methane can an elephant make?
5. The average daily caloric intake of 40 countries from the poorest (Bangladesh) to the richest (U.S.) is 2,571 Cal., of which meat provides 205 and vegetables only 41. Even in Guatemala meat provides 49 of 2,020 Cal/day, vegetables only 18. The bulk is from maize (corn) which probably is high in carcinogenic mycotoxins.
6. Water pollution by animal wastes is often cited as evidence that livestock are less desirable environmentally than crops, but we have already seen how the latter can pollute through chemical fertilizers.

There is mounting evidence that world food supplies are declining at a time when the population is exploding. These two factors appear to be on a collision course. The production of cereal grains seems to be declining faster than the production of meat, but fish stocks are also being depleted. A prime example of this latter point is the decline in North Atlantic cod stocks. Overfishing has been singled out as the major culprit, but competition from an expanding seal herd for caplin, a herring-like fish that is the principal food source for both species, has also been incriminated. Surveys indicate that all ground species, including noncommercial ones like monkfish and eelpouts, are in decline from waters off northern Newfoundland all the way up the coast of Labrador, suggesting that natural phenomena may be contributing to the decline. Levels of pollutants such as dioxins and heavy metals are negligible in Atlantic cod, making it unlikely that pollution is the culprit. A possible explanation may be the 0.5–1.0°C cooling of northern spawning waters, which may reduce dramatically the survival of cod fingerlings. El Niño, which involves a massive pooling of warm water in the mid-Pacific, has caused marked reductions in the fish catches off South America.

With regard to cultivated foodstuffs, there is now great concern that the development of special strains of food grains like rice and wheat, with increased resistance to specific diseases, may render them more susceptible to other diseases, some of which may not have emerged as yet. There is now an effort to collect and preserve the wild strains of important food sources such as rice, corn, potatoes and fruits to constitute a genetic library that can be called upon in the future when it is needed. A more recent concern over the environmental impact human-created species relates to the development of genetically engineered species of macro- and microorganisms. Herbicide resistance has been conferred genetically on cultivated plants, resulting in concern that their survival advantage might lead to their invasion of inappropriate habitats. There is concern that rDNA-modified species might interpollenate with wild ones (in the case of plants), attack benign insects (in the case of predatory or parasitic pest control species) or protect undesirable species (as in the case of the anti-frost bacteria *Pseudomonas syringae*). None of these scenarios has yet been identified as a practical problem, but there have been calls for improved methods of risk assessment of genetically engineered species before they are turned loose in the world.

Not all authorities are in agreement regarding the degree of crisis in food production. J. Ausubel, writing in *The Sciences*, predicts that the Earth will hold 8 billion people by the year 2020. While others believe that the current 5.5 billion already is overstressing our food-producing ability, his premise is that advanced technology will cope with the larger numbers. He believes that we are currently in a period of "creative destruction" brought on by the flagrant abuse of credit, both public and private, in the 1980s. The cycle of bankruptcies, fiscal crises and unemployment that followed will in turn be followed by a period of sustained growth that will foster the emergence of new technologies to solve the food crisis. His recommendations, however, focus mainly on the need to foster cooperation among nations to promote sustainable development and reduce the competitive nature that dominates most current international affairs. Apart from stressing the need to reduce our dependence on polluting energy sources, as by substituting natural gas for coal and oil, no specific, innovative measures are offered. Not surprisingly, his article generated a flurry of letters in support of the conventional wisdom that zero population growth must be achieved within the span of the present generation if we are to have any chance of reversing the pollution and starvation generated by overpopulation. There was no discussion, however, of the fact that much of the world's food-producing capacity is underutilized because of market forces that prevent its redistribution to areas of need. In 1992, farmers in Prince Edward Island buried millions of kilos of potatoes because there was no market for them and no agency could be found to convert them to potato flakes for shipment to war-torn areas such as Bosnia, even if they were donated at no cost. American-Canadian trade disputes over wheat exports further illustrate the fact that cost is often the limiting factor in getting food to those who need it.

In his reply to his critics, Ausubel uses a quotation that bears repeating. It goes as follows:

> The most convincing examinations of the phenomenon of overpopulation hold that we humans have by this time become a weight on the earth, that the fruits of nature are hardly sufficient for our needs, and that a general scarcity of provisions exists, which carries with it dissatisfaction and protests, given that the earth is no more able to guarantee the sustenance of all. We thus ought not to be astonished that plagues and famines, wars and earthquakes come to be considered as remedies, with the task, held necessary, of reordering and limiting the excess population.

These words were written by Tertullian, a priest, circa 200 A.D. Were they prophetic or merely alarmist?

■ THE ENVIRONMENT AND CANCER

The public fear that anthropogenic chemicals in the environment may be contributing to cancer incidence has already been noted, as has the existence of natural carcinogens such as certain mycotoxins and radon gas. Statements in the press that a high percentage of cancers are "environmentally produced" are usually taken to mean "anthropogenically produced", with no reference to the existence of natural carcinogens. Indeed, anthropogenic carcinogens may be of greater significance to aquatic organisms than they are to human beings. Cancer statistics for North America indicate that, in fact, the incidence of most cancers is declining. There are some noteworthy exceptions, such as smoking-related lung cancer. In contrast, the apparent "epidemic" of breast cancer in North America has been discounted by both the American Cancer Society and the National Cancer Institute as being a statistical abberation due to better detection and a population bulge in the 20–40-year-old age group of women. Despite the generally encouraging statistics regarding cancer incidence, the (U.S.) National Academy of Sciences recently issued a report "Pesticides in the Diets of Infants and Children" in which it is stated that allowable levels of pesticides may be several hundred times too high for these age groups because of their age-related susceptibility (they consume more food per unit weight and may not be efficient detoxifiers) and because their eating habits may lead them to consume many times more of a food than the amounts used to calculate allowable levels. The economic need for pesticides in agriculture may not be as great as previously thought, and the pressure is increasing to limit their use.

Evidence has emerged recently in Utah of an epidemic of bone cancer, including chondrosarcoma (cartilaginous tumors) which is believed to be associated with an environmental carcinogen. The victims were dinosaurs. They lived 135 million years ago. The problem of cancer obviously is not a new one.

■ FURTHER READING

Ausubel, J.H., 2020 Vision. *The Sciences*, 33, 14, 1993; see also comments and response. *The Sciences*, 34, 7 and 51, 1994.
Fisher, D., *Fire and Ice; The Greenhouse Effect, Ozone Depletion and Nuclear Winter*, Harper and Row, New York, 1990.

Hantavirus infection: southwestern United States: interim recommendations for risk reduction. *Morbid. Mortal. Weekly Rep.*, 42, RR-11, 1993.

Hantavirus pulmonary syndrome — United States. *Morbid. Mortal. Weekly Rep.* 42, 816, 1993.

Huberman, B.A., An ecology of machines: how chaos arises in computer networks. *The Sciences*, July/Aug., 39, 1989.

Joseph, L.E., *Gaia: The Growth of an Idea*, St. Martin's Press, New York, 1990.

Lloyd, S., The calculus of intricacy; can the complexity of a forest be compared with that of Finnegan's wake? *The Sciences*, Sept/Oct., 38, 1990.

Lovelock, J., *Gaia: A New Look at Life*, Oxford University Press, Oxford, 1979.

Lovelock, J., *The Ages of Gaia*, W.W. Norton, London, 1988.

Margulis, L., Gaia in science (letter). *Science*, 259, 745, 1993.

Margulis, L. and Dobb, E., Untimely requiem. *The Sciences*, Jan./Feb., 44, 1990.

Marshall, E., Hantavirus outbreak yields to PCR. *Science*, 262, 832, 1993.

Mazumder, A., Ripple effects; how lake dwellers control the temperature and clarity of their habitat. *The Sciences*, Nov./Dec., 39, 1990.

Miller, H.I. and Gunary, D., Serious flaws in the horizontal approach to biotechnology risk. *Science*, 261, 1500, 1993.

Morse, S.S., Stirring up trouble; environmental disruption can divert animal viruses into people. *The Sciences*, Sept./Oct., 16, 1990.

News and Comment: Experts clash over cancer data. *Science*, 250, 900, 1990.

Ramade, E., *Ecotoxicology*, John Wiley & Sons, Chichester, 1987.

Seielstad, G., *At the Heart of the Web*, Harcourt Brace Jovanovich, New York, 1989.

Tsonis, A.A., *Chaos: From Theory to Applications*, Plenum Press, New York, 1992.

What we eat. *Geo; the Earth Diary*, 3, 139, 1981.

Yanko, D., Are animal disease patterns changing because of global warming? *Vet. Mag.*, 2, 18, 1990.

14 CASE STUDY REVIEWS*

■ CASE STUDY #1

This population of laboring miners has an incidence of a rare cancer many times that of the general population, and those who live near the mine, as well as family members of miners, have a lower, but still elevated, incidence of the same cancer. The incidence of lung cancer in smoking miners is 60× that in nonsmoking miners, and several times higher than in smokers who are not miners.

This is actually the situation in asbestos miners who were first employed 30 years ago or more. The cancer is mesothelioma. The miners carried home the asbestos fibers on their clothing, and these were inhaled by family members, especially the wives who did the family laundry. Living in close proximity to the mine also was a risk factor, as particles were airborne over short distances.

Cigarette smoke and asbestos fibers were acting as cocarcinogens and/or promoters to greatly increase the risk of lung cancer.

Socioeconomic factors that might increase cancer incidence, in addition to smoking, would include diet (high saturated fats and low fiber), high alcohol consumption, lack of exercise, and other workplace hazards (e.g., benzene). Miners would likely be 18–65 years of age and predominantly male, so that the elderly and the very young, as well as women, would be excluded from the group at risk.

* All case studies are based on actual occurrences.

CASE STUDY #2

Working in a confined space always constitutes a risk whenever volatile solvents or gases, explosive or otherwise, are in use. In this particular case, explosion-proof electrical devices (lights, ventilating fans) should have been available. Failing that, the ventilation fan could have been placed at a distance from the site and flexible, large-bore conduit used to conduct air into the tank.

The tank should have been pumped dry before attempting to work in it. A safety person should have been left at the surface to summon help (not to enter the tank alone), and safety lines should have been attached to the workers in the tank so that they could be recovered if they lost consciousness. Breathing apparatus could have been used if available. Ideally, air quality should have been tested before any workers were allowed into the tank (O_2 >19.5%, flammable substances <10%, toxic chemicals to meet published standards).

This accident took place in the Philippines, in a remote area. Epoxy paint solvents may contain a variety of volatile substances that are sedating and heavier than air. In this case, volatile glyceridyl ether was identified.

CASE STUDY #3

This 4-year-old boy had acrodynia, a rare form of childhood mercury poisoning. His 24-hr urine mercury level was 65 µg/L as were those of his mother and siblings. The house was air-conditioned and therefore sealed. The interior latex paint used on the walls of the house was identified as containing about 950 ppm mercury. Seventeen gallons had been applied to the interior walls. The U.S. EPA allows the addition of 300 ppm mercury as phenylmercuric acetate to interior latex paint as a fungicide and bactericide to prolong shelf-life. This limit was exceeded by more than threefold.

CASE STUDY #4

The manure pit is another example of the dangers of working in a confined space. In this case, methane gas from the decomposition of the manure displaced O_2 so that the workers died of asphyxiation. All of the safety measures noted for CS #2 apply here as well. It is almost unbelievable that four men would consecutively enter the confined space, but it is a testimony to how panic can overcome training and common sense. Over 25 people have died in this type of accident in the last 5 years in the U.S.

■ CASE STUDY #5

Yet a further example of a "confined space" problem, these workers was using a chlorofluorocarbon (Freon-113) as a degreaser in a pit beneath a large piece of machinery when they were overcome and one died on route to hospital. The portal of entry was obviously by inhalation, and the target organs were the brain and the heart.

CFCs are CNS depressants, causing narcosis, stupor and loss of consciousness. Asphyxiation may occur from respiratory depression as well as from displacement of O_2. These agents, like the related anesthetics halothane and chloroform, sensitize the heart to adrenaline, with resulting arrhythmias and ventricular fibrillation. A number of years ago, sniffing CFCs in aerosol propellants became a cheap way of becoming "high" in some adolescent cultures. A number of deaths resulted from cardiac arrest. From 1983 to 1990, 12 deaths were recorded in the U.S. from the industrial use of CFCs.

■ CASE STUDY #6

These Inuit soapstone carvers are exposed to the dust of their carving stones. This can cause pneumoconiosis similar to silicosis. Like the asbestos miners, they could carry the dust home on their clothing and thus put their families at risk as well. The use of respirators capable of filtering out the dust, and of special work clothes left at the site, are necessary safeguards.

■ CASE STUDY #7

The variety of symptoms, not all of which are infectious, in these agricultural workers do not suggest that the primary cause is a microorganism but rather an inhaled pollutant. Potential airborne causative agents in this environment would include animal dandruff, dried feces, grain dust, dust from feed additives (antibiotics, sulfa drugs, minerals, etc.), parasites, bacteria, bacterial endotoxins, fungi, ammonia (adsorbed to dust particles) and methane. Many of these agents are allergens, and an allergic component may be present in some individuals.

Corrective measures could include improved ventilation, the use of filter masks or respirators, damping down floors to control dust and removal from this work environment of anyone suffering from allergies or respiratory disease. In many jurisdictions, agricultural workers are not protected by workplace legislation.

CASE STUDY #8

The array of respiratory symptoms is strongly suggestive of an inhaled toxicant. Since ice surfacing machines use internal combustion engines, CO poisoning should be suspected. Carboxyhemoglobin levels in blood should be measured. This was done, and values ranging from 10% to 20% were found (normal levels are <2% for nonsmokers, 5–9% for smokers). CO would account for the headache, nausea and dizziness, but not for the difficulty in breathing nor the coughing up of blood. This is a strong indication of nitrogen dioxide NO_2 poisoning.

The following day, tests were conducted with an ice resurfacing machine equipped with an internal combustion engine. The machine had not been serviced for some time. Levels of NO_2 reached 1.5 ppm. The standard set by the U.S. National Safety and Health Administration (OSHA) is 1 ppm for the Short-Term Exposure Level (STEL) which has a 15 min limit. The CO reached 150 ppm and may have been higher the previous evening. The maximum recommended for ice arenas is 30 ppm.

CASE STUDY #9

These family members became dizzy and nauseated after eating a snack food purchased at a local convenience store. Two of them suffered convulsive seizures. The brain is obviously the primary organ of toxicity. The rapid onset (<1 hr) indicates a preexisting toxin rather than an infectious agent. Since common bacterial toxins do not produce this type of reaction, and since the preparation did not involve seafood or mushrooms, the likely offender is a chemical contaminant.

It is difficult to identify the source of the contaminant since all cases involved the same retail outlet and the same manufacturer. The fact that the snacks were prepackaged would suggest the manufacturer as the source, but the nature of the package was not identified, and it is not impossible that some chemicals could penetrate paper or polyethylene bags.

Endrin is an organochlorine, cyclodiene insecticide. No traces were found in the store or the manufacturing plant. Another possibility is that the flour from which the taquito snacks were made was contaminated. There have been previous incidents of such contamination.

Signs of acute organochlorine intoxication include headache, nausea, vomiting, dizziness, clonic jerking and epileptiform seizures.

■ CASE STUDY #10

This is another case of a worker dying from exposure to fumes from a volatile chlorinated hydrocarbon solvent being used as a degreaser. Trichloroethethane acts like CFCs to sensitize the heart and depress the CNS. Inadequate ventilation is the critical factor in all these cases (refer to CS #2). Chronic intoxication would have probably involved central necrosis of the hepatic lobules.

■ CASE STUDY #11

These demolition workers are cutting up an old iron bridge. The organ systems involved in the toxic reaction are the musculoskeletal system (joint and muscle pain), the CNS (headache) and the gastrointestinal system (nausea). This last symptom could also be central in origin, through stimulation of the chemoreceptor trigger zone. The most likely portal of entry is the lungs. These torch cutters would be wearing coveralls and gauntlets to protect against sparks. The source of the toxicant is thus likely to be vapors from the cutting process.

An old bridge of this nature is unquestionably going to be coated in many layers of lead-based paint. Blood lead levels were performed and ranged from 60–160 µg/dL. U.S. regulations require that workers having levels >60 µg/dL be removed from the work site. The highest level was detected in the barge worker who did not benefit from the nearly constant breeze encountered up on the superstructure of the bridge. The paint from the bridge was found to contain 30% lead by weight.

Treatment was initiated with chelation therapy (EDTA), and substantial amounts of lead were excreted in the urine with an accompanying reduction in symptoms.

The employer was fined for not providing appropriate respirators, clean work clothing and facilities for washing up before lunch and at the end of the day.

■ CASE STUDY #12

This is another case of lead poisoning. Blood lead level was 70 µg/dL in the primary patient and elevated in other family members. In chronic gastric pain that cannot be attributed to ulcer or cancer, a blood lead determination is useful, as gastric distress is a common symptom of lead poisoning. The portal of entry in this case is almost certain to be oral since no activities were in place that could have caused lead vapors. The ceramic jug was suspect because it was imported from Mexico where lead glazing still is used. The jug had been used to store a fermenting beverage, so that considerable time was available for leaching to occur. The entire family partook of this beverage over the course of the summer.

■ CASE STUDY #14

An investigation of the home environment revealed that the house had been built after the banning of lead paints. Water was obtained from a well by means of a galvanized pipe system, eliminating solder joints as a potential source of the lead. No suspicious ceramic or pewter utensils were used for food storage or preparation. All of this pointed to the work environment as the source of contamination.

A 19-day course of treatment was begun with dimercaptosuccinic acid (DMSA), an orally administered lead chelating agent. The patient was instructed to remain off work during the treatment period. At the end of this period, his blood lead level (BLL) fell to 13 µg/dL. One month after returning to work, it was back up to 53 µg/dL, confirming this environment as the source of the lead.

Further careful questioning elicited the information that he habitually chewed on bits of insulation cut from the ends of electrical wires. Analysis of this colored plastic (white, blue and yellow) indicated that it contained 10,000–39,000 µg of lead/G of plastic. He was instructed to desist from this habit and within four months his BLL fell to 24 µg/dL and he reported a subjective improvement in his symptoms.

Lead, usually as lead chromate, is used in pigments employed in the manufacture of colored plastics. Lead salts are used in the manufacture of polyvinyl chloride (PVC) plastics. Thus, any colored plastic may contain lead and should not be chewed or ingested.

Lead sometimes appears as an unexpected contaminant in unusual circumstances. Ethnic health remedies are such an example. Lead, in amounts up to 90% by weight, has been detected in "Azarcon" from Mexico (used as a digestive aid), "Greta", also from Mexico (same use), "Paylooah" from Southeast Asia (applied to inner lower eyelid to improve vision) and a substance from Tibet given to improve development. All of these have resulted in raised blood lead levels in children (20–80 µg/dL) and symptoms of lead poisoning.

■ CASE STUDY #15

These five steam press operators became ill with a variety of signs and symptoms suggesting involvement of the CNS, the blood and the heart. The use of a solvent-borne adhesive raises questions about ventilation of the workplace, the nature of the solvent and its toxicity, whether respirators were in use and, if so were they approved for the task, and whether protective gloves and clothing were in use. Since the steam presses had been in use for many years without incident, it is important to probe carefully to identify any changes in work habits or materials that had taken place recently.

A wide variety of aromatic nitro compounds is capable of causing methemoglobinemia. Remember that nitrites are administered deliberately in cases of cyanide poisoning to form methemoglobin which has a higher affinity for CN than does reduced hemoglobin.

The top ten aromatic nitro compounds implicated in methemoglobin formation include, in order of decreasing potency, ortho-chloraniline, dinitrobenzene, meta-nitroaniline, para-toluidine, nitrobenzene, meta-toluidine, ortho-nitrochlorobenzene, aniline, para-dinitrosobenzene and ortho-toluidine.

Since analysis of air samples was not helpful, analysis of the actual solvent should be undertaken if possible. Analysis of the adhesive in use at the time of the toxic event was undertaken and the results compared with a newly delivered lot. Samples were extracted with carbon disulfide-methanol and analyzed by gas chromatography with flame ionization detection. The "old" sample was found to contain 1% by weight para-dinitrobenzene (pDNB) versus 0.03% in the new sample. Tracing the product back to the manufacturer, it was found that a proprietary solvent used in the preparation of the adhesive was contaminated with pBNB. MetHb levels were monitored after the removal of the contaminated lot and found to be normal. Periodic monitoring was instituted, and workers were required to wear butyl rubber gloves as it was felt that significant skin absorption may have been occurring.

CASE STUDIES #15 AND #16

The rapid onset of symptoms in both cases, the hemodialysis patient and the people who attended the picnic, makes it unlikely that an infectious agent is involved. There are no water-borne bacterial toxins that could account for the symptoms. The CNS appears to be the organ system to which most symptoms are related (sleepiness, dizziness, etc.).

In both of these cases, it emerged that the potable water system was cross-connected with a chilled water system used for air-conditioning. Ethylene glycol was used as an antifreeze. The ethylene glycol diffused into the tap water and was consumed by those at the picnic and diffused across the dialysis membrane into the bloodstream of the patient with renal disease. Ethylene glycol acts first as an intoxicant similar to alcohol. Subsequently, the formation of oxalate occurs. This chelates calcium to form calcium oxalate which precipitates in the kidneys and other tissues to cause renal failure and other tissue damage. Blood calcium levels are low also because of the binding of calcium to oxalate. Ethanol is administered because alcohol dehydrogenase is the metabolic enzyme for ethylene glycol, but ethanol is the preferred substrate.

■ CASE STUDY #17

These patrons of a Chinese restaurant most likely took in something orally. Again, the rapid onset of symptoms suggests a preexisting substance rather than an infectious agent. The signs and symptoms relate to the CNS. The symptoms are not compatible with known pesticide toxicity.

The physician inquired about fish and mushroom consumption because a type of scalefish poisoning produces similar symptoms. Toxic mushrooms (toadstools) also might have been responsible.

The fact that only these three patrons consumed a particular dish is helpful. Inquiry revealed that this particular dish, pork chow yuk, had been prepared by three different chefs. Each had added monosodium glutamate to the full amount, resulting in concentrations three time normal. This is a classical case of "Chinese restaurant syndrome".

■ CASE STUDY #18

This "cropduster" pilot is displaying characteristic signs of excessive cholinergic activity. The signs and symptoms are opposite to those of atropine overdose and include pupillary constriction, visual disturbances, mental confusion and other mental disturbances, profuse sweating, dizziness, weakness and diarrhea. Parathion is an organophosphate insecticide. Questioning revealed that the pilot had been careless in handling the concentrated stock solution, had not worn protective clothing, including gloves, and had not worn a respirator. The signs and symptoms of organophosphate poisoning result from its irreversible inhibition of cholinesterase, and they are frequently delayed. Treatment would consist of atropine to block the effects of acetylcholine overload and pralidoxime to reactivate the acetylcholinesterase enzyme by removing the phosphate group.

Pralidoxime would be contraindicated if a nonorganophosphate inhibitor of acetylcholinesterase had been used, such as a carbamate like Sevin. Since no phosphate is involved, no antidotal response occurs, and the situation could be worsened because pralidoxime has some anticholinesterase activity in its own right.

Plasma cholinesterase levels and erythrocyte acetylcholinesterase would be depressed in this individual.

■ CASE STUDY #19

The signs and symptoms of this outbreak are characteristic of poisoning with a cholinesterase inhibitor. Questioning revealed that 23 of the men had been involved in the mixing, loading or application of mevinphos, an organophosphate insecticide being used for aphid control. It has high toxicity (EPA Class 1). The remaining men had entered and worked in orchards within 24 hr of spraying.

Questioning of the orchard operators revealed that protective clothing or equipment either was not available or had not been used. Respirators, gloves, goggles, coveralls and rubber footwear are recommended and, in some jurisdictions, required by law.

Seven individuals required hospitalization. Their plasma and red blood cell cholinesterase levels were depressed 75–95%. Other workers were treated as outpatients and had levels depressed by 15–25%. Atropine was given to a total of 11 patients.

■ CASE STUDY #20

This restaurant-related syndrome involved a small number of patrons (six) out of four dozen who had eaten lunch there. Careful inquiry should be undertaken to determine whether these six had eaten anything different from the others. It emerged, after questioning these and other available patrons, that the six had consumed yellow-fin tuna. On the surface, this resembles a monosodium glutamate reaction, but the chefs adamantly denied using this flavor enhancer in the fish dish. The persistence of the symptoms, up to nine hr, also argues against MSG as the cause of the reaction.

Three of the affected individuals reported that the tuna had a distinctly peppery or "Cajun" taste, but again, the chefs claimed not to have used any such spicing.

This is a typical case of scrombroid fish poisoning. Analysis of the yellow-fin tuna revealed histamine levels of 50–160 mg/100 g (normal <1 mg/100 g). A telephone survey of hospital emergency departments in the city uncovered nine more cases over a period of 2 days. All had eaten yellow-fin tuna either in restaurants or in the home. Investigation did not elicit any evidence of a serious lapse in refrigeration or handling. The fish were cleaned and packed in ice on board the fishing boat, delivered by refrigerated vehicle to a distributor where they were repacked in ice in smaller lots for delivery to retailers and restaurants. It was extremely hot during the period of the outbreak, however.

All cases were treated successfully with oral antihistamines. In asthmatics and cardiac patients, the condition can be life-threatening, requiring more intensive emergency treatment.

■ CASE STUDY #21

One would want to inquire carefully as to recent ingestion of food, especially seafood, given the environment. These six sport fishermen consumed blue mussels that had been collected in deep water far offshore by commercial fishing boats. This makes bacterial contamination from sewage effluent or other human sources unlikely. The toxin is obviously a neurotoxin. Shellfish toxins that must be considered include domoic acid (amnesic shellfish poisoning), okadaic acid (diarrhetic shellfish poisoning) and saxitoxin (paralytic shellfish poisoning). Tingling and loss of sensation are strongly indicative of saxitoxin poisoning. Significantly, the episode occurred in June, when red tides would be more common in the northern hemisphere.

The symptomatology and mechanism of action of saxitoxin are identical to those of tetrodotoxin from fugu (puffer fish).

Analysis of the remaining blue mussels revealed saxitoxin levels of 25,000 µg/100 g in uncooked ones and 4,300 µg/100 g in cooked mussels. The boiling destroyed much of the saxitoxin; otherwise, the attacks would have been fatal, as they were in a similar incident in which clams were steamed but not boiled. The raw clams contained up to 13,000 µg/100 g of saxitoxin, most of which survived the steaming.

■ CASE STUDY #22

In this case, the differential diagnosis must include food poisoning (some of the symptomatology could be due to *Salmonella* infection or *Staphylococcus* toxin), a chemical contaminant such as a pesticide, or some other cause. The time to onset is not suggestive of poisoning by a preexisting toxin as in Staph food poisoning. Careful questioning might reveal whether a possible source of *Salmonella* existed in the diet, e.g., undercooked eggs, chicken or beef. The symptomatology is reminiscent of poisoning by an organophosphate insecticide (see Case Study #16), but it is unlikely that spraying to control insects would occur so close to harvesting the tobacco.

This is a condition known as Green Tobacco Disease. It is due to the absorption across the skin of nicotine, and it characteristically occurs in periods of wet weather. The signs and symptoms are those of stimulation of nicotinic receptors in the ganglia of the autonomic nervous system and of the neuromuscular junction. Some of the symptoms will relate to parasympathetic stimulation, as do those of organophosphate poisoning.

■ CASE STUDY #23

This poisoning obviously involves the gastrointestinal tract and the peripheral nervous system with different times-to-onset of signs and symptoms. Without knowing the common denominator (amberjack), one would initially have to consider an infectious agent, a contamination occurring in the restaurant and a preexisting toxin. Stool and vomitus cultures were negative for all common bacterial causes of gastroenteritis (*Salmonella, Shigella, Campylobacter and Yersinia*). Monosodium glutamate has a different array of symptoms and a more rapid time-to-onset. Seafoods that could be responsible could be shellfish (diarrhetic shellfish poisoning, paralytic shellfish poisoning) and scalefish (scrombroid and ciguatera poisoning). Shellfish poisoning is ruled out by the commonality of amberjack consumption. The rapid onset of histamine-related symptoms of scrombroid poisoning are lacking. The combination of gastrointestinal and neurological symptoms is characteristic of ciguatera poisoning. Analysis of samples of the amberjack by mouse bioassay was positive for ciguatera-type biotoxins.

■ CASE STUDY #24

The rapidity of onset of the symptoms after ingesting the root points clearly to this as the source of the toxin. There is also a clear dose-dependency as the man who consumed three bites died, whereas his brother, who ate only one bite, survived. This is a classical case of water hemlock poisoning, with cicutoxin as the causative agent. The involvement of the gastrointestinal tract and the central nervous system is typical.

INDEX

A

Abiotic modifiers, 68, 69–71, see also specific types
Absorption, 4–5, 21, 47, 99, 141
Acceptable daily intake (ADI), 53
Acclimation, 71, 72
Acesulfame potassium, 172
Acetaldehyde, 184
Acetaminophen, 26
Acetic acid, 10
Acetone, 151, 154
Acetylation, 24
Acetylcholine (ACh), 13, 28, 195
Acetylcholinesterase (AChE), 28, 77, 195, 196
N-Acetylcysteine, 26
Acetylene torches, 96
Acetylsalicylic acid (ASA, aspirin), 14, 23
N-Acetyltransferase, 24
ACh, see Acetylcholine
AChE, see Acetylcholinesterase
Acidic drugs, 5, see also specific types
Acidifiers, 166, see also specific types
Acidity of water, 78–79
Acidosis, 154
Acid rain, 78–79, 83, 98–99, 252
Acids, 5, 10, 91, 92, 99, see also specific types
Acidulants, 166, see also specific types
Acquired immunodeficiency syndrome (AIDS), 56, 178
Acrolein, 58
Acrylonitrile, 52
Actin, 32
Actinomycin D, 219
Activated charcoal, 26, 154
Active sediment, 68
Acute effects of air pollution, 95
Acute exposure, 2, 42
Acute toxicity, 20, 27–29, 28, 197, 211
Additive interactions, 26
Additive risk factors, 54
Adenocarcinomas, 180, see also Cancer; specific types
Adenylcyclase, 13
Adhesives, 151, 160, see also specific types
ADI, see Acceptable daily intake
Adjuvants, 166, see also specific types
Adrenal gland cancer, 159
Adrenaline, 125, 211

Adriamycin, 219
Aerating agents, 166, see also specific types
Aerosols, 91–92, 100
Aflatoxins, 8, 118, 211–212
Age, 21–22, 47–48, 71, 159
Agent Orange, 77, 113
Agricultural runoff, 73, 74
Agricultural wastes, 75
AHH, see Aryl hydrocarbon hydroxylase
AIDS, see Acquired immunodeficiency syndrome
Air pollution, 91–111
  acute effects of, 95
  case studies on, 109–111, 274–276
  chemical impact of, 98–104
  chronic effects of, 95
  climate change and, 104–105
  gaseous, 91, 94
  health effects of, 95
  natural sources of, 92, 104–105
  remedies for, 106–107
  review questions on, 107–109
  sources of, 92–93, 104–105
  transport of, 94
  types of, 91–92, 94–95
  in workplace, 95–98
Alachlor, 203
Alar, 60, 76, 83
Albedo factor, 103
Albumin, 5, 22
Alcohol consumption, 86
Alcohol dehydrogenases, 27, 154, 155
Alcohols, 11, 27, 28, 153–154, 155, see also specific types
Aldehydes, 58, 155, see also specific types
Aldicarb, 77, 196
Aldrin, 26, 76, 84, 194, 201
Aleukia, 211
Algae, 86, 143, 219, 266
Aliphatic alcohols, 153–154, see also specific types
Aliphatic hydrocarbons, 152–153, see also specific types
Alkalies, 166
Alkaline drugs, 5, see also specific types
Alkaloids, 27, 28, 233, see also specific types
Alkyalting agents, 157, see also specific types
Alkylbenzenes, 156
Allergies, 179

285

Allyl isothiocyanate, 185
Alternative fuels, 106
Alum, 78
Aluminum, 78, 79, 99, 132, 143
Ames test, 49
Amines, 24, 91, see also specific types
4-Aminobiphenyl, 59, 93, 118
Aminoglycosides, 175, see also specific types
Aminomercuric chloride, 138
Aminopterin, 34
Ammonia, 91
Amygdalin, 234
Analgesics, 13, see also specific types
Androgenic hormones, 34, see also specific types
Anemia, 32, 78, 184, see also specific types
   aplastic, 29, 30, 61, 156
   hemolytic, 183
   lead and, 133
   sickle cell, 25
Anesthetics, 6, 11, 125, 151, 152, see also specific types
Aneuploidization, 30
Angiosarcomas, 96, see also Cancer; specific types
Aniline, 184
Animal studies, 2, 42, 43, 48–50, 126
Anitrole, 199
Antagonistic interactions, 26
Anthelmintics, 199, see also specific types
Anthropogenicity, 3, 72
Antibacterial agents, 47–48, see also Antibiotics; specific types
Antibacterial disinfectants, 114–115, see also specific types
Antibiotics, 174–179, see also specific types
   in cancer treatment, 219
   as food additives, 166
   as growth promotants, 174, 175
   microorganism-produced, 223
   resistance to, 25, 177, 178
   teratogenicity of, 219
   toxicity of, 219
Antibrowning agents, 167
Anticaking agents, 167
Anticancer drugs, 12, 25, 30, 32, 33, 34, see also specific types
Anticholinesterase, 198
Anticoagulant protease, 230
Antidotal remedies, 26
Antifoaming agents, 167, see also specific types
Antifreeze, 155
Antifungals, 175, see also specific types
Antiinflammatory agents, 145, see also specific types

Antimalarials, 25, see also specific types
Antimetabolites, see Anticancer drugs
Antimold agents, 167, see also specific types
Antimony, 144
Antineoplastic drugs, see Anticancer drugs
Antioxidants, 167, 170–171, 185, see also specific types
Antipyrine, 22
Antistaling agents, 167
Apamin, 231, 236
Aplastic anemia, 29, 30, 61, 156
Appliances, 254
Aquatic species, 67, 68, 72–73, see also specific types
Aquifers, 73, 94
Aromatic hydrocarbon receptors, 119
Aromatic hydrocarbons, 13, 85, 155–156, see also specific types
   polycyclic, see Polycyclic aromatic hydrocarbons (PAHs)
Arsenate, 29
Arsenic, 52
   in air, 144
   carcinogenicity of, 141, 144
   environmenal effects of, 143
   in herbicides, 142
   history of, 131
   intake of, 142
   in pesticides, 142
   pharmacokinetics of, 142
   poisoning from, 142
   in tobacco, 142
   toxicity of, 141–143
   in water, 78
   in workplace, 95
Arsenic acid, 141
Arsenical compounds, 192, see also specific types
Arsenic pentoxide, 141
Arsenic trichloride, 141
Arsenic trioxide, 141, 142, 192
Arsine, 142
Arsphenamine, 192
Arthropods, 230–232
Artificial flavors, 167
Artificial food colors, 167, 168–170
Artificial sweeteners, 168, 171–173
Aryl amines, 24, see also specific types
Aryl compounds, 24, see also specific types
Aryl hydrocarbon hydroxylase (AHH), 119
Aryl hydrocarbon receptors, 119
Aryl hydrocarbons, 13, 85, see also specific types
ASA, see Acetylsalicylic acid (aspirin)
Asbestos, 36, 52, 93, 96–97, 273

Asbestosis (white lung syndrome), 96
A-solanin, 27
Aspartame, 172
Aspartic acid, 172
Aspirin (acetylsalicylic acid), 14, 23
Astringents, 233, see also specific types
Atmospheric distribution, 93–94
Atomic bombs, 157, 247
Atrazine, 199, 202
Atropine, 26, 28, 198, 233, 236
Automobile accidents, 55
Automobile emissions, 92, 94, 103–104
Autonomic agents, 233, see also specific types
Azathioprine, 24

B

Bacteria, 25, 92, 119, 137, 262, 269
Bacterial conjugation, 177
Bacterial resistance, 178
BaP, see Benzo(a)pyrene
Barbiturates, 26, see also specific types
Basement membranes, 140
Bases, 5, 10, see also specific types
Batteries, 140
Bauxite miners, 78
Bayley scale, 86
BCME, see Bis(chloromethyl) ether
Belladonna alkaloids, 28, see also specific types
Benzene, 27, 36, 52, 59, 273
  carcinogenicity of, 156–157
  in solvents, 155, 159, 160
  in water, 78, 80, 81
Benzene hexachloride, 195, 201
Benzene leukemia, 156
Benzo(a)pyrene (BaP), 8, 26, 36, 72, 120, 185, 233
BHA, see Butylated hydroxyanisole
Bhopal, India accident, 28
BHT, see Butylated hydroxytoluene
Bicarbonate, 78
Bifurcation theory, 263
Binding agents, 167, 216, see also specific types
Binding sites, 5, 15
Bioaccumulation, 72, 191
Bioconcentration, 72
Biodegradation, 123
Biological control methods, 199–200
Biological half-life, 19, 42
Biological oxygen demand (BOD), 75
Biological variation, 14–18
Biomagnification, 67, 72, 76, 191
Biosphere, 67

Biotic modifiers, 68, 71
Biotransformation, 7–10, 12, 15, 20
  age and, 21, 22
  altered, 26
  of chloroform, 125
  genetics and, 23–24, 25
  hepatic, 153
  interactions and, 26
  placenta and, 34
Biotransforming enzymes, 33, 34, see also specific types
Biphenyls, see also specific types
  biodegradation of, 123
  carcinogenicity of, 123
  chlorinated, 85
  disposal of, 124
  exposure to, 123–124
  metabolism of, 123
  pharmacokinetics of, 123
  physicochemical characteristics of, 122–124
  polybrominated, 26, 36, 85, 124
  polychlorinated, see Polychlorinated biphenyls (PCBs)
  toxicity of, 123
Bipyridyls, 76, 198–199, see also specific types
Birth defects, 30, 115, 252
Bis(chloromethyl) ether (BCME), 153, 157–158
Bladder cancer, 93, 118, 168, 172, 185
Bleaching agents, 167
Blood-brain barrier, 6, 34
Blood flow, 6
BOD, see Biological oxygen demand
Body composition, 21, 22
Bone cancer, 270
Bone marrow toxicity, 155
Botanical insecticides, 197, see also specific types
Botulinum toxin, 28, 42, 171, 219
Bradykininogen, 230
Brain cancer, 159
Brake fluid, 77
Brake linings, 97
Brass, 95
Brazelton scale, 86
Breast cancer, 118, 121, 181, 182, 270
Bromides, 12
Brominated vegetable oil (BVO), 170
Bromine, 113, 125, 152
Bromism, 114
Bromodichloromethane, 126
Bromoform, 126
Brush cleaners, 151
Buffers, 167
Bungarotoxin, 229, 236
Butadiene, 50, 51

Butanol, 154
2-Butoxyethanol, 158
Butylated hydroxyanisole (BHA), 170
Butylated hydroxytoluene (BHT), 170
BVO, see Brominated vegetable oil

C

Cadmium, 2, 29, 46, 99
  absorption of, 141
  carcinogenicity of, 141, 144
  intake of, 140
  pharmacokinetics of, 140
  specific gravity of, 132
  toxicity of, 140–141
  in urine, 145
  in water, 69, 78
  in workplace, 95
Calcium, 5, 27, 144
  chelation of, 155
  deficiency of, 141
  in eggshells, 194
  metabolism of, 76
  uptake of, 254
Calcium arsenate, 141
Calcium-binding proteins, 141, see also specific types
Calcium channels, 144, 227
Calcium chloride, 74
Calcium oxalate, 155, 232
Calcium/sodium ethylene diamine tetra-acetate, 135
Camphor, 120
Canadian Shield lakes, 70
Cancer, 30, 84, see also Carcinogenicity; specific types
  adrenal gland, 159
  antibiotics in treatment of, 219
  bladder, 93, 118, 168, 172, 185
  bone, 270
  brain, 159
  breast, 118, 121, 181, 182, 270
  cervical, 35, 56, 180, 181
  Chernobyl and, 248
  cigarette smoking and, 270
  colon, 27, 255
  endometrial, 118
  environment and, 270
  fear of, 55, 60
  gastrointestinal, 96
  kidney, 159
  liver, 81, 117, 118, 121, 125, 211, 233
  lung, 81, 96, 118, 125, 143, 144, 159, 255, 270
  mesothelioma, 96, 273
  nasal, 57, 117
  pancreatic, 126
  prostate, 144
  radiation and, 248, 254, 255
  rectal, 126
  respiratory, 144
  risk of, 2, 43, 45, 46–49
  skin, 99, 144, 159
  socioeconomic factors in, 273
  solvent-related, 156–159
  spleen, 159
  squamous cell, 57
  statistics on, 270
  stomach, 117, 126, 159
  testicular, 158, 182
  uterine, 180, 181
  vaginal, 35, 180
Cancer registries, 81
Capacitors, 122
Caplin, 268
Capsaicin, 236
Captan, 199, 203
Car accidents, 55
Carbamates, 28, 76, 77–78, 193, 196, 199, see also specific types
Carbamic acid, 76
Carbaryl, 196
Carbon, 70, 106
Carbonates, 69, see also specific types
Carbon-based compounds, 91, 92, see also specific types
Carbon dioxide, 101–102, 105, 106, 261, 266
Carbon dioxide sinks, 103
Carbon monoxide, 27, 94, 97–98, 104, 125, 153
Carbon tetrachloride, 29, 36, 78, 113, 124, 152, 153
Carbon trichloride, 113
Carboxyhemoglobin, 104, 125
Carcinogenicity, 1, 2, 3, 8, 9, 13, 24, 29–32, 43
  of aflatoxins, 211
  of arsenic, 141, 144
  of artificial sweeteners, 172
  of benzene, 27, 156–157
  of benzo(a)pyrene, 185, 233
  of biphenyls, 123
  of bis(chloromethyl) ether, 157–158
  of cadmium, 141, 144
  of chromium, 141, 144
  co-, 10, 48, 172, 233
  of copper, 144
  of DDT, 122
  of dimethylformamide, 158
  of dioxins, 42
  of estrogens, 180
  of ethylene oxide, 158–159

INDEX   289

of food preservatives, 170, 171
of fungicides, 199
genetics and, 25
of glycol ethers, 158
interactions and, 26
of lanthanum, 144
of lead, 134, 141, 144
of metals, 134, 141, 144
multistage, 119
of mycotoxins, 185, 268
natural vs. anthropogenic, 50
of nickel, 144
of nitrosamines, 171
of organochlorines, 194
of PCBs, 185
of pesticides, 203
of phorbol esters, 213, 233
of preservatives, 170, 171
of radiation, 253, 254, 255, 256
reliability of tests for, 50–51
risk of, 43
risk assessment and, 43–51
of solvents, 156–159
sources of, 35, 36
steps in, 30–31
of sweeteners, 172
of TCDD, 59, 117–119
tests for, 49, 50–51
of THMs, 126
transplacental, 35–36
Carcinomas, 35, 57, 96, 121, 180, 181, see also Cancer; specific types
Cardiac glycosides, 232, see also specific types
Car emissions, 92, 94, 103–104
Carotenoids, 185
Carrageenin, 170
Case studies, 273–283
  on air pollution, 109–111, 274–276
  on food additives, 188, 280
  on halogenated hydrocarbons, 128–129, 276, 277
  on metal toxicity, 147–148, 277–278
  on pesticides, 205–206, 280–281
  on risk analysis, 64, 273, 274
  on solvents, 161–162, 274, 278–280
  on toxic animals, 238–240, 281–283
  on toxic plants, 238–240, 281–283
Catalases, 137
Catecholamines, 125, 173, see also specific types
Ceiling exposure value (CEV), 53, 54
Cell regeneration, 32–33
Cell repair, 32–33
Cellular toxicity, 49, 133–134
Cervical cancer, 35, 56, 180, 181

Cesium, 2, 245, 248, 249
CEV, see Ceiling exposure value
CFCs, see Chlorofluorocarbons
Chaconine, 233–234
Chaos theory, 262–263
Charcoal, 26, 154
Charybdotoxin, 236
Chelating agents, 26, 135, 138, 141, 142, 145, 155, 167, see also specific types
Chemical warfare, 223
Chernobyl disaster, 2, 57, 245, 248
Chloracne, 12, 58, 77, 114, 115, 116, 117, 123, 197
Chloralkali plants, 138, 139
Chloramphenicol, 21, 30, 61, 177, 178
Chlordane, 26, 194, 203
Chlordane heptachlor, 201
Chlordecone, 194, 202
Chlordiazepoxide, 25
Chlordibromomethane, 126
Chlorides, 12, see also specific types
Chlorinated biphenyls, 85, see also specific types
Chlorinated dibenzofurans, 36, 84, see also specific types
Chlorinated ethylenes, 152
Chlorinated hydrocarbons, see Organochlorines
Chlorinated organics, 83, see also specific types
Chlorine, 5, 99, 100–101, 113, 125
  mercury in manufacture of, 136
  substituents of, 152
Chlorine compounds, 100, see also specific types
Chlorine oxides, 100
Chlorodibromomethane, 126
Chlorofluorocarbons (CFCs), 94, 100, 153, see also specific types
Chloroform, 78, 113, 124, 125–126, 152, 153
  biotransformation of, 125
  in drinking water, 124
  in water, 81
Chloromethanol, 125
Chlorphenoxy acids, 76, 77, 197, see also specific types
Chlortetracycline, 174
Choline esters, 223
Cholinesterase-inhibiting insecticides, 42, see also specific types
Cholinesterases, 230
Chondrosarcoma, 270
Chromium, 69, 132, 141, 143, 144
Chromosomal abnormalities, 30, 34, 142, 156, 247

Chronic effects of air pollution, 95
Chronic exposure, 43, 125
Chronic toxicity, 20, 27–29
Chrysotile, 96
Cigarette smoke, 10, 26, 58, 81, 140, 160
Cigarette smoking, 27, 48, 56, 61, 86, 97, 117, 118
 asbestos and, 273
 cancer and, 270
Ciguatoxin, 224–225
Citric acid cycle, 29
Clastogenesis, 30
Cleaning agents, 1, 151, 159, 160, see also specific types
Clearance, 20
Clear-cell adenocarcinoma, 180
Climate change, 104–105
Coagulant protease, 230
Coal, 103
Coal-fired electrical generating stations, 92
Coal mining, 106, 252
Coating agents, 167
Cocarcinogenicity, 10, 48, 172, 233
Coelenterate toxins, 227
Coke oven emissions, 52
Colchicine, 233, 236
Colon cancer, 27, 255
Colors, 167, 168–170
Combustion, 104
Concentration gradients, 11
Conjugating enzymes, 27, see also specific types
Conjugation, 10, 176, 177
Conotoxins, 227, 235, 236
Convulsants, 235–236, see also specific types
Copper, 69, 70, 95, 132, 144
Copper arsenite, 192
Core samples, 99
Corticosteroids, 145, see also specific types
Crocidolite, 96
Cross resistance, 201
Cumene, 156
Cumulative effects, 18–20
Cyanide, 28
Cyanobacteria, 262
Cyanocobalamin, 137
Cyanogenic glycosides, 234–236, see also specific types
Cyclamates, 172
Cyclodienes, 193, 194, 202, see also specific types
Cyclo-oxygenase, 14
Cyclopropane, 125
Cyhexatin, 203
Cysteine, 137, 140, 144

Cytochalasins, 236
Cytochrome a, 234
Cytochrome P-448, 119, 123
Cytochrome P-450, 8, 27, 29, 48, 120, 125, 126, 153
 DDT and, 193, 194
 glycosides and, 234
 organochlorines and, 193, 194
Cytochrome P-450 oxidases, 152
Cytochromes, 28, see also specific types
Cytosolic enzymes, 12, see also specific types
Cytosolic receptors, 13, 119
Cytotoxicity, 49, 133–134

D

2,4-D, see 2,4-Dichlorophenoxyacetic acid
Daminozide, 60
Data manipulation, 14–18
DDE, see Dichlorodiphenyldichloroethane
DDT, see Dichlorodiphenyltrichloroethane
DDVP, 76
Dealkylation, 8
Debrisoquine, 24
Dechlorination, 123
Defoaming agents, 167
Delaney Amendment, 59–60, 182
Dementia, 78, 99
De minimis concept, 59
Deoxynivalenol (vomitoxin, DON), 215, 216, 217–218
Derosene, 152
DES, see Diethylstilbestrol
Desferoxime, 143
"Designated substances", 52
Desulfuration, 8
Detergents, 75, see also specific types
Detoxification, 84, 215–218, 234–235
Dexamethasone, 144
Dhurrin, 234
Diabetes, 11–12, 34
Diacetoxyscirpenol, 214, 215
Diacyl glycerol, 233
Diatoms, 86, 226
Diazepam, 25, 144
Diazinon, 195
Dibenzofurans, 36, see also specific types
Dicarboximides, 199, see also specific types
Dichlorodiphenyldichloroethane (DDE), 122, 194
Dichlorodiphenyltrichloroethane (DDT), 6, 76, 192
 in Antarctic snow, 94

carcinogenicity of, 122
chemical structure of, 193
as enzyme inducer, 26
in fish, 194
in Great Lakes, 84
history of, 192, 193
neurotoxicity of, 76, 122, 193
resistance to, 25, 201
toxicity of, 122, 193
in water, 84
1,2-Dichloroethylene, 80
Dichloromethane (methylene chloride), 81, 124, 125, 153
2,4-Dichlorophenoxyacetic acid (2,4-D), 29, 76, 77, 115, 197
Dichlorvos, 195
Dieffenbachia, 232
Dieldrin, 26, 76, 84, 194
Diesel locomotives, 92
Diethylstilbestrol (DES), 8, 35, 56, 179–183
Diffusion, 5, 11
Digoxin, 236
Dihalocarbonyl compounds, 126, see also specific types
Dihydrodiol, 118
Dimercaprol, 135, 138, 142
2,3-Dimercaptosuccinic acid (DMSA), 135, 145
1,1′-Dimethyl-4,4′-bipyridinium, see Paraquat
Dimethylformamide (DMF), 158
Dimethylhydrazine, 60
Dimethylmercury, 29
Dinitrobenzene, 156
4,6-Dinitro-o-cresol (DNOC), 197, 198
Dinitrophenols, 197–198, see also specific types
Dinoflagellates, 224, 226
Dinosaurs, 270
Dinoseb, 197, 198
Dioxides, 91, 94, 97, 98–99, 103, see also specific types
Dioxins, 13, 58–59, 83, 197, 268, see also Tetrachlorodibenzo-p-dioxin (TCDD); specific types
carcinogenicity of, 42, 46
as enzyme inducers, 26
in Great Lakes, 84, 202
physicochemical characteristics of, 115, 116–119
toxicity of, 42
in water, 76, 77, 84
Diquat, 198
Discharge, 67, 68, 74, 92
Disinfectants, 114–115, see also specific types
Disposal, 80, 124

Dissolved organic carbon, 70
Distribution, 4, 5–6, 15, 44, 93–94
Disulfide groups, 142, see also specific types
Dithiocarbamates, 199, see also specific types
Diuretics, 138, 144, see also specific types
DMF, see Dimethylformamide
DMSA, see 2,3-Dimercaptosuccinic acid
DNA, 13, 29, 30, 48
DNA adducts, 118
DNA genotyping, 24
DNA-modified species, 268
DNOC, see 4,6-Dinitro-o-cresol
Domoic acid, 226
DON, see Deoxynivalenol
Dopamine, 13, 211
Dosage, 1
defined, 16
estimated maximum tolerated, 50
lethal, 16, 69–70, 77, 84, 123, 197
toxic, 16
Dose response, 15–17
Dow Chemical, 138
Doxorubicin, 219
Drainage, 73
Drinking water, 67, 99, 124, 125–126, see also Water pollution
Drug residues, 174–183, see also specific types
review questions on, 186–188
Drugs, 1, 3, see also specific types
acidic, 5
adverse reactions to, 26
age and response to, 71
alkaline, 5
anticancer, 12, 25, 30, 32, 33, 34
antimalarial, 25
antineoplastic, see Anticancer drugs
biotransformation of, 23–24
breakdown of, 10
dose response to, 15–17
elimination of, 10
as enzyme inducers, 26
genetic aspects of response to, 23–24, 71
illegal sales of, 144
interactions of, 12, 26–27
metabolism of, 119
poisoning from, 26
resistance to, 25, 174–179
response to, 23–24, 71
sulfa, 24
teratogenicity induced by, 33, 34
in workplace, 95
Dust, 91, 92, 95, 97
Dust storms, 92

E

Earth's orbit, 103
Earth's temperature, 103
Echinoderm venoms, 227–228
Economic toxicology, 3
Effluents, 68, 75, 178
Electrical generating stations, 54, 92
Electrical lines, 254
Electrical transformers, 76, 122
Electrical workers, 254
Electric blankets, 254
Electrolyte balance, 10
Electromagnetic radiation, 254–255
Electromagnetic spectrum, 243
Elemental mercury, 135, 136
ELF, see Extra-low frequency
Elimination, 10, 101–102
ELISA, see Enzyme-linked immunosorbant assays
El Niño, 99, 105, 261, 268
Embryotoxicity, 213
Emetics, 26, see also specific types
EMTD, see Estimated maximum tolerated dose
Emulsifiers, 167, 170, see also specific types
Endometrial cancer, 118
Endorphins, 13
Energy measures, 245–246
Enkephalins, 13
Enterotoxins, 219, see also specific types
Environmental monitoring, 51
Environmental risk, 42–43, 54–57
Environmental toxicology, 3
Enzymatic hydrolysis, 12
Enzyme inducers, 10, 26, 85, 119–121, see also specific types
Enzyme inhibition, 30
Enzyme-linked immunosorbant assays (ELISA), 179
Enzymes, 12, 13, see also specific types
    biotransforming, 33, 34
    conjugating, 27
    cytosolic, 12
    hepatic, 10
    induction of, 77
    microsomal, 77
    protein-synthesizing, 113
    serum, 58, 116
EPI, see Exposure/Potency Index
Epichlorhydrin, 46
Epinephrine, 211
Epoxide hydrolases, 8
Epoxides, 8, 9, 51, 158, see also specific types
Equilibrium, 4, 5, 6

Ergot, 209–211
Ergotism, 209–211
Erosion, 73, 94
Erythromycin, 175
Estimated maximum tolerated dose (EMTD), 50
Estrogenic effects, 77
Estrogen receptors, 118, 121, 181
Estrogens, 180
Ethanol (ethyl alcohol), 27, 151, 153, 154, 159
Ethnic background, 24
2-Ethoxyethanol, 158
2-Ethoxyethyl acetate, 158
Ethyl acetate, 151
Ethyl acrylate, 46
Ethyl alcohol (ethanol), 27, 151, 153, 154, 159
Ethylbenzene, 156
Ethylenediamine dihydroiodide, 182
Ethylene glycol, 155
Ethylene oxide, 52, 158–159
Ethylenes, 152, see also specific types
Eutrophication, 68
Evaporation, 94
Exchange of toxicants, 68
Excretion, 12, 21, 22, 34
Exposure
    acute, 2, 42
    to biphenyls, 123–124
    ceiling, 53, 54
    chronic, 43, 125
    frequency of, 20
    long-term, 43
    low-level, 43
    to metals, 144–145
    short-term, 53, 54, 159
    to solvents, 156–160
    time-weighted average, 53, 54
    very low-level, long-term, 43
Exposure/Potency Index (EPI), 45
Extenders, 167
Extraction, 151
Extra-low frequency (ELF) radiation, 254–255
Extrapolation of data, 1–2, 42, 43, 49–50

F

Facilitated diffusion, 5
FAS, see Fetal alcohol syndrome
Fat, 6, 123, 170
Fate, 47
Fat solubility, 7, 8
Fatty acid anilides, 184
Fatty acids, 8
Favism, 183
Feedback loops, 265–266
Fermentation, 102

Fertility disorders, 30
Fertilizers, 74, 75, 140, 171
Fetal alcohol syndrome (FAS), 35
Fetal hemoglobin, 93
Fetal toxicity, 2, 33–36, 134, 137, 181
Fiber, 27, 185
Fiberglass, 93
Figitalis, 232
Fillers, 167
Fire detectors, 56
Firefighters, 96
Fires, 92
First-order kinetics, 20
Fixatives, 167
Flavor adjuvants, 166
Flavor enhancers, 167, 173–174, see also specific types
Flavors, 167
Fluoride, 55
Fluorine, 113, 125, 152
Fluoroacetates, 29
Foam insulation, 58
Fog, 92
Folic acid antagonists, 34, see also specific types
Folpet, 199
Food
  carcinogens in, 183–185
  colors for, 167, 168–170
  defined, 165
  irradiation of, 256–257
  packaging materials for, 166
  production of, 266–270
Food additives, 163–174, see also specific types
  case studies on, 188, 280
  defined, 165–166
  regulations on, 165–168
  uses of, 166–168
Food preservatives, 1, 170–171, see also specific types
Forensic toxicology, 3
Forest fires, 92
Formaldehyde, 57–58, 93, 154, 158
Formic acid, 154
Fossil fuel, 54, 92
Free radicals, 71, 153
Freon, 94
Frequency of exposure, 20
Fukuoka region, Japan, 123
Fumigants, 158, 167, 191, see also Pesticides; specific types
Fumonisins, 212–213, see also specific types
Fungal infections, 264, 265
Fungi, 92

Fungicides, 140, 168, 191, 199, see also Pesticides; specific types
Furazolidone, 175
Furcelleran, 170

G

GABA, see Gamma-aminobutyric acid
Gaia hypothesis, 106, 261–262
Gamma-aminobutyric acid (GABA), 13
Gamma radiation, 245
Gaseous pollutants, 91, 94, see also specific types
Gasoline, 80, 103, 151, 152, 159
Gasoline sniffing, 139, 159
Gastrointestinal cancer, 96
Gene-locus mutation, 30
Genetic factors, 22–25, 30, 71
Genetic polymorphism, 24
Genotyping, 24
Gentamicin, 25
Global cooling, 103
Global warming, 99, 264–265
Glucose-6-phosphate dehydrogenase, 25, 183
Glucuronic acid, 10
Glucuronides, 12, 21
Glues, 159
Glue sniffing, 159
Glutamate, 173, 174
Glutamic acid, 13
Glutathione, 10, 125
Glutathione-S-transferase (GST), 84, 234
Glycine, 10, 13
Glycogenic alkaloids, 233, see also specific types
Glycolate, 155
Glycol ethers, 155, 158
Glycols, 155
Glycosides, 175, 223, 232, 234–236, see also specific types
Gold, 132
G-protein receptors, 13
G proteins, 13
Grain detoxification, 215–218
Grain production, 267–270
Granite, 79
Grassy Narrows reserve, Canada, 29, 138–139
Great Lakes, 68, 71, 74, 81, 82, 83–86
  mercury in, 138, 139
  pesticides in, 202
Greenhouse effect, 101, 102–103
Greenhouse gases, 100, see also specific types
Growth factors, 32, see also specific types
Growth hormone, 77, 183

Growth promotants, 174, 175, 183, see also specific types
GST, see Glutathione-S-transferase
GTP, see Guanosine triphosphate
Guanosine triphosphate (GTP), 13
Guantal response, 15
Guatemala, 203
Gaussian distribution, 14

H

Hairspray, 159
Half-life, 2, 19, 42
Halides, 91, see also specific types
Haloalkane, 153
Halogenated anesthetics, 125, see also specific types
Halogenated hydrocarbons, 13, 36, 86, 113–129, see also specific types
　case studies on, 128–129, 276, 277
　hepatotoxicity of, 154
　physicochemical characteristics of, 114–126
　as solvents, 152–153, 154
　toxicity of, 113–114, 199
Halogenated phenols, 29, see also specific types
Halogenated substances, 91, see also specific types
Halogens, 12, 113, see also specific types
　organo-, see Halogenated hydrocarbons
Halothane, 113, 125, 153
Hanford, Washington radiation release, 249
Hazard, defined, 42
Hazardous sports, 61
Haze, 91, 92
HCFCs, see Hydrochlorofluorocarbons
Heavy metals, see Metals
Helmets, 55, 61
Hemlock, 233, 235
Hemodialysis, 154
Hemoglobin, 28, 74, 93, 104
Hemolytic anemia, 183
Hepatic biotransformation, 153
Hepatic enzymes, 10, see also specific types
Hepatotoxicity, 2, 116, 219
　of halogenated hydrocarbons, 154
　of hydrocarbons, 152–153
　of solvents, 125, 152–153, 154
　of THMs, 126
Heptachlor, 76, 194, 201
Herbicides, 76, 191, 197–199, see also Pesticides; specific types
　acute toxicity of, 197
　arsenic in, 142
　bipyridyl, 76, 198–199

carbamate, 76, 199
chlorphenoxy, 76, 77, 197
　physicochemical characteristics of, 115–121
　resistance to, 268
　toxicity of, 197, 198, 199
Hereditary disease, 30, see also specific types
Hexachlorbiphenyls, 85, see also specific types
Hexachlorobenzene, 84, 116, 199, 203
Hexachlorophene, 114, 115
Hexane, 151, 152, 159
Hexane-2,5-dione, 152
2-Hexanone, 152
HFCs, see Hydrofluorocarbons
Hippuric acids, 156
Hiroshima, 157, 247
Histamine, 13, 225
Histidine, 225
Hodgkin's disease, 117
Hooker Chemical Company, 80, 81
Hormones, 8, 13, see also specific types
　androgenic, 34
　growth, 77, 183
　plant growth, 77
　sex steroid, 22
　steroid, 13, 22, 26–27
Household appliances, 254
Household products, 159, see also specific types
HT-2 toxin, 214, 215
Humectants, 168
Hyaluronidase, 230
Hydraulic fluids, 77
Hydrocarbons, 91, 94, see also specific types
　aliphatic, 152–153
　aromatic, see Aromatic hydrocarbons
　aryl, 13, 85
　chlorinated, see Organochlorines
　halogenated, see Halogenated hydrocarbons
　hepatotoxicity of, 152–153
　nephrotoxicity of, 152–153
　polycyclic aromatic, see Polycyclic aromatic hydrocarbons (PAHs)
　polyhalogenated aromatic, 85
　toxicity of, 199
Hydrochlorofluorocarbons (HCFCs), 100, 101, see also specific types
Hydrochlorthiazide, 144
Hydroelectric dams, 106
Hydrofluorocarbons (HFCs), 100, see also specific types
Hydrogen, 106
Hydrogen bonds, 13
Hydrogen cyanide, 234–235
Hydrogen sulfide, 94

Hydrolysis, 7, 8, 12
Hydroscopic agents, 168
Hydroxylation, 8
Hyoscine, 233
Hyoscyamine, 233
Hyperplasia, 33

I

Ice Age, 103, 105
Icthyotoxin, 226
Immunosuppression, 77
Immunotoxicity, 199
Imposed risk, 55–56
Incineration, 80, 124
Indoles, 185
Indomethacin, 144
Industrial accidents, 57–59, see also specific accidents; specific types
Industrial chemicals, 122–124, see also Xenobiotics; specific types
Industrial discharge, 67, 92
Industrial hygiene, 3
Industrial particulates, 92
Infections, 34, see also specific types
Infectious drug resistance, 175
Infrared, 99
Inherited disorders, 25, see also specific types
Inorganic material, 70, see also specific types
Insecticides, 191, see also Pesticides; specific types
  botanical, 197
  carbamate, 76, 196
  cholinesterase-inhibiting, 42
  classes of, 193–197
  as enzyme inducers, 26
  in Great Lakes, 84
  organochlorine, see Organochlorines
  organophosphate, 13, 26, 28, 76, 77, 193, 195–196
  physicochemical characteristics of, 122
  thiocarbamate, 204
  toxicity of, 4
  in water, 76, 84
Insulation, 58
Intake, 53, 140, 142
Interconnected systems, 264–266
Iodine, 2, 113, 125, 248, 249
Ion channels, 13
Ion exchangers, 13
Ionic bonds, 13
Ionization, 5
Ionizing radiation, 93, 243
Ion sensors, 245
Iraq, 137, 203

Iron, 132
Iron-containing vitamins, 143
Isocyanates, 52
Isoniazid, 23, 24
Isopropyl alcohol, 154
Isopropylbenzene, 156
Isothiocyanates, 185
Itai-Itai disease, 141

K

Ketoacidosis, 11–12
Ketones, 151, see also specific types
Kidney cancer, 159
Kidney disease, 99
Kinetics, 20

L

Lacquers, 160
Lactate, 155
Lactation, 6
Lake St. Clair, 138, 139
Landfills, 80
Lanthanum, 144
Law of mass action, 4
Lead, 52, 99
  in automobile exhaust, 103–104
  carcinogenicity of, 134, 141, 144
  cellular toxicity of, 133–134
  fetal toxicity of, 134
  in Great Lakes, 84
  in health supplements, 145
  history of, 131
  from plumbing, 78, 104
  poisoning from, 26, 132
  from solder, 78, 104, 133
  sources of, 133
  specific gravity of, 132
  toxicity of, 27, 132–135
  toxicokinetics of, 133
  in water, 78, 84
  in workplace, 95
Lead arsenate, 141
Lead chelators, 135
Lead smelting, 103
Legal aspects of risk, 59–60
Lethal dose, 16, 69–70, 77, 84, 123, 197
Leukemia, 59, 81, 118, 156, 251
  atomic bomb and, 157, 247
  benzene, 156
  radiation and, 254, 255
Leukemia viruses, 51
Ligand binding, 12–14
Light stress, 70–71

Limestone, 70, 79
Linamarin, 234
Lincomycin, 175
Lindane, 26, 76, 194, 195
Linearized, Multistage Assessment Technique, 44
Lipid solubility, 4, 5, 6, 7, 33
Lipophilicity, 76
Liquid pollutants, 95
Little Ice Age, 105
Liver angiosarcoma, 96
Liver cancer, 81, 117, 118, 121, 125, 211, 233
Liver cell carcinoma, 121
Lotaustralin, 234
Love Canal, 80–83
Lung cancer, 81, 96, 118, 125, 143, 144, 159, 255, 270
Lyme disease, 178–179
Lymphomas, 59, 81, 117, 118, 255, see also Cancer; specific types
Lymphoreticulosarcoma, 118
Lysergic acid, 211

M

Malarial parasites, 25
Malathion, 76
Manganese, 95, 132
Marine species, 72–73, 224–228
Mass action, law of, 4
Material Safety Data Sheets, 52
Maturing agents, 168
Meat production, 267–270
Mechanistic models, 44
Medicinals, 1, 3, see also specific types
Mefanemic acid, 144
Melanomas, 255
Membrane receptors, 12
Membranes, 4, 5, 6, 10, 33
Mercaptides, 137
6-Mercaptopurine, 24
Mercuric chloride, 136
Mercury, 2, 29, 52
 elemental, 135, 136
 fetal toxicity of, 137
 in fish, 138
 forms of, 135
 in Great Lakes, 84
 history of, 131
 methylation of, 69
 organic, 203
 poisoning from, 137, 138
 specific gravity of, 132
 toxicity of, 135–139
 in water, 70, 78, 84

 in workplace, 95
Mercury cell process, 136
Mercury compounds, 135, 138, 199, see also specific types
Mercury nitrate, 136
Mesothelioma, 96, 273
Metabolism, 24, 29, 34, 76, 119, 123
Metal chelators, 26, 135, 138, 141, 142, see also specific types
Metal-fume fever, 96
Metallic oxides, 92, see also specific types
Metallothionein, 140, 141
Metal plating, 140
Metals, 2, 99, 268, see also specific types
 absorption of, 21
 carcinogenicity of, 134, 141, 144
 cellular toxicity of, 133–134
 exposure to, 144–145
 fetal toxicity of, 134, 137
 in food, 184
 in Great Lakes, 84
 half-lives of, 2
 as "Miracle Herb", 145
 pharmacokinetics of, 140, 142
 poisoning from, 26, 29
 sources of exposure to, 144–145
 specific gravity of, 132
 toxicity of, 78–79, 131–148, see also under specific metals
  case studies on, 147–148, 277–278
  cellular, 133–134
  fetal, 134, 137
  mechanism of, 137–138
  review questions on, 145–147
  treatment for, 135, 141, 142
 toxicokinetics of, 133
 transport of, 144
 in water, 69, 70, 78–79, 84
 in workplace, 95
Methane, 94, 102, 104, 106, 267, 268
Methanol, 154, 155
Methemoglobin, 74, 104, 156, 171
Methemoglobinemia, 184
Methemoglobin reductase, 74
Methionine, 144
Methoxychlor, 194, 195, 201
Methoxyflurane, 153
Methylation, 24, 69
3-Methylcholanthrene, 26
Methylene chloride (dichloromethane), 81, 124, 125, 153
Methylene dichloride, 152
Methylethyl ketones, 151, see also specific types
Methylmercuric chloride, 137
Methylmercury, 2, 29, 35, 70, 74, 78, 137, 138

Methylmercury-cysteine complex, 144
4-Methyl pyrazole, 154
Methylsergide, 211
Metoprolol, 24
Microorganisms, 92, 223, 262, see also specific types
Microsomal enzymes, 77, see also specific types
Microtubule dissolvers, 233, see also specific types
Microwaves, 252–253
Minamata Bay, Japan, 29, 137
Minamata Disease, 137, 138
Mineral turpentine, 152
Mirex, 81, 84, 194
Mist, 91
Mitosis, 30
Mitotic activity, 12
Mixed function oxidases, 8, 126
Moana Loa observatory, 105
Models, 44, 46, see also specific types
Modifiers, 68–71, see also specific types
Mollusc venoms, 227
Molybdenum, 132
Monitoring, 51
Monoethyl ethylene glycol, 155
Monomethyl glycol, 155
Monomethylmercury, 29
Monooxygenases, 8, 27, 29, 234
Monosodium glutamate (MSG), 173–174
Monsanto, 117, 119
Motor vehicle accidents, 55
Motor vehicle emissions, 92, 103–104
Mount Pinatubo, Philippines, 92, 99, 102
Mount St. Helen volcanic explosion, 43, 92
MSG, see Monosodium glutamate
Mu conotoxin, 236
"Multiple Drug Resistance", 175
Multiple myeloma, 118
Multiple resistance, 175, 201–202
Multistage models, 44, 46
Municipal sewage discharge, 74
Murine leukemia virus, 51
Muscarine, 28
Muscarinic blockers, 28, see also specific types
Mushrooms, 232–236
Mutagenicity, 29–32, 49, 199, see also Mutations
Mutations, 12, 30, 119, see also Mutagenicity
Mycotoxins, 185, 209–221, 270, see also specific types
  carcinogenicity of, 268
  defined, 209
  economic impact of, 214–215
  embryotoxicity of, 213
  health problems from, 209–214
  review questions on, 219–221
  teratogenicity of, 213
  toxicity of, 213
Myeloid leukemia, 118
Myosin, 32

N

Nagasaki, 157, 247
Naloxone, 26
Narcotic analgesics, 13, see also specific types
Nasal cancer, 57, 117
Natural carcinogens, 184–185, 270
Natural factors in air pollution, 92, 104–105
Natural gas, 106
Natural radiation, 249–251
Natural sources of water pollution, 74
Natural toxicants, 183–185
NDMA, see N-Nitrosodimethylamine
NEDs, see Normal equivalent deviations
Nematodes, 75
Neomycin, 175
Neonatal toxicity, 21
Nephrotoxicity, 2, 152–153
Nerve gases, 28
Neurotoxicity, 2, 28, 76, 119, 122, 193, 223, 229
Neurotransmitters, 13, 173, 211, see also specific types
New Ice Age, 103
Niagara Falls, 81
Nickel, 95, 144
Nickel-cadmium batteries, 140
Nicotine, 197
Nicotine sulfate, 192
Nicotinic blockers, 230
Nitrate fertilizers, 74
Nitrates, 48, 74, 104, 171, see also specific types
Nitric oxide, 94, 100
Nitriles, 234, see also specific types
Nitrites, 48, 171, see also specific types
Nitrofurans, 175, see also specific types
Nitrofurazone, 175
Nitrogen, 91
Nitrogen compounds, 91, see also specific types
Nitrogen dioxide, 94, 97
Nitrogen monoxide, 94
Nitrophenols, 29, see also specific types

Nitrosamines, 171, 185
N-Nitrosodimethylamine (NDMA), 36
NMBY, see Not in My Back Yard
NOEL, see No Observable (Adverse) Effect
Non-Hodgkin's lymphoma, 117, 118
Nonionizing radiation, 254
No Observable (Adverse) Effect (NOEL), 16–17, 44, 53
Noradrenaline, 13, 125
Normal equivalent deviations (NEDs), 17
Not In My Back Yard (NMBY) response, 124
Nuclear decay, 245
Nuclear disasters, 2, 45, 57, 157, 245, 247–249, see also specific disasters; specific types
Nuclear energy, 106
Nuclear fuel, 54, 92
Nuclear inclusion bodies, 134
Nuclear plant workers, 252
Nucleic acids, 12, see also specific types
Nutrition, 15, 71
Nystatin, 175

O

Occidental Petroleum, 81
Occupational exposure, see Workplace
Ochratoxin A, 213, 214
Oil combustion, 103
Okadatic acid, 226
Oleander, 232
Oncogenes, 31
Oncogenic viruses, 31
Opioates, 26, see also specific types
Opioid receptors, 13
Oral contraceptives, 179, see also specific types
Organic carbon, 70
Organic materials, 70, 78, 92, see also specific types
Organic solvents, see Solvents
"Organized complexity", 263
Organochlorines, 2, 76, 122, 193–195, see also specific types
  carcinogenicity of, 194
  health hazards from, 76–77
  resistance to, 201
  in water, 80
Organohalogens, see Halogenated hydrocarbons
Organophosphates, 13, 26, 28, 76, 77, 193, 195–196, see also specific types
Orthophosphate, 75
Osteomalacia, 78
Overcrowding, 71

Overpopulation, 36, 268
Oxalates, 155, 232
Oxathizinondioxides, 172, see also specific types
Oxidation, 7, 8, 71, 143
Oxidative metabolism, 24
Oxidative phosphorylation, 28, 133, 142
Oxide hydrolase, 51
Oxides, 91, 92, 95, see also specific types
Oxygen, 70, 261, 262
Ozone, 94, 99–100

P

PAHs, see Polycyclic aromatic hydrocarbons
Paint removers, 154, 159
Paints, 159, 160
Paint strippers, 151
Paint thinners, 154, 159
Pakistan, 203
Pancreatic cancer, 126
Paper mills, 75, 138, 139
Paraoxon, 195
Paraquat, 43, 76, 121, 198
Parathion, 76, 77, 195, 203
Particle feeders, 68
Particulates, 91–92, 95, 98, see also specific types
Partition coefficients, 4, 6, 42
Passive diffusion, 11
Pathology, 2, 25–26
Patulin, 60, 213
PBBs, see Polybrominated biphenyls
PCBs, see Polychlorinated biphenyls
PCDDs, see Polychlorinated dibenzodioxins
PCDFs, see Polychlorinated dibenzofurans
PCR, see Polymerase chain reaction
Penicillamine, 135, 138, 142
Penicillin, 25
Penicillin G, 174, 179
Pentachlorophenol, 47, 69, 199
Pentapeptides, 219, see also specific types
Peptides, 223, 227, see also specific types
Percolation, 68
Peroxidases, 199
Persistence, 2
Pesticides, 1, 28, 191–206, see also Insecticides; specific types
  arsenic in, 142
  benefits of, 202–203
  bioaccumulation of, 191
  biological control methods for, 199–200
  biomagnification of, 191
  carcinogenicity of, 203
  case studies on, 205–206, 280–281

chemical classification of, 76
classes of, 193–197
defined, 191
in food, 270
in Great Lakes, 202
neurotoxicity of, 28
organochlorine, see Organochlorines
problems associated with, 201–202
regulation of, 200–201
resistance to, 201
review questions on, 204–205
risks of, 202–203
teratogenicity of, 203
toxicity of, 203–204
in water, 75–76
in workplace, 95
pH, 6, 10, 69, 70
Pharmacodynamics, 12–20
Pharmacogenetics, 23–25
Pharmacokinetics, 3–12
of arsenicals, 142
of biphenyls, 123
of cadmium, 140
defined, 3
of metals, 140, 142
species differences in, 48
Pharmacology, 3, 4
Phenacetin, 10
Phenobarbital, 26
Phenols, 29, 156, see also specific types
Phenylalanine, 172
Phenylalanine hydroxylase, 173
Phenylbutazone, 30
Phenylketonuria, 173
Phorbol esters, 31, 213, 233, 234
Phorbol myristate, 236
Phorbol myristate acetate (PMA), 233
Phosgene, 125, 153
Phosphatase, 230
Phosphate fertilizers, 75
Phosphates, 68, 75, see also specific types
aluminum-bound, 79
organo-, 13, 26, 28, 76, 77, 193, 195–196
transport of, 142
in water, 75
Phospholipase A, 229, 230
Phospholipase C, 13
Phospholipids, 4
Phosphorus, 79
Phosphorylation, 28, 29, 133, 142
Photochemical reactions, 92, see also specific types
Photochemical transformations, 70–71
Photographic chemicals, 140
Photolysis, 123

Photo-oxidation, 71
Photosynthesis, 101, 262
Physicochemical characteristics, 114–126
of antibacterial disinfectants, 114–115
of biphenyls, 122–124
of disinfectants, 114–115
of herbicides, 115–121
of industrial chemicals, 122–124
of insecticides, 122
of solvents, 124–125
Pickering Nuclear Plant, Ontario, 251
Pigments, 140
Pinocytosis, 5
PKU, see Phenylketonuria
Placenta, 33
Plankton, 224
Plant growth hormones, 77
Plasma albumin, 22
Plasma membrane proteins, 12, see also specific types
Plasmids, 175, 176
Plasticizers, 168
Plastics, 97, see also specific types
Plastic stabilizers, 140
Platinum, 132
Plumbing, 78, 104
Plutonium, 132
PMA, see Phorbol myristate acetate
Pneumoconiosis, 96
Podophylotoxin, 233
Poikilotherms, 70
Point mutations, 30
Poisoning, see also Toxicity
acetaminophen, 26
arsenic, 142
carbon monoxide, 125
cyanide, 234
drug, 26
lead, 26, 132
mercury, 137, 138
metal, 26, 29
methanol, 155
methylmercury, 137
from nitrates, 171
organophosphate, 26
scrombroid, 225
shellfish, 226
from solvents, 151
Polishes, 159
Pollen, 92
Polybrominated biphenyls (PBBs), 26, 36, 85, 124, see also specific types
Polychlorinated biphenyls (PCBs), 36, 76, 77, 82, see also specific types
carcinogenicity of, 185

dechlorination of, 123
disposal of, 124
as enzyme inducers, 26
in Great Lakes, 84, 202
physicochemical characteristics of, 122
in water, 76, 81, 84, 86
Polychlorinated dibenzodioxins (PCDDs), 84, 85, 119, see also specific types
Polychlorinated dibenzofurans (PCDFs), 85, 119, see also specific types
Polycyclic aromatic hydrocarbons (PAHs), 8, 36, 71, 72, 233, see also specific types
detoxification of, 84
as energy source, 120
in Great Lakes, 84
in water, 84, 85
Polyhalogenated aromatic hydrocarbons, 85, see also specific types
Polymerase chain reaction (PCR), 24
Polymorphism, 24
Polyphosphates, 75, see also specific types
Polyurethane, 97
Porphyria, 116
Porphyrins, 116
Portal of entry, 4, 46–47
Potassium, 5
Potassium bromide, 114
Potassium nitrate, 171
Potassium transport, 122
Power generating stations, 54, 92
Pralidoxime, 196
Preservatives, 1, 170–171, see also specific types
Probit analysis, 17–18
Procarcinogens, 8
Progesterone, 180
Promoters, 30–31, 48, 77
Propylene glycol, 155
Prostate cancer, 144
Proteases, 230, see also specific types
Protein kinase C, 234
Proteins, 12, 13, 29, 141, see also specific types
Protein-synthesizing enzymes, 113, see also specific types
Protozoa, 143
Prunasin, 234
Pseudomonas, 120
Psychological impact, 54–56, 82, 182, 248
Public perceptions of risk, 41–64
carcinogenesis and, 43–51
environmental monitoring and, 51
environment risk and, 42–43, 54–57
legal issues and, 59–60
statistical problems and, 60–63

workplace vs. environment and, 42–43
workplace limit setting and, 51–54
Pulp mills, 75, 138, 139
Pyrethrins, 197, 201
Pyrogallol tannins, 233, see also specific types
Pyrolysis, 97
Pyruvate dehydrogenase, 142

Q

Quantal data, 17–18
Quantal response, 16

R

Race, 24
Radar, 252
Radiation, 32, 34, 57, 243–258, see also specific types
carcinogenicity of, 253, 254, 255, 256
causes of, 244
damage from, 246
disasters involving, 247–249
electromagnetic, 254–255
extra-low frequency, 254–255
fear of, 243, 256
of food, 256–257
gamma, 245
ionizing, 93, 243
measurement of, 245–246
medical uses of, 253
microwave, 252–253
natural, 249–251
nonionizing, 254
sources of, 244–245
sterilization by, 200
tissue sensitivity to, 251–252
toxicity of, 246
types of, 244–245
ultraviolet, 70–71, 99, 100, 103, 123, 253
wavelengths of, 244
X-, 245
Radioactive energy, 245, see also Radiation
Radioactive isotopes, 94
Radioisotopes, 245
Radon gas, 93, 245, 249–251, 270
Rain, 68, 73
Rainforests, 101, 261
Randomness, 262–263
Receptors, 5, 12–14, 15, see also specific types
aromatic hydrocarbon, 119
aryl hydrocarbon, 119
competition for, 26
cytosolic, 13, 119

estrogen, 118, 121, 181
G-protein, 13
membrane, 12
opioid, 13
steroid, 13
Rectal cancer, 126
Reduction, 7, 8
Refrigerants, 152, see also specific types
Regulatory proteins, 29, see also specific types
Reproductive toxicity, 2, 119
Resistance
  to antibiotics, 25, 177, 178
  bacterial, 178
  cross, 201
  to DDT, 25
  to drugs, 25, 174–179
  to herbicides, 268
  infectious drug, 175
  multiple, 175, 201–202
  to oxygen, 262
  to pesticides, 201
  to toxicity, 25
Respiratory cancer, 144
Retention time, 98
Retroviruses, 31, see also specific types
Risk
  age and, 47–48
  avoidance of, 56–57
  of cancer, 2, 43, 45, 46–49
  costs of avoidance of, 56–57
  environmental, 42–43, 54–57
  imposed, 55–56
  legal aspects of, 59–60
  of pesticides, 202–203
  prediction of, 42–43
  psychological impact of, 54–56
  public perceptions of, 41–64
    carcinogenesis and, 43–51
    environmental monitoring and, 51
    environment risk and, 42–43, 54–57
    legal issues and, 59–60
    statistical problems and, 60–63
    workplace vs. environment and, 42–43
    workplace limit setting and, 51–54
  species differences in, 48–49
  voluntary, 55–56, 61
  workplace, 42–43
Risk analysis, 41–64
  carcinogenesis and, 43–51
  case studies on, 64
  environmental monitoring and, 51
  environmental risk and, 42–43, 54–57
  legal issues and, 59–60
  statistical problems with, 60–63
  workplace vs. environment and, 42–43

workplace limit setting and, 51–54
Risk assessment, 42, see also Risk analysis
  carcinogenicity and, 43–51
  statistical problems with, 60–63
Risk-benefit analysis, 54, 191
Risk factors, 54, see also specific types
RNA, 29
RNA polymerase, 213
RNA retroviruses, 31, see also specific types
Rodenticides, 191, see also Pesticides; specific types
Rotenone, 192, 197
Runoff, 68, 73, 74, 94
Russell's viper venom, 236
Rutin, 27

S

Saccharin, 168, 171, 172
Safety equipment, 55, 56, see also specific types
Safrol, 233
Salinity, 74
Sarcomas, 117, 118, 255, see also Cancer; specific types
Saurine, 225
Saxitoxin, 226
Scalefish toxins, 224–226
Scopolamine, 28, 233
Scrombroid poisoning, 225
Seat belts, 55, 61
Sediment, 68, 70, 73
Seepage, 94
Selenium, 78, 132
Semipermeable membranes, 4, 10, 33
SER, see Smooth endoplasmic reticulum
Serotonin, 13, 211
Serum albumin, 5
Serum enzymes, 58, 116
Serum glutamic-oxaloacetic transaminase (SGOT), 116
Serum glutamic-pyruvic transaminase (SGPT), 116
Service station storage tanks, 80
Seveso, Italy, 58, 115, 116, 118, 121
Sewage, 67, 74
Sewage sludge fertilizers, 140
Sewer effluents, 178
Sex, 22
Sex steroids, 22
SGOT, see Serum glutamic-oxaloacetic transaminase
SGPT, see Serum glutamic-pyruvic transaminase
Shellfish poisoning, 226

Shellfish toxins, 226–227
Short-term exposure limit (STEL), 54, 159
Short-term exposure value (STEV), 53
Sickle cell anemia, 25
Silica, 52
Silicosis, 97
Skin cancer, 99, 144, 159
Smog, 92
Smoke, 91
Smoke detectors, 56, 245
Smoking, see Cigarette smoking
Smooth endoplasmic reticulum (SER), 8, 10
Snakes, 228–230
Sodium, 5
Sodium arsenite, 141, 142
Sodium bromide, 114
Sodium channels, 227
Sodium dodecylbenzenesulfonate, 75
Sodium hydroxide, 136
Sodium nitrate, 171
Sodium nitrite, 171
Sodium/potassium ATPase, 79
Sodium pump, 79
Sodium saccharin, 171
Sodium transport, 122
Sodium tripolyphosphate, 75
Softeners, 168
Soft tissue tumors, 117, 118
Soil erosion, 73, 94
Soil transport, 94
Solanine, 233–234
Solar activity, 105
Solar energy, 106
Solar photolysis, 123
Solder, 78, 104, 133
Solid tumors, 156
Solubility, 4, 5, 6, 7, 8, 33
Solvents, 11, 26, 27, 151–162, see also specific types
  aliphatic alcohols as, 153–154
  aliphatic hydrocarbons as, 152–153
  aromatic hydrocarbons as, 155–156
  bone marrow toxicity of, 155
  carcinogenicity of, 156–159
  case studies on, 161–162, 274, 278–280
  classes of, 152–156
  exposure to, 156–160
  halogenated hydrocarbons as, 152–153, 154
  hepatotoxicity of, 125, 152–153, 154
  nephrotoxicity of, 152–153
  nonoccupational exposure to, 159–160
  physicochemical characteristics of, 124–125
  poisoning from, 151
  toxicity of, 125, 155, 159
  in workplace, 156–159

Sorbitol, 172
Species differences, 48–49, 71
Specific gravity of metals, 132
Sphingosine, 213
Spices, 166
Spleen cancer, 159
Sports, 61
Squamous cell carcinoma, 57
St. Basile-le-Grand PCB fire, 82, 124
St. Clair River chemical spills, 83
St. Thomas, 80
Standard deviation, 14
Statistical problems, 60–63
STEL, see Short-term exposure limit
Sterilants, 158
Steroidal alkaloids, 27, see also specific types
Steroid glycosides, 223
Steroid receptors, 13
Steroids, 13, 22, 26–27, see also specific types
STEV, see Short-term exposure value
Stilbestrol, 8
Stinging fishes, 226–227
Stomach cancer, 117, 126, 159
Storage tanks, 80
Storm drains, 74
Stratosphere, 93–94
Streptomycin, 175, 178
Strontium 90, 94, 143, 256
Subacute toxicity, 20
Subchronic toxicity, 20
Sugar substitutes, 168, 171–173
Sulfa drugs, 24, 175, 179, see also specific types
Sulfamethazine, 174
Sulfhydryl groups, 137, 142, 213, see also specific types
Sulfides, 91, see also specific types
Sulfonamides, 25, see also specific types
Sulfoxides, 8, see also specific types
Sulfur, 79, 91
Sulfur compounds, 91, 151, see also specific types
Sulfur dioxide, 94, 98–99, 103
Sulfuric acid, 10, 98
Sulfurous acid, 98
Sulfur trioxide, 94, 98
Sunspots, 105
Suppressor genes, 32
Surface evaporation, 94
Surface impoundments, 80
Surfactants, 75, 167, see also specific types
Suspending agents, 168
Sweeteners, 168, 171–173
Symbiosis, 262
Synergistic interactions, 26

Synergistic risk factors, 54
Synthetic chemicals, 1, see also specific types
Syphilis, 56

T

T-2 toxin, 213, 214, 215
2,4,5-T, see 2,4,5-Trichlorophenoxyacetic acid
Tambora volcano, 102
Tamoxifen, 121
Tannins, 233, 236, see also specific types
TCDD, see Tetrachlorodibenzo-p-dioxin
TEFs, see Toxicity equivalency factors
Temperature, 15, 70, 103, 105
Teratogenicity, 13, 30, 33–35, 119
    of alcohols, 155
    of antibiotics, 219
    causes of, 34
    defined, 33
    drug-induced, 33, 34
    of mycotoxins, 213
    of pesticides, 203
Termites, 102
Terpenes, 26, see also specific types
Testicular cancer, 158, 182
Tetrachlorethylene, 80
Tetrachlorodibenzo-p-dioxin (TCDD), 28, 36, 58–59, 204, see also Dioxins
    carcinogenicity of, 59, 117–119
    cardiovascular effects of, 117
    as enzyme inducer, 119–121
    estrogen receptors and, 121
    in Great Lakes, 84
    hepatotoxicity of, 116
    neurotoxicity of, 119
    physicochemical characteristics of, 115, 116–119
    reproductive toxicity of, 119
    toxicity of, 116–119, 197
    in water, 77, 84
Tetracyclines, 174, 175, 177, 178
Tetraethyl lead, 103, 133
Tetrodotoxin, 28, 225, 235
Thalidomide, 33, 34, 115, 199
Theophylline, 10
Therapeutic Index, 16
Thiabendazole, 199
Thiocarbamate insecticides, 204
Thiocyanates, 234
Thiol groups, 134, 144
Thiopurine S-methyltransferase (TPMT), 24
THMs, see Trihalomethanes
Three Mile Island, 57, 248–249

Threshold limit values (TLVs), 54
TI, see Therapeutic Index
Time-weighted average exposure (TWAEV), 53, 54
Tin, 95
Tire-dump fire, Hagersville, 82
Tissue regeneration, 32–33
TLVs, see Threshold limit values
Toad-licking, 223
Tobacco, 142, see also Cigarette smoking
Toluene, 27, 81, 151, 156
Tomatin, 27
Toxaphene, 84, 194, 195
Toxic animals, 224–232
    case studies on, 238–240, 281–283
    review questions on, 236–238
Toxic dose, 16
Toxicity, 17–18, see also Poisoning; specific types
    acute, 20, 27–29, 197, 211
    of aflatoxins, 211
    age and, 21–22, 159
    of alcohols, 27
    of aluminum, 143
    of aniline, 184
    of antibiotics, 219
    in aquatic species, 72–73
    of arsenic, 141–143
    of biphenyls, 123
    body composition and, 21, 22
    bone marrow, 155
    of cadmium, 140–141
    of carbamates, 199
    cell regeneration and, 32–33
    cell repair and, 32–33
    cellular, 49, 133–134
    of chromium, 143
    chronic, 20, 27–29
    cyto-, 49, 133–134
    of DDT, 122, 193
    of dioxins, 42
    embryo-, 213
    ethnic background and, 24
    factors influencing, 20–27
    fetal, 2, 33–36, 137, 181
    of fumonisins, 212–213
    of fungicides, 199
    genetic factors in, 22–25
    of halogenated hydrocarbons, 113–114, 199
    hepato-, see Hepatotoxicity
    of herbicides, 197, 198, 199
    of hydrocarbons, 199
    immuno-, 199
    of insecticides, 4

of lead, 27, 132–135
in marine species, 72–73
mechanism of, 125, 137–138
of mercury, 135–139
of metals, 78–79, 131–148, see also under specific metals
   case studies on, 147–148, 277–278
   cellular, 133–134
   fetal, 134, 137
   mechanism of, 137–138
   treatment for, 135, 141, 142
of monosodium glutamate, 173–174
of mushrooms, 232–236
neonatal, 21
nephro-, 2, 152–153
neuro-, 2, 28, 76, 119, 122, 193, 223, 229
of paraquat, 121
pathology and, 25–26
of pesticides, 203–204
of phorbol esters, 233
of plants, 232–236
race and, 24
of radiation, 246
reproductive, 2, 119
resistance to, 25
sex and, 22
of solvents, 125, 155, 159
subacute, 20
subchronic, 20
of TCDD, 197
tissue regeneration and, 32–33
of uranium, 143
Toxicity assessment, 42
Toxicity equivalency factors (TEFs), 85
Toxicity testing, 72–73
Toxic oil syndrome, 183–184
Toxicokinetics, 133
Toxicology, 2, 3, 4, 27–29, see also Toxicity
defined, 3
fetal, 33–36
Toxic plants, 232–236
   case studies on, 238–240, 281–283
Toxins, 223, see also specific types
afla-, 8, 118, 211–212
botulinum, 28, 42, 171, 219
coelenterate, 227
cono-, 227, 235, 236
entero-, 219
HT-2, 214, 215
myco-, see Mycotoxins
podophylo-, 233
scalefish, 224–226
shellfish, 226–227
T-2, 213, 214, 215
tetrodo-, 28, 225, 235

TPMT, see Thiopurine S-methyltransferase
Tranquilizers, 25, 144, see also specific types
Trans-1,2-dichloroethylene, 80
Transduction, 176
Transformation, 176
Transplacental carcinogenicity, 35–36
Transplacental transfer, 22
Transport, 6, 94, 122, 142, 144
Transposons, 176
Tremolite, 96
Triazines, 199, see also specific types
Trichlorethane, 80
Trichlorethylene, 27, 46, 80, 113
1,1,1-Trichloroethane, 151
Trichloroethylene, 113, 151, 152, 153
2,4,5-Trichlorophenoxyacetic acid (2,4,5-T), 76, 77, 115, 197
Trichothecenes, 215, see also specific types
Trihalomethanes (THMs), 125–126, see also specific types
Trioxides, 91, 94, see also specific types
Troposphere, 93, 98
Trypanosomiasis, 141
Tryptophan, 184
T-2 toxin, 213, 214, 215
Tuberculosis, 178
Tubocurarine, 28
Tumors, 2, see also Cancer; specific types
bladder, 168
breast, 121
formation of, 27, 28
initiation of, 30
progression of, 31
promotion of, 30–31, 48, 77
soft tissue, 117, 118
solid, 156
uterine, 121, 180, 181
Tungsten, 132
Tunnel projects, 80
Turkey, 203
Turpentine, 152
TWAEV, see Time-weighted average exposure
Tyrosine, 173

U

UDMH, see Unsymmetrical dimethylhydrazine
Ultraviolet radiation, 70–71, 99, 100, 103, 123, 253
Uncoupling agents, 29, see also specific types
Unsymmetrical dimethylhydrazine (UDMH), 60